Science on the Ropes

Carlos Elías

Science on the Ropes

Decline of Scientific Culture in the Era of Fake News

Springer

Copernicus Books is a brand of Springer

Carlos Elías
Department of Journalism and Film Studies
Carlos III University of Madrid
Madrid, Spain

ISBN 978-3-030-12977-4   ISBN 978-3-030-12978-1   (eBook)
https://doi.org/10.1007/978-3-030-12978-1

This Copernicus imprint is published by the registered company Springer Nature Switzerland AG
The registered company address is: Gewerbestrasse 11, 6330 Cham, Switzerland

Acknowledgements

Writing this book has been a goal pursued for years. I have been making progress little by little in my free time, since the lack of STEM vocations and their relationship with the media and the rise of anti-scientific culture in Western universities are not topics to be found in financial, scientific, or engineering discourses and belong more to the social sciences. However, since the approach I wanted to explore implied "blaming" the social sciences and the humanities in some way, it was not easy for those responsible for sponsoring projects in these areas to support my work. While researching more orthodox issues in media communication to fulfill the requirements of my academic career, I still tried to find time to investigate on my own this fascinating issue, namely how the war between Snow's two cultures—science and humanities— was still alive and was influencing not only the decline in STEM vocations but also the rise of post-truth, alternative facts, and fake news. And, in short, in a decline of Western culture related to rationality, empiricism and facts.

I have not had financial help, but I have had many encouragements to write an informative—and controversial—essay analyzing this problem in such a way that it is accessible not only for university experts but also, and above all, for secondary school teachers, scientists, and engineers who are reluctant to read reports from the social sciences. My purpose is to encourage debate, not to "be right."

In this adventure, I owe an enormous debt to Springer Nature Editor Alexis Vizcaíno, who, from the very beginning, was not only enthusiastic about the idea, but whose input has also greatly enriched this book. I also thank him for his infinite patience, as my other professional duties prevented me from making good progress in this adventure. I am also indebted to Alison Williamson, who put enormous effort into the English edition of this text.

I am grateful to the two institutions that welcomed me for, among other endeavors, also advancing this one: the London School of Economics (specifically Martin Bauer from the Department of Social Psychology) and The Department of History of Science at Harvard University (its Director, Janet Browne, my Faculty Sponsor Everett Mendelsohn, and Jimena Canales). The talks I gave on this subject at these institutions also helped me to improve certain approaches.

Finally, I would like to thank my colleagues at the Universidad Carlos III de Madrid (Daniel Peña, Juan Romo, Fernando Broncano, Francisco Marcellán, and Juan José Dolado) and those at the Universidad Nacional de Educación a Distancia (David Teira and Jesús Zamora) for their role in getting me to explore politically incorrect terrain. But only in this way, through critical dissidence, has Western thought advanced.

Contents

1

Introduction

Two worrisome themes are currently running through the West: a disturbing rise in fake news and an alarming loss of vocations in science, technology, engineering and mathematics, the so-called STEM subjects. These two streams run parallel, threatening to erode the culture of searching for truth through rationality and the scientific method. The cultural particularity that has favoured this type of European thought, rationality, originated in ancient Greece and was rediscovered and improved in Renaissance Europe. This led to the Enlightenment and, with it, an unprecedented improvement in the understanding of how the world works.

Each of the two streams has its analysts, yet few have stopped to reflect that, perhaps, each arises from the same source: that rationality has been increasingly questioned in a process that began in the 1960s. What began as a departmental culture war in the university has spread to society, because for the most part the anti-scientific departments are the ones that prepare those who have taken over the discourse on science policy and politics itself in the mass media.

The university and its influence on the media have fostered a celebrity culture that puts the emotional before the rational. They have established university degrees in which literature about magic is studied more than physics and chemistry. These two problems, which as yet nobody connects but in which I, modestly, claim to see a relationship, have led on the one hand to the rise of post-truth, fake news or alternative facts in Western public communication, and on the other hand to an alarming drop in young people taking up a vocation in STEM-related professions.

Without science and technology creators, there will be a loss of cultural and economic influence. With the rise of fake news, democracy will be restricted. To the latter is added an increase in the content of magic at the expense of

© Springer Nature Switzerland AG 2019
C. Elías, *Science on the Ropes*, https://doi.org/10.1007/978-3-030-12978-1_1

science in the media, as well as extolling irrational thinking and, in some ways, criticizing scientific method thus, ultimately, scientists and technologists.

The result of the first trend includes, among others, the referendum by which the United Kingdom opted to leave the European Union (Brexit) and the election, against all rational odds, of Donald Trump as the president of the United States. In 2017, the *Oxford English Dictionary* considered 'fake news' to be the word of the year. It was reported that its use increased by 365%, and in January 2018 the French president Emmanuel Macron announced in the traditional presidential speech on his wishes for the press that France would legislate against false news.

The change had begun in 2016 Germany. The Christian Democratic Union had suffered a heavy defeat and its president, Angela Merkel, had uttered a phrase that would define the era: 'Lately,' she had said, 'we live in post-factual times. This means that people are no longer interested in facts, but in feelings' (Schaarschidt 2017).[1] After Merkel's reflection, the term 'post-factual' became so important that the German Language Society unanimously chose it as its word of the year in 2016. Also in 2016, the *Oxford English Dictionary* selected the US/UK 'post-factual approach'—that is, post-truth—as its word of the year.

The issue of STEM vocations is less noticed by the general public and the media, but has for years been the focus of intellectual debate by—and a main concern of—scientists and university and secondary school teachers around the world: why are young people not interested in studying science in Western and Westernized countries? In 2011, the President of the United States, Barack Obama, and the country's leading business leaders had launched the STEM—Science, Technology, Engineering and Mathematics—initiative, with the aim of graduating 10,000 more scientists and engineers each year in the United States and obtaining 100,000 new STEM teachers. The proposal had been picked up by the media.[2] The goal was that the United States—and the West in general—should not lag behind the emerging Asian countries in science and technology.

The disturbing question is why this situation had arisen. Why do young Americans—and Europeans—not decide to study science, technology, engineering and mathematics (STEM), but aim for social sciences or media careers? Does mainstream popular culture have anything to do with it? What about the social networking culture? Or does it have something to do with the fact that

---

[1]T. Schaarschmidt. (2017). 'La era de la posverdad', *Mente y Cerebro [Spanish edition of Mind Scientific American]*, 87, 22–28.

[2]Among others, *New York Times*, 4 November 2011: 'Why science majors change their minds (it's just so darn hard)'. http://www.nytimes.com/2011/11/06/education/edlife/why-science-majors-change-their-mind-its-just-so-darn-hard.html?pagewanted=all&_r=0.

these youngsters' high school teachers of social sciences and humanities have studied at universities where the practice of questioning science has become a legitimate intellectual option? It is less easy for a high school student than a scientist to refute an anti-scientific sociologist, historian, journalist or philosopher. Many believe that science is what their humanity and social science teachers have told them about science.

According to scholars, the STEM vocation crisis may cause the West to fall behind the Asian countries, leading to significant adverse economic effects for Europe, the United States, Australia, and so on. Above all, it may be the beginning of the decline of the West's cultural hegemony, which is based on rationality in decision-making (the decisions that Trump is making—about climate change, for example—are not rational).

This book is the result of trying to answer two questions that, I believe, are related: why irrationality is emerging in the media and in society; and why there is a decrease in the number of vocations using the most rational degrees: science, technology, engineering and mathematics.

This work is the by-product, not the main research objective, of my two stays at very different universities to find out firstly 'the state of the art' and secondly its evolution over time. I say it is a by-product because I've never had the funding to undertake it. In the natural sciences and engineering, academics believe that this problem is not one that they should fund research into since it concerns the social sciences and the humanities. On the other hand, in social sciences and humanities the academics regard it as a problem for science and engineering, areas that have far more economic resources. There is a huge lack of knowledge on the part of those from the social sciences and the humanities about how science works (and even what science is) and its importance in Western culture.

As an anecdote, when I applied for a start-up grant from the European Research Council to study the phenomenon of pseudoscience's and the media, one of the evaluators wrote, without blushing, that one of the problems with the project that I had proposed was that it put disciplines such as palmistry and tarot reading on the same level as homeopathy. Apparently this evaluator, who was from the social sciences, considered homeopathy to be a valid study. I keep in my mind the (negative) resolution of this project, because it shows that the level of natural science knowledge among the elite of European social scientists and those from the humanities is rather lower than that of mediaeval intellectuals about nature.

But there is something even worse: academics without the slightest scientific knowledge or scientific method have more and more power. Basically, this is because their disciplines are attracting more and more students and, in many

Western universities, if you have students, you have power (enrolments, votes, money…).

The first phase of this investigation was the result of my year-long stay at the London School of Economics (LSE) from 2005 to 2006. From that period a book emerged—*The Strangled Reason*—on the crisis of science in contemporary society (Penguin/Random House 2008, 2014, 2015), which had considerable impact on Spanish media because, among other elements, it warned that the same thing could be happening to the West as happened in Spain during the seventeenth century in the Counter-Reformation. The contempt and intolerance of science by the elite was one of the factors behind the decline of Spanish imperial power and the rise of those countries that did embrace the new scientific ideas and, from them, the newly created technology, the seed of the industrial revolution.

As Margaret Jacob, a professor of the history of science, so eloquently says:[3]

> The writings of Galileo, and his subsequent trial and condemnation, moved the new science into the forefront of learned discourse throughout Europe. Anyone attracted by the ideas of Copernicus, if living in Catholic as opposed to Protestant Europe, now had to think very carefully about how to announce that support. In France, for instance, the clerical opponents of papal intervention in the affairs of the French Church saw in Copernicanism a new weapon in their struggle; the Jesuits, with their strongly ultramontane (propapal) conception of religious authority, sided with the Inquisition's condemnation. In Protestant countries, on the other hand, support for Copernicanism could now be constructed as antipapal and hostile to the power of the Catholic clergy. What a splendid incentive for its adoption. The ideological linkage was to prove critical in creating the alliance between Protestantism and the new science. (Jacob 1988, 24–25)

The role that Spain, the Jesuits and the Catholic clergy played against science could be taken on these days by many humanities academics and social scientists, who are highly critical of science (like the Catholic clergy of the time), while countries such as Great Britain, Germany and the United States had the key to scientific and technological progress. As a result, the latter obtained scientific, and also cultural and linguistic, hegemony.

In fact, as we shall see below, young people in the West now no longer see science and technology as the best way to obtain real answers to the eternal questions of philosophy, about who we are and where we come from, but as a threat. The scarcity of science graduates in the West also means that these subjects are taught less and less every day and, furthermore, that those who teach them may not have been the best prepared. There is a lack of enthusiasm

---

[3] Margaret Jacob. (1988). *The Cultural Meaning of the Scientific Revolution*. New York: Alfred Knopf.

to proclaim that science and technology are the most elaborate and sublime cultural products of humanity. Science, for many young Westerners, is no longer the way to solve problems—from environmental to health—or seek the truth, but a cultural tool that is imperialist and totalizing. These ideas, if not similar, are as opposed to scientific thought as those that were put forward by the Spanish Inquisition to reject science in Hispanic culture until almost the twentieth century.

If Latin is no longer a relevant language, it is because scientists, starting with Galileo, preferred to write in their native language. In the nineteenth and early twentieth centuries, the best science was written in German and English. From World War II onwards, it has been English. Scientists' revenge on the clergy, who were so reluctant to accept science, was to stop using Latin. But the anti-scientific clergy, now converted into 'secular humanists'—that is, academics from the humanities and social scientists who are opposed to science and technology—have not yet lost the war.

What is more, in the West, they are winning battles. Their most important wins have been the establishment of an anti-scientific prejudice in parts of the university (in the departments of arts and social sciences), the loss of vocations, the domination of the media, praise for the irrational and mystical in the face of scientific fact and evidence and, above all, influence over the minds of young people. These youngsters will one day have power, yet their scientific training is at a lower level than that of educated people in the pre-scientific era.

The level of mathematics and physics possessed by cardinals, policy-makers and artists in the pre-scientific era is hard to find in such groupings in the West in our twenty-first century. It is hard enough to follow the discussions in the Renaissance about the fall of bodies, optics or planetary orbits. Nonetheless, many of today's academics from the humanities have dug the trench of Snow's 'two cultures' and consider that one can indeed be educated without knowing what matter is like, as if it had somehow been demonstrated that human beings and the whole environment around them are something other than matter.

Moreover, the influence of the media has made the pernicious culture of the celebrity endemic in the West (and in Westernized countries such as Japan, South Korea and Australia), wanting an immediate attainment of goals, disdaining studies with high academic standards and, above all, preferring disciplines in which, according to the postmodern philosopher Paul Feyerabend, 'anything goes'. Under the auspices of the French Theory of 'Deconstructivism' (Foucault, Derrida, Deleuze & Co.), teachers of humanities and social sciences are obliged to pass almost all of their students. They are afraid of students denouncing them, saying that they feel discriminated against because the points of view that they defend in their essays are not approved of by their teachers.

Chemical reactions or physical equations are just what they are. They are not discussed, and this implies that you have to study to know and approve them. In the humanities and social sciences, many interpretations and points of view are equally valid. And this makes them careers where almost everyone passes, as opposed to science, mathematics or engineering where there are no multiple answers or creative points of view. This implies that STEM careers require a culture of effort that in social or cultural studies is less vital, and many of which, such as communication, tend also to be very playful. This enables those groups that have effort as part of their cultural identity—such as Asians—to be more successful in universities and STEM studies.

My grant proposal for that first study was to document how, in two paradigmatic countries of Western culture with scientific traditions as disparate as Spain and Great Britain, there is a similar phenomenon of a lack of vocations. On that basis, using figures from other countries, it may be concluded that this claim is well founded. My proposal was also to examine another curious fact: while the number of science students is decreasing, the number of journalism, film or media studies students is increasing. And not only do the numbers increase. In these disciplines, thought is dominated by the postmodern philosophy that holds that everything has the same validity, that science is just another ideology (but with the added sin of being dangerous for society and totalitarian for thought) or that alternative facts are equally valid. What is even worse than their increasing enrolment is that those who study them do not later work on an assembly line or in a supermarket, restocking the shelves: on the contrary, they have infiltrated all key positions in Western society.

While scientists are locked up in their laboratories, the non-scientists of cultural studies or media studies work in the media, indoctrinating the masses, or are advisors to politicians—or themselves politicians—or even policy-makers in science and technology management agencies. That someone who believes that homeopathy deserves to be regarded differently from palmistry is now evaluating research projects in the European Union and putting their ideas into writing in their evaluations, in my opinion, is the clearest indicator that Europe is in decline and that rigorous thinking has disappeared from an important part of its academy.

The second part of this research was carried out at Harvard, in its History of Science Department. There I spent the academic year 2013–2014 researching this phenomenon in the United States, the leading country (so far) in science and technology. There I discovered that, in a way, in academia it was taboo or at least politically incorrect to address the decline of scientific vocations.

There were several factors. The first was like the elephant in the corner that nobody wanted to see: an excessive number of some minority races, such as Asians, in STEM compared to other minorities, such as African Americans and Hispanics. There was also concern about the shortage of women in science and engineering, but this was not a feature of social studies or the humanities, where they were in the majority.

I was very interested in the underground debate in the departments of history, philosophy or sociology of science in the United States about whether their academics should come from the natural sciences, or whether they need not have any idea about them yet should talk on them. The influence of French Theory—French postmodern philosophers such as Derrida, Lyotard, Foucault, and so on—which maintains that scientific interpretation is just another cultural form and should not be hegemonic, has had such a bad influence that it has led to Donald Trump being chosen as president, with his questioning of science and, above all, his defence of 'alternative facts' or alternative realities, beginning with his questioning of climate change. There are more and more researchers who approach science from the theory of literature or cultural studies and who are highly critical of scientific work.

Donald Trump is a great follower—perhaps unwittingly—of the French postmodern philosophers, who point out that everyone has their own reality and that no point of view—not even a scientific one or one of facts or data—should prevail over others. Authentic mediaeval mysticism and obscurantism were elevated to epic status by another of the great destroyers of Western science: the film and television industry. Fiction, in general, despises science and yet defends not only mediaeval aesthetics but magic in arguments that are presumed to be coherent. One of the latest examples is the television series, *Game of Thrones.*

My proposal is a question of exploring the hypothesis that the only feature that these Western and Westernized countries share is a culture that is dominated by the media. It proposes a tour of all the fields—which are possibly responsible for this decline in rationality—that connect science and the media. These range from the image of science and scientists that is portrayed by film, television and the press, to how media culture influences scientific journals themselves, which are also media, and how science is produced. Nor does it forget other channels of communication that are less massive yet highly influential in the media, such as literature and art, while at the same time asking why irrational thinking is advancing so much in our digital-age society.

This book aims to explore something that I consider very important: what humanities and social science graduates think about science. There is an academic struggle between science and the arts, and the sciences are suffering in certain areas. These areas include the paucity of students, politicians, intellectuals, actors, media experts, academic and social leaders with a solid scientific background. This is not the case, as we shall see, in countries such as China.

Why does cinema prefer the archetype of a mad, cold scientist? Why does Western television give more space to a sorcerer than a physicist? Why does a newspaper pay more attention to an athlete or actor than a researcher? Why does a graduate with a Master's degree in business earn more than a Doctor of Science? Why does *Nature* publish studies on topics such as how most of us turn to the right to kiss? Why do some people hate science so much? Why does the mystical *X-Files* triumph over portrayal of the scientific? Why don't law students study genetics, or film students mathematics or the physics of light?

An important difference from 2007, when I started the research, to my update for this book in 2018 is the presence of social networks. Trump's victory in 2016 shocked political scientists, sociologists and, above all, journalists: in university departments, how do we explain that a candidate won despite having all the influential media against him—from the *New York Times* to the *Washington Post*, the *Huffington Post*, CNN, NBC, ABC, MSNBC, *USA Today* and *Atlantic*, among many others? Rarely have Right, Left, and centre media united in an 'anti-someone' campaign such as this; yet that enemy won. Are the 'influential' media no longer influential?

Once he was elected, it was easy to argue why Trump had won, although some of us had been afraid of him for a while. We had warned that the decline of the traditional press and the rise of social networks, of search engines such as Google or of phenomena like Wikipedia—a real ideological battleground—have led to a fragmentation of the majority opinion states, which we now call media tribes. These often feed on what Damian Thompson calls counter-knowledge and lead to a credulity pandemic. 'Ideas that in their original and gross form flourished only in the outskirts of society today are seriously considered by even educated people in the Western world,'[4] says Thompson. This is the strategy of Trump's populist parties, who slipped into the campaign that American Muslims applauded the 9/11 attack or that Clinton was a follower of Satan. The same applies to Venezuelan Chavism, whose social networks and state media—VIVE, the channel of popular power—promoted the idea that the United States caused the 2010 Haiti earthquake.

These hoaxes and uncontested news are sent and forwarded via Facebook, Twitter and WhatsApp. But they also appear on Google, if we look for 'Haiti-Chávez earthquake' or 'Clinton-Satan'. Some entries confirm them and others deny them, and readers will always click on those that reinforce their previous

---

[4] Damian Thompson. (2008). *Counterknowledge: How we surrendered to conspiracy theories, quack medicine, bogus science and fake history.* London: Atlantic Books.

belief, thus they will no longer be alone in their extravagant conjecture. What social psychologists call 'confirmation bias', whereby we humans tend to consider only ideas that confirm our previous beliefs, has now become a major problem: while in the 1960s it was virtually impossible to find anti-vaccine information, it is now within everyone's reach.

Social media create echo chambers in which you receive only information that confirms your ideas, so that you become more radical (in favour of magic and against science and technology, for example). 'Many of us seem to feel trapped in a filter bubble created by the personalization algorithms owned by Facebook, Twitter, and Google. Echo chambers are obviously problematic; social discourse suffers when people have a narrow information base with little in common with one another (Hosangar 2016)'.[5] This confirmation bias also flourishes in Western universities, which, in theory, should prepare people to avoid such bias. At the moment, those from the humanities and social sciences who are critical of science and who analyse only its negative side are more likely to prosper in their academic careers, in which they will teach those students who do not learn chemistry, physics or biology at university level. These students may well end up as jurists, politicians, journalists, filmmakers, political scientists and philosophers.

Causing an intoxication with counter-knowledge, alternative facts or fake news is an unethical yet effective strategy: it is about having an army of trolls that sends hoaxes to the media tribes. These tribes send them back to their relatives and internet news portals—blogs, websites, and so on—without any deontological control, and well designed and linked to the search engines. We know of the best techniques for a hoax to work on the net: counter-hegemonic narratives; small distortions of real facts; contempt for scientific method; use of scientific terminology to support the story; and, in general, a panic to obtain the truth. This is based on the idea that it is impossible to lie if the truth is not known. Suddenly, the alternative not only has a good image but has found a place to flourish. And anti-science has become fashionable in the West.

Journalism was an 'invention' to fight hoaxes. Its aim—the same as that of modern science—has been to seek the truth and make it public, but both systems have been perverted. The search for truth leads us to a better knowledge of reality, which, in a system of freedoms, that is to say of decision-making—political, business, labour, medical, and so on—offers us a great competitive advantage. Along with the scientific method, journalism defines contemporary Western culture. To be more precise, even if this seems contradictory, it is necessary to generalize: it is not only journalism that defines

---

[5] Kartik Hosangar (*Wired*, 25 November 2016). 'Blame the echo chamber on Facebook. But blame yourself, too', https://www.wired.com/2016/11/facebook-echo-chamber/.

contemporary Western society but the entire system of mass communication, which has included cinema, radio and television since the twentieth century. Its result has been the rise of our celebrity culture and the ethics of the easy and the achievable with little effort. Until the emergence of mass media, public opinion was shaped in the churches.

In the twenty-first century, in addition to traditional media, the influence of social networks has to be reckoned with. A human is, above all, a social being who cannot stand isolation. Therefore, apart from a small group of recalcitrants, most of us will accept dominant thinking in order to be socially tolerated. This is what Noelle Neumann called the 'spiral of silence'. She studied this when television (mass media) already existed and stated that it helped to consolidate a climate of opinion. In these environments, polls can work, yet if there is even the slightest loophole to confirm our ideas, even if they are absurd, so as not to feel lonely they will be consolidated and we will form our own ideological tribe, now with social media support.

The theory of the spiral of silence also states that the socio-economic and cultural elite—academics, artists, journalists, politicians, writers, bankers—dares to speak out with non-dominant thoughts and that, ultimately, these permeate society. This has happened since the acceptance of science versus religion, right up to the defence of civil rights. The traditional and serious media have played an important role in social progress: it is true that they are not anti-capitalist, as some would like, yet neither are they racist, nor do they defend dictatorships or pseudosciences. Up to now, they have been of great value.

The media—and the establishment among those who were politicians and, in a prominent place, the university—established what is 'politically correct': globalization is positive, immigration is valuable, xenophobia is repugnant… And those who do not agree, according to the laws of the spiral of silence, are silent. Hillary Clinton had the support of dominant thinking opinion-formers: university professors, artists and the media. That's why, in classical theory, it was unthinkable for Trump to win.

But that changed. Now, controversial television stars such as Trump, and politically incorrect ones, can also be mass media. On the day that Trump won the election, he had 13.5 million followers on Twitter. That was more than the *Wall Street Journal*, considered to be the most influential newspaper in the world, or the *Washington Post* had at the time. So what, then, is Trump, or Lady Gaga: a source or a means of mass communication, in itself? Each is both things, and this dismantles much of the current communication theory. But Trump had an advantage over Clinton: he handled Twitter (although his team forbade him to, in the final few days), while Hillary used community managers.

On the internet, there are tribes for all tastes: those who believe that aliens have manipulated our DNA, that vaccines are dangerous or that AIDS was created by the CIA. The information has been tribalized, and is no longer mass. And those tribes that are informed by the algorithms that decide on the news that will make them feel good (Facebook, Twitter, and so on) live in parallel worlds, in which they do not listen to other but where their xenophobic, anti-scientific or anti-system ideas are the dominant ideas in their microcosm. That's why the results—from Brexit to Trump—seem incredible to many (those who don't belong to that tribe). They don't dare to confess their ideas in opinion polls, but they do in a secret vote.

Very interesting research published in *Science* has shown that fake news spreads faster on the internet and social networks than real news.[6] After analysing a data set of rumour cascades on Twitter from 2006 to 2017 (about 126,000 rumours were spread by ~3 million people), it was determined that: 'falsehood diffused significantly farther, faster, deeper, and more broadly than the truth in all categories of information, and the effects were more pronounced for false political news than for false news about terrorism, natural disasters, science, urban legends, or financial information' (Vosoughi et al. 2018). Researchers found that 'false news was more novel than true news, which suggests that people were more likely to share novel information. Whereas false stories inspired fear, disgust, and surprise in replies, true stories inspired anticipation, sadness, joy, and trust' (Vosoughi et al. 2018). It continued: 'contrary to conventional wisdom, robots accelerated the spread of true and false news at the same rate, implying that false news spreads more than the truth because humans, not robots, are more likely to spread it.'

Although it was good to prove this empirically, we have known since ancient times that rumour is more effective than the truth. The book *Fama, la historia del rumor* ('Fame, Rumour's History') by the German journalist Hans-Joachim Neubauer (2013, Madrid: Siruela), the role and power of rumour throughout history explains this perfectly. It would seem that a technological and scientific society is safe from rumour, but Neubauer warns that loose talk has found shelter in new forms of communication.[7] These tools contribute to its dissemination, understanding it as a voice 'as relevant as it is impossible to corroborate', which spreads the content autonomously and rapidly.

---

[6] Soroush Vosoughi, Deb Roy and Sinan Ara. (2018). 'The spread of true and false news online'. *Science*, 359, 6380, 1146-1151.

[7] Sergio Delgado. (2013). 'Internet, la nueva era del rumor. Hans Joachim Neubauer aborda la problemática de la habladuría a lo largo de los siglos. [Internet, the new era of rumor. Hans Joachim Neubauer addresses the problem of gossip over the centuries]' *El País* 23 March 2013).

In Neubauer's opinion, the internet and new technologies promote two vital aspects: reaching a large group of people; and appealing to strong feelings such as fear, hatred or uncertainty—something always topical in the face of a crisis. 'The Internet is very fast and any denial is always late. We are facing a new era of rumor,' Neubauer said in an interview with the Spanish newspaper *El País*.[8] In a Shakespeare extract quoted by Neubauer, rumour is likened to as 'a flute. Guesswork, suspicion, and speculation are the breath that makes it sound, and it's so easy to play that even the common masses—that dim monster with innumerable heads, forever clamoring and wavering—can play it'.

In *The Aeneid*, the Roman poet Virgil (70–19 BC) describes fame as a Greek divinity, a messenger of Jupiter, yet in heaven considered infamous. It is the 'public voice' or rumour that conveys all sorts of truth and slander, and spreads around the world at great speed. The Greeks worshipped fame because they knew the power of this goddess, 'who grows as she spreads'.

Neubauer concluded his interview in *El País* with a highly relevant statement:

> (rumour) is a question of power, it allows anyone to be part of a moral discussion without being the person who gives his opinion. The talk focuses on the secret, on the hidden, which is usually a negative thing. People hide their dark side from others (…) Telling something puts you in the role of someone who knows what's behind it, you've discovered something. Rumours like to find out something, it's sexy and everybody wants to have it.

Some journalism professors celebrated the emergence of the internet of alternative media, as opposed to the traditional: they believed that the new forms were going to be all Leftist. They did not foresee that they could also be both Right wing and irrational. Alternative means that it is not massive, and it does not mean a specific ideological or political orientation. Trump won thanks to the Alt-Right. Mainstream media aspire to have the widest possible audience: therefore, they will be neither anti-capitalist nor xenophobic. It is true that, as we will see in this book, from the 1960s onwards mainstream media in the West began to be highly critical of science. To this is now added social networking and tribalization, or ideological hooliganism.

With the tribalization of media and audiences, the spiral of silence no longer works: there are audiences for all tastes, and these that feed off each other, increasing uncertainty. The internet has favoured unscientific information: depending on how you ask Google about vaccines, you will be proffered sites that are to a greater or lesser extent pro- or anti-vaccine, not to mention the

---

[8]Delgado. (2013). Op. cit.

effect of having 'friends' whom you have accepted on Facebook from your anti-vaccine group. If this can happen with something that is scientifically proven, what could happen with something that is ideologically questionable?

If we Google 'Donald Trump', the first entry will be from Wikipedia. For neutral terms—such as 'proton' or 'Romanesque art'—its definitions may be valid, but for controversial ones such as transgenics or homeopathy, let alone biographies on such as Hugo Chávez or Donald Trump, an army of defenders and detractors are continually editing and republishing. As we journalists have known for a long time, neutrality is a misrepresentation: one cannot give two antagonistic versions of something. One can give them the one that is true—and, while the entry for radioactivity in the *Encyclopaedia Britannica* was written by Marie Curie (Nobel Prize winner for discovering this physical property), we don't know who wrote the Wikipedia one or what moved them to write it.

Traditional media were not overtly unscientific, but they were highly critical of science, technology and the image of scientists, especially since the 1960s, when this image awakened an anti-enlightened spirit in many humanities departments that trained students who later were to be become communicators. To this, we must add diffusion by the internet, where truth coexists with hoax, and neither science, anti-science nor reason can triumph—rather, what succeeds is what is most sent, most linked or most 'liked'. The West—liberalism—has always defended online freedom, in contrast to countries such as China. Behind Brexit or Trump, there is debate about the freedom to broadcast fake news and whether this is protected by freedom of expression.

I will try to relate these approaches to the crisis of scientific vocations. I am referring not only to a shortage of scientists or engineers—particularly worrying in the case of women—but to something much more serious: an absence of even a minimal scientific culture in many in power in the West, whether political, economic, media, legal.

This did not happen before World War II. It is not a question of wanting more scientists, engineers or mathematicians to work in their respective areas, which is also necessary, but of them occupying many more positions of responsibility in the political, economic, media and social spheres, as happened in the Enlightenment when scientists imposed their criteria on those who wanted religious, magical, mystical or mythological thought.

I have chosen to present this research in the form of an informative essay, because I believe that the issue is important enough not to be simply left in the drawer of some educational leader or university evaluator. I would like to create debate, even controversy, to make this issue a priority in social and political discussion. It's a rehearsal. I do not seek to be right, but to prompt another look at what is happening in science and in Western culture in general. In my view, the first step in solving a problem is to make more and more people aware of it.

This book cannot move scientists to lead society, nor even science alone. It cannot make science faculties full again and arrange for young people to study science rather than journalism, sociology or film. But it can try to provide an explanation of why this trend occurs and where it may lead in today's civilization. It can ease the anxiety of scientists by at least providing an explanation of why things happen, which is the basis of science.

Science, the rational explanation of facts, is the most fascinating journey that the human species has ever made. It is an unnatural journey, because the human brain likes emotions and rumours. But science works. The economy, polls, political scientists and sociologists do not always manage that. If the severity with which philosophers or sociologists have scrutinized physics were to be applied to economics or sociology, let alone media studies or cultural studies, these degrees would not be taught at university, just as astrology or palmistry are no longer taught—or, maybe they could be taught, provided that the students contribute.

Scientific method does not tell us the truth, but it is the best way to get to the truth. The rest is just opinion, story or myth. These are important, no doubt, but not comparable to arriving at the truth. Is this way of finding answers in crisis? Is rationality in crisis? I will discuss these issues in the following pages.

2

Science in the Twenty-First Century

The paradox that defines today's world is that every day we depend more and more on science and technology, every day science knows more and more and explains the world to us better, but also every day people feel less and less appreciation for it. Every kid knows every player in the English football league. They read and reread their biographies. The Western media regard them as heroes. But no teenager knows anything about a scientist. They despise them: they think their lives are boring and disappointing compared to those of singers or footballers. Scientific vocations are alarmingly being lost in the West. The gap between what science knows and what 'educated' people know about science is widening to such an extent that the population considers it impossible to follow their progress and literally turns their backs on it. There are people in high places in today's society who still think that it is the Sun that revolves around the Earth or that dinosaurs and man lived at the same time. This contempt for science is a relatively new and unstoppable phenomenon in the Western world, which we do not know where it will lead us. But in any case we must urgently find the origin.

The above ideas are not mine; they are a paraphrased summary of ideas on two similar concerns that were imparted to me by two contrasting characters in interview during my time as a journalist: Sir Martin Rees and David Filkin. They had a common bond: a passion for science.[1]

Their reflections on the decline of science could not be published at the time due to restricted space and current events. Journalistically, perhaps, they were

---

[1] The views expressed here are also broadly presented in my book, *La ciencia a través del periodismo*. (2013). Madrid: Nivola. This book is a summary of the articles published during my time as a science reporter, such as those in the scientific part of *El Mundo*.

© Springer Nature Switzerland AG 2019
C. Elías, *Science on the Ropes*, https://doi.org/10.1007/978-3-030-12978-1_2

irrelevant, but the matter seemed to me to be of great importance to Western culture and, at least, worthy of in-depth investigation.

Sir Martin Rees (UK, 1942) is one of our most eminent contemporary scientists. A Royal Astronomer, Knight of the British Empire and Professor Emeritus of Astrophysics at the University of Cambridge, he has suggested interesting theories such as parallel universes. Rees is also an excellent scientific disseminator, and in 2005 he was elected President of the Royal Society (2005–2010).

David Filkin (UK, 1942–2018) was the director of the BBC's science programmes for forty years. He is considered one of the leading figures in the dissemination of science in the media, and these programmes revolutionized the way in which science was transmitted, winning audiences of millions.

When I spoke to them in 2001 and 2002, both had a long history behind them and both were concerned at the growing contempt for science. I must admit that this had passed me by: I had thought that the contempt was restricted to Latin countries, countries like Spain that have historically turned their backs on science. Spain failed to be part of the scientific revolution of the seventeenth, eighteenth and nineteenth centuries, and this still weighs it down like a millstone around its neck. The fact that this reflection came from two Brits such as Rees and Filkin was worrying. Both belong to the country, the United Kingdom, that has contributed the most in the history of science, a country in which the industrial revolution began, founded on technological and scientific elements; to a society that, for centuries, has hosted the most prestigious scientific institution in the world: the Royal Society, presided over, among others, by Isaac Newton.

How could this phenomenon of contempt for science also occur in Great Britain? I could not understand it, because in Great Britain, unlike in Spain, everything that had to do with science and its dissemination had been done well. I kept thinking about the Royal Society. It had been the engine that turned something that was no more than a gentleman's pastime, science, into the profession that would change the world. It was founded in London in 1662, although since 1645 there had already been meetings of a group of people interested in science.

One of the Royal Society's successes was that it introduced a groundbreaking custom: the publication of periodic journals to communicate all scientific results. Faced with the hermeticism of the Egyptian 'wise men' or mediaeval alchemists, everything suddenly changed and the greatest value of t was now to communicate it, to share it and to discuss it. This was an impressive advance. Those Royal Society publications were the forerunners of the current 'impact journals' that we will discuss later, and with this background it is no coincidence that the scientific (generalist) journal with the greatest impact is the British journal *Nature*, which we will also analyse in this book. The Royal Society also favoured conferences on a particular subject—what today would be called congresses—as well as conferences and information books. If, in the country that produced the Royal Society, science was losing prestige, what was like in the rest of the world?

For the reader to understand the contrast with Spain, I will mention that a similar scientific society, the Royal Academy of Sciences, was not created in Spain until well into the nineteenth century, 1847, during the reign of Isabella II of Spain. This meant two centuries of delay at a time that was foundational to the history of science.

When the Royal Society is mentioned in England, everyone understands that we are talking about science; by contrast, in Spain, when the Royal Academy is mentioned everyone understands the opposite: that we are talking about literature and the arts. It is necessary to specify if one is speaking of the Royal Academy of Sciences, because the Real Academia, without any other term, in Spain refers to the Royal Academy of Spanish Language. Perhaps Spanish is a language only of literature, while English is a language of both literature and science, so perhaps English is more important than Spanish. In any case, the differences between the two countries, in terms of science, are obvious.

It should be noted that during the seventeenth, eighteenth and early nineteenth centuries in Spain the crown opposed the creation of an academy of sciences, although an academy of humanities, thus Royal National Academies of language, history and fine arts were constituted, but not of science. The science historian José Manuel Sánchez Ron stresses: 'Here, too, the arts and literature defeated the sciences.'[2] Numerous examples can be cited of this triumph in Spain of literature and arts over the natural sciences. One of the most symbolic reminders is that the current Prado Museum building was actually designed to house an academy of sciences. Then the crown decided that it was more convenient to use it to house something 'less revolutionary' for the people and for the monarchy, so the collection of portraits of the country's kings was placed there.

Sánchez Ron recalls that the now well-developed Newtonian science, developed and highly mathematicized, now stripped of its dark initial mathematical clothing, was compelled to enter Spain through its navy, not through any scientists or natural philosophers who were determined to understand, in the final analysis, simply why nature works in the way it does. 'It was not the best way, but it was a way', he adds. In my opinion, the fact that scientific advances were known in Spain only through the military would deprive civil society of

---

[2]José Manuel Sánchez Ron. (1999). *Cincel, martillo y piedra*. Madrid: Taurus (p. 44).

knowledge, and this would later be translated into a lack of appreciation of scientific disciplines in Latin countries. It is condensed in the famous phrase of the Spanish philosopher Miguel de Unamuno: 'Let them invent others.' In July 1906, Unamuno, Rector of the University of Salamanca, the oldest in Spain at 800 years, published an essay, '*El portico del templo*' ('The Door to the Temple'), where, in a dialogue, its characters reflect:

> ROMAN: What have we [the Spanish] invented? And what does that do to us? In this way we have spared ourselves the effort and eagerness of having to invent, and the spirit is fresher and fresher…. So, let them invent, and we will take advantage of their inventions. Well, I trust and hope that you will be convinced, as I am, that electric light shines as well here as it did where it was invented. SABINO: Perhaps better.

This attitude was widely criticized by the only openly pro-scientific Spanish philosopher, Ortega y Gasset (1883–1955). Ortega was convinced that the greatness of Europe was due to science. At the beginning of the twentieth century (1908) he wrote:

> Europe = science; everything else is common to it with the rest of the planet. (…) If China travels, it exists and grows today as it did ten centuries or twenty centuries ago, if it soon reached a higher degree of civilization than Greece and stopped there, it was because it lacked science. (…) If we believe that Europe is 'science', we will have to symbolize Spain in the 'unconsciousness', a terrible secret disease that when it infects a people turns it into one of the slums of the world. (Ortega, *El Imparcial*, 27/07/1908, in Ortega y Gasset 1966, 99–104)[3]

Likewise, Ortega observed numerous obstacles to Spain becoming a truly European and twentieth-century country. Perhaps the most serious, in his opinion, was its backwards education system, from the configuration of the university right through to its curricula. With regard to these, Ortega noted that Spanish culture and education lack a training in science—in reality, it is not that this training is deficient but that it is totally absent—and that this is a burden on other cultural and educational areas:

> The Spanish problem is certainly a pedagogical problem, but the genuine thing, the characteristic of our pedagogical problem, is that we first need to educate a few men of science, to raise even a shadow of scientific concerns. (Ortega, *El Imparcial* 27 July 1908, in Ortega y Gasset 1966, 103).

---

[3] J. Ortega y Gasset (1966). *Obras Completas*, vol. I (1902–1916). Revista de Occidente (7th edn). Madrid. All translation are the author's own.

Unfortunately, the cultural influence that Spain exerted during the sixteenth, seventeenth, eighteenth and nineteenth centuries in America caused scientific activity and the appreciation of science to be also weak in those countries of that continent where Spain exercised its political and cultural power, giving rise to the current technological and economic gap. The German sociologist Max Weber had already demonstrated in 1905, in his book, *The Protestant Ethic and the Spirit of Capitalism* ('Die Protestantische Ethik und der "Geist" des Kapitalismus'), that societies that embraced modern science developed two phenomena that helped them to succeed economically: a process of demagnification (the extolling of matter, the transformation of matter (chemistry) and the material against the mystical and spiritual), together with a rational method (of mathematical logic) in decision-making (political, mercantile, intellectual).

Weber observed that in countries like Germany or France, where Catholics and Protestants coexisted, the latter were not only richer but made more rational decisions and had professions that were more related to the study and transformation of matter: chemistry, engineering, physics. These new professionals, according to Weber, were what made a country powerful, not the contemplative life of Catholic friars and nuns. This is not the case today. Many of the elite Protestant youth prefer art or social and humanities disciplines to science and technology. In this sense, there are hardly any differences between Catholic and Protestant countries.

Still at the end of the nineteenth century, Spanish intellectuals defended the Inquisition. Menéndez Pelayo (1856–1912), with a Spanish public university named after him and a statue that presides over the National Library of Spain, wrote in 1880 in his book *La ciencia española* ('Spanish Science'):[4]

It is a great honour for me that De la Revilla[5] calls me a neo-Catholic, inquisitorial, defender of barbaric institutions and other nice things. I am Catholic, not new or old, but Catholic to the bone, like my parents and grandparents, and like all of historic Spain, fertile in saints, heroes and sages far more than modern Spain. I am a Catholic, a Roman apostolic without deviation or subterfuge, without making any concession to impiety or heterodoxy.... I consider it a most honourable coat of arms for our country if it did not have heresy rooted in it during the 16th century, and I understand and applaud it, and I even bless the Inquisition as a formula of thought of unity that rules and governs national life throughout the centuries, as a product of the genuine spirit of the Spanish people, and not as an oppressor of it but in very few individuals and very rare occasions. I deny those supposed persecutions of science, that annulment of intellectual activ-

---

[4]M. Menéndez Pelayo. (1880). *La ciencia española* (vol. LVIII, national edition of the *Complete Works of Menéndez Pelayo* (ed. Rafael de Balbín). Santander: CSIC, 1953.

[5]Manuel de la Revilla (1846–1881) was a Spanish journalist and philosopher who was a supporter of the sciences. He translated Descartes into Spanish.

ity, and all those atrocities that routinely and without foundation are repeated, and I have bad taste, fashionable backwardness and write laborious cogitations like those of De la Revilla… I believe that true civilization is within Catholicism. (Menéndez Pelayo 1880, in 1953, 200–201).

And in the middle of the twentieth century, in the opening ceremony of the Consejo Superior de Investigaciones Científicas[6] (the most important scientific institution in Spain, then and now), it was stated:

> We want a Catholic science, that is, a science which, because it is subject to reason, supreme of the universe, because it is harmonized with faith, in the true light which illuminates every man who comes into this world, reaches its purest universal note. Therefore, at this time, we are liquidating all the scientific heresies that have dried up and exhausted the channels of our national genius and have plunged us into atony and decadence. Our science, the Spanish science of our Empire, the one that wishes to promote the new Spain with maximum vigour, repudiates the Kantian thesis of absolute rationalism. (…) The imperial tree of Spanish science grew lush in the garden of Catholicity and was not disdained to be the essential fiber and nerve of its trunk, the sacred and divine science, from whose juice all the thick branches were nourished in unison. (…) Our present science – in connection with that which in the past centuries defined us as a nation and as an Empire – wants to be above all Catholic. (…) In vain is the science that does not aspire to God. (Ibáñez 1940. Consejo Superior de Investigación Científicas (CSIC) 1942, 32–33)

In Spain, the Counter-Reformation—that is, the Catholic fundamentalism initiated after the Council of Trent in 1545—stifled the promising intellectual development that had begun in Toledo, with its School of Translators (in the twelfth and thirteenth centuries). It is revealing that here not only were mathematics and medicine translated but the astronomical data used by Copernicus (1473–1543), from the Alphonsine tables, the leading astronomy funded by the King of Castile, Alfonso X the Wise (1221–1284), the first tables since Ptolemy (second century AD).

The University of Salamanca was founded in the thirteenth century, yet it has not made any relevant contribution to the history of science, because it turned its back on it. It preferred to defend—and teach—Aristotle against Galileo. And that has been the downfall of Spanish culture. It has not produced Nobel Prize scientists, even among its ex-students and professors. This is still taboo in Spain, and I wonder if the mystical and anti-scientific philosophy from Salamanca is not attacking the entirety of Western universities.

---

[6]Spanish National Research Council.

Let's compare the University of Salamanca to the University of Cambridge. This is hard for a Spanish academic, such as me, but maybe that is why I'm in the best position to do it. In 2018, the University of Salamanca celebrated its 800 years with great pomp. More than 800 rectors attended from all over Latin America. The Rector of Salamanca, Ricardo Rivero, affirmed that 'the roots of the university system of Latin America are in Salamanca, because with our statutes and following the Salamanca model, the first American universities were founded, such as Mexico, Santo Domingo, Lima, Cordoba...'.[7] Zero self-criticism. But, of course, Rivero was a professor of administrative law, a Spanish specialty whose object is to produce complex bureaucratic regulations, not of chemistry or biology. The worst thing that could have happened to Latin America was to have had universities based on the Salamanca model, whereby more importance is given to law than to physics or chemistry. That is basis of why North America is rich and South America, where the elite study more law than engineering, is poor.

The University of Cambridge celebrated its 800th anniversary in 2009. What would have happened if the University of Salamanca had not been founded? We would live in exactly the same way: we would know the same physical laws, synthetic chemistry, electricity, magnetism, antibiotics, drugs, mobile phones, the internet or genetics. What if there had been no University of Cambridge? It was here that Francis Bacon, the father of empiricism, and Gilbert who, with Galileo, was considered to be the first modern scientist for his studies of magnetism, were formed. If only for Newton and his law of universal gravitation, Cambridge would have been worth it by now. A poet, Alexander Pope (1688–1744), wrote the most sublime epitaph dedicated to a human:

Nature and nature's laws lay hid in night;
God said, 'Let Newton be!' and all was light.

In Cambridge, Darwin studied, who discovered how we evolved; and Rutherford, who gave us an atomic model built from empiricism; Watson and Crick, who discovered DNA and, with it, modern genetics; and Turing, the father of computer science, from an immense list.

Why was this not in Spain, which in the seventeenth century was more powerful than England? Because in Spain it was believed that culture lay in scholastic philosophy, literature, law, history or theology, while at Cam-

---

[7] 'The fourth International Meeting of University Rectors will make Salamanca the world capital of higher education', ran the headline of the press release distributed to the media by the press office of the University of Salamanca. It can be consulted on the website of the University of Salamanca (accessed 15 September 2018 at Saladeprensa.usal.es/node/112753.

bridge chemistry, physics, mathematics and biology were able to compete—and win—against studies in theology or law.

In 800 years, no one with any affiliation to Salamanca, either as a teacher or student, has won a Nobel Prize. According to its website in 2018, the members of the University of Cambridge had won 98 Nobel Prizes: 32 in physics, 24 in chemistry, 26 in medicine and 11 in economics (in addition to three in literature and two in peace, which are more political awards and do not count towards university rankings). Trinity College alone had 32 Nobel Prizes in 2017.

Spain's adherence to the Council of Trent and its religious orthodoxy is an excellent example of how a dominant country (Spain had the world's most important empire in the sixteenth century) can be scientifically and culturally retarded and, in the process, lose economic, political and intellectual power. Let us not forget that, in Trent, under the auspices of Spain's King Philip II, it was decreed that the Bible was a source of scientific data and that any statement contained in it should be taken as scientifically true. To differentiate itself from Rome—and from Trent—the Protestant religion embraced science. According to the science sociologist Robert Merton, this correlation of deep interest in religion and science—which might have been incongruous in later times—was entirely consistent with the incisive Protestant ethic.[8]

This can be seen, for example, in England during the reign of Elizabeth I (1558–1603), when Elizabeth I, the daughter of Edward VIII and his second wife Anne Boleyn, became an icon: 'The Virgin Queen' (Philip II of Spain offered his hand in marriage but Elizabeth I refused). Elizabeth I distanced herself from Rome, consolidating the Church of England and promoting culture: the golden age of drama, from Shakespeare to Christopher Marlowe. Above all else, she encouraged the study of nature (considered a pagan act by some Catholic theorists).

During her reign, she named as her personal physician William Gilbert (1544–1603), who rivals Galileo in being considered the first modern scientist. Gilbert was the first to study the nature of electricity and magnetism, publishing his findings in *De Magnete* (1600). Elizabeth knighted him. Another of her councillors was John Dee (1527–1608), the great mathematician and astronomer, who also practised alchemy at a time when science and magic were still one. In fact, the subsequent generation—people such as Newton (1642–1727)—also practised both science and alchemy. Knighting a scientist was a true statement of intent. In comparison, during the reign of Philip II, the great physician Vesalius (author of *De Humani Corporis Fabrica*) was forced to

---

[8]Merton, Robert K (1938). *Science, Technology & Society in Seventeenth-Century England.* Bruges: The Saint Catherine Press.

flee Spain in order to escape the Inquisition, which wanted to burn him at the stake for practising dissection.

Newton's funeral on 25 December 1642 at Westminster Abbey (where royalty is buried) was a mass event with the rank of a state funeral. And this was 1642! The French philosopher Voltaire (1694–1778), one of the promoters of the Enlightenment, was then in exile in London. He attended the funeral and was amazed. His words were: 'England honoured a mathematician in the same way that subjects of other nations honoured a king.' And England, not as an empire but as the source of universal culture and language, took off.

There are other scientists buried in Westminster, from Darwin to Lord Kelvin. For a visitor such as myself, from a Catholic country, it is admirable that here are outstanding figures of modern atomic physics. They include scientists such as Thompson (discoverer of the electron and with his own atomic model) and Rutherford (who discovered that an atom's positive charge and almost all its mass are in its nucleus, and whose model was almost definitive—until quantum mechanics). Faraday has a commemorative plaque and, in 2018, Stephen Hawking's ashes were deposited here. There are also poets, musicians, painters and novelists. Westminster Abbey is not arranged by culture but by profession, and there are scientists who are honoured above writers, musicians and painters.

To put it very simply, it could be said that, after the Counter-Reformation, if you were Catholic you could not defend science. The Spanish example is revealing for those countries that now wish to embrace religious fundamentalism or ideas opposed to science and technology. The links between emerging modern science, the first industrial revolution and the market economy have been widely analysed: from Weber in 1905 to Merton in 1938, to Jacob in 1988.

Weber, in his influential *The Protestant Ethic and the Spirit of Capitalism* (1905), dwells on the figure of Benjamin Franklin (1706–1790), a prominent scientist who is considered to be one of the founding fathers of the United States and one of the ideologues of its Constitution. The fact that a country's constitution and foundation are the work of a scientist is of the greatest importance to its technological and economic development. The fact that there are few scientists among today's US politicians says a great deal about where that country is headed.

Spain suffered enormous ravages by following Catholic orthodoxy: Philip II forbade Castilians, both religious and lay, from going to study or teach outside Spain (1559) and barred French professors from teaching in Catalonia (1568) (Sánchez Ron 1999). The Royal Academy of Sciences was not founded in Spain until 1847, under the reign of Isabella II who, only a few years earlier, in 1834, had abolished the Spanish Inquisition. By contrast, its British equivalent, the Royal Society, was founded earlier, in 1662. That's almost a 200-year delay, two centuries that were foundational to the history of the natural sciences and their consolidation in the countries where they flourished. In Great Britain in the eighteenth and early nineteenth centuries, chemists such as Joseph Priestley (1733–1804), Humphry Davy (1778–1829) and Faraday (1791–1867) were almost public idols (Golinski 1992).[9] They were portrayed by the best painters. The eighteenth century was crucial to science, because it saw the emergence of the concept of the professional scientist, the popularization of science and the difference between science for the elite (researchers) and science to be explained, taught and communicated to the general public (Holmes 2010).[10]

Many of the fathers of modern science are British: not only geniuses such as Newton or Darwin but the more modest yet indispensable scientists (who taught secondary education) such as Faraday, Maxwell, Thompson, Dalton and Hooke, among many others. Another way to see clearly the differences between Spain and Great Britain is the number of Nobel Prize winners. Spain—in 2017—did not have a Nobel Prize in basic sciences (physics and chemistry). However, Great Britain had 53 (26 in physics and 27 in chemistry). In an applied science such as medicine, Spain has only one Nobel Prize winner, Ramón y Cajal, whose centenary was celebrated in 2006. Severo Ochoa was awarded one in 1959, but he was a US citizen and worked there with US resources, so he works for that country. He never wanted to retain his Spanish nationality. Britain has 32 Nobel Prizes in medicine. As I have mentioned, their trend—of hatred/love for science—is not only intrinsically Spanish or British: both countries have exported their trend to their colonies.

It is telling that, apart from the United States—the former British colony, and now with more than 200 scientific Nobel laureates—the former British colonies that are still relatively small countries, such as Australia, with 24 million inhabitants, have nine scientific Nobel laureates. On the Hispanic side, Mexico (a former Spanish colony and therefore a recipient of its anti-scientific culture) has 128 million inhabitants yet only one Nobel Prize winner in science—who, by the way, works for the United States.

The historian Eric Hobsbawm wrote in 1994 (in his book *The Age of Extremes: the short twentieth century, 1914–1991*[11])

[9]J. Golinski. (1992). *Science as Public Culture: Chemistry and Enlightenment in Britain, 1760–1820.* Cambridge University Press. New York.

[10]R. Holmes. (2010). *The Age of Wonders.* Vintage Books. New York.

[11]Eric Hobsbawm (1994, edition of 1995). *The Age of Extremes: the short twentieth century, 1914–1991.* London: Abacus.

Euro-centric science ended in the twentieth century. The Age of Catastrophe, and especially the temporary triumph of fascism, transferred its centre of gravity to the USA, where it has remained. Between 1900 arid 1933 only seven science Nobel Prizes were awarded to the USA, but between 1933 and 1970 seventy-seven. (…) At the same time the rise of non-European scientists, especially those from East Asia and the Indian subcontinent, was striking. Before the end of the Second World War only one Asian had won a science Nobel prize (C. Raman in physics, 1930). Since 1946 such prizes have been awarded to more than ten workers with obviously Japanese, Chinese, Indian and Pakistani names, and this dearly under-estimates the rise of Asian science as much as the pre-1933 record under-estimated the rise of US science. However, at the end of the century [XX] there were still parts of the world which generated notably few scientists in absolute terms and even more markedly in relative terms, e.g. most of Africa and Latin America. (Hobsbawm 1994).

The eternal question that the former Spanish colonies ask themselves, and that we in Spain dare not ask, is always the same: if Hispanic America had been conquered by Britain and, therefore, adopted its culture of love for and interest in the natural sciences, would the Latin American countries now be as rich and influential as their neighbours, the United States and Canada? If the Philippines had been conquered by Britain rather than Spain, would their standard of living now be similar to that of New Zealand? Why is India, a former British colony, so interested in computers, robotics or chemistry?

The answers are complex and are not the subject of this book. However, the British and the North Americans are very clear that the success of their culture from the eighteenth century onwards has much to do with their view that science is hugely important in their intellectual preparation. They believe that the fall of the Spanish Empire and of Latin culture globally were due in large part to the lack of appreciation and knowledge of the natural sciences among its ruling classes—whether political, economic or intellectual. And the current alarm—which is also present in Germany, together with Great Britain and the United States the country with the most Nobel scientists—is whether the English-speaking world is 'Spanishizing'. This is what happened in the Spanish Counter-Reformation, where scholasticism and criticism of science were widespread until well into the nineteenth and even twentieth centuries. Is postmodern philosophy descended from scholasticism, which even after centuries of reflection has arrived at nothing productive? Could it be that the sociologists, philosophers, political scientists and US/UK communicators, the new versions of the Jesuits who did so much damage to Spain, could now be doing it to the US/UK world?

After my interviews with Rees and Filkin in 2001, what I did not understand was that, according to them, at the beginning of the twenty-first century in English-speaking countries, too, there was a phenomenon of disinterest in—even contempt for—the natural sciences. If we started from totally different cultural conditions with respect to science, why had we reached a similar situation in both countries? Could it be a trend across the entire West?

Science and Technology with a Magic Wand

Perhaps because of my professional training as a journalist, I came to the conclusion that the only 'cultural' product that united these Latin and English-speaking countries was the existence of the media, along with a related element, their influence on society. However, I was not sure about the hypothesis: that the loss of the cultural impact of science in the West in recent years is due to the media.

It was reading an article by Eco (1932–2016) that convinced me.[12] Eco was an Italian who was deeply familiar with the Middle Ages and also with Western culture as a whole and is, above all, cosmopolitan. He argued that, curiously enough, magic is more and more present in the media: 'It may seem strange that this magical mentality survives in our era, but if we look around us, it appears triumphant everywhere. Today we are witnessing the rebirth of satanic sects, of syncretistic rites that we cultural anthropologists used to study in the Brazilian *favelas*' (Eco 2002, 13). According to Eco, in the twenty-first century the magic wand has been transformed into the buttons on our devices. Technology has replaced magic, as people do not understand or care about how, for example, a photocopier works. All they know is that they press the button and the miracle—or the magic fact—takes place, the instantaneous appearance of an exact reproduction of the original:

> What was magic, what has it been for centuries and what is it still today, even if under a false appearance? The presumption that we can go directly from the cause to an effect by means of a short circuit, without completing the intermediate steps. For example, you stick a pin in the doll of an enemy and get his death; you pronounce a formula and are all of a sudden able to convert iron into gold; you call the angels and send a message through them. Magic ignores the long chain of causes and effects and, especially, does not bother to find out, trial after trial, if there is any relation between cause and effect. (Eco 2002)

---

[12]Umberto Eco. (2002). 'El mago y el científico' ('The magician and the scientist'). *El País,* 15 December 2002, 13.

And, as Umberto Eco concludes, the confidence, the hope in magic, has not faded in the slightest with the arrival of experimental science: 'The desire for simultaneity between cause and effect has been transferred to technology' (Eco 2002, 13).

Rational thinking, of which the Greeks were the greatest exponents, led us to wonder about the relationship between cause and effect. Socratic debate (or the Socratic method) is the best tool to fight fundamentalism. The Renaissance took up this attitude of the classics; the Enlightenment consecrated it. But, today, few people know how the device works that allows you to send emails from Spain to Australia in a second, how the music on a flash drive plays or how air conditioning cools down the air. We caress our new magic wand—we press the button—we invoke who knows what or whom, so that nothing untoward happens, and that's it: we just wait briefly for the miracle to take place. Some argue that this mentality is being transferred to other spheres, such as politics or economics, and that nobody is now interested in the background to the news.

Be that as it may, the media are not in the business of explaining cause and effect: 'It is useless,' said Eco, 'to ask the media to abandon the magical mentality: they are condemned to do so not only for what we would today call audience reasons, but also because the nature of the relationship they are obliged to put daily between cause and effect is magical' (Eco 2002, 13).

Umberto Eco argues that it is technology and disinterest in cause and effect that have led to society's lack of appreciation for science. It is difficult to communicate to the public that research is made up of hypotheses, of controlled experiments, of evidence of falsification. It is true that journalists are looking for a substance that cures cancer rather than an intermediate step, because they know that is not of interest to today's society. A person with a background in journalism will never be fascinated by the method by which a science has come to fruition. He or she is only interested—and that's if you're lucky—in the outcome. But for a scientist, the method is as important as the result, or even more. Just as, for a traveller, not a common tourist, the journey can be more attractive than the destination.

In the West, the social model itself has isolated us from the natural world. It is true that science today understands how nature works better than at any other time in history. Yet it is also true that the common man has never known less about the subject. Any child from the lost tribes of the Amazon knows much more about how nature works than a pupil from even a magnificent Western school.

Most of the Western population born shortly after World War II still had the opportunity to practise traditional agriculture, fishing or cattle raising. This implies the need to understand nature in order to survive. But that requirement has now disappeared. I do not defend that kind of life: the living conditions of my grandparents in post-war Spanish might have been natural, yet they were appalling. However, I believe that scientific and technological development could be combined with a greater contact with nature. For example, those of us who were born in the late 1960s or early 1970s were still lucky enough to have grandparents who kept animals that they raised at home. Their houses then had all the comforts of the twentieth century and also enough space to have a vegetable garden with crops and animals. This was true even in the city: on the roofs of city houses in there were cages with chickens or rabbits. Today, that is little short of forbidden by law for the majority of the Western population.

Like a large part of my generation—Generation X, which has, for example, full mastery of all the new technologies—I remember how, with our grand-parents, we were also able to enjoy waiting impatiently for the chicks to break the shell of their egg to hatch. After killing a chicken to eat it, we had to pluck it and so had the opportunity to see how the feathers emerged from the skin or how the viscera were arranged inside, what the muscles were like and where the fat had settled.

In December 2004, just as I was preparing reports to investigate this issue of scientific decline, *Nature* published the chicken genome. Shortly afterwards, I asked my journalism students—who were born in the late 1980s—to write a story based on the press release sent by *Nature*. Most of them did so quickly. I asked how many had seen a chick hatching from the egg in real life. Only two out of 63 students. How many had observed how the different viscera or muscles were arranged in a chicken? None of them had even seen the death of an animal, of which, they confessed, they often ate as meat. How then could they understand what the chicken genome meant and write about it for a newspaper?

The public communication of science obviously goes beyond journalism. In fact, the main means of communication for scientists is still scientific jour-nals. Here, we will discuss how these have evolved since the early days of the Royal Society. Above all, we will analyse how journalism has influenced these publications. At the end of this book, I will take up the question posed by Rees and Filkin: where does this lack of appreciation for science lead us?

Objective Science and Subjective Journalism

Kant, in his *Critique of Pure Reason*, argued that the knowledge that we have of things does not reflect true reality but a subjective reality sifted through human cognitive faculties (sensitivity, understanding and reason). While I do not entirely agree with Kant's perception of the natural sciences, journalism has adapted this idea into a slogan that states, unequivocally, that 'objectivity does not exist'. This is because all journalists are people with a particular background that conditions the way in which they see the world. Unlike the natural sciences, which aspire—and, in my opinion, usually succeed—in being objective, the journalist is a person—not a computer or an equation—and should only aspire to be 'honestly subjective'. For this reason, faculties of journalism insist on knowing the subject in order to understand what one is writing.

In this book I will try to be 'objective', understanding objectivity as the sum of many subjectivities, coming from different directions. My professional life has taken me in these directions, which are in principle very far apart yet which I believe can add to, rather than subtract from, each other. First of all, my life journey led me to study a degree in chemistry and to develop research in that discipline. I then completed a graduate program in advanced inorganic chemistry and published the results of my research in three impact journals, including one published by the Royal Society of Chemistry. I was a high school physics and chemistry teacher and was able to see for myself the STEM vocation crisis. Yet I could also see the difficulty for students (and humans in general) in understanding science and mathematics. Physics, chemistry and mathematics are the most complex disciplines that the human mind has developed. Then I studied journalism and took a doctorate programme in social sciences.

It's funny: in neither my journalism degree nor my doctorate training was I ever validated as a chemist, which gives some idea of how far apart these disciplines are in a Spanish university. I finally defended a thesis that united the two fields and addressed the use of scientists as elements of political manipulation. They awarded it a grade of 'outstanding' in the social sciences: even though the topic was the natural sciences, I couldn't present it in the field of science.

I gave up my research in chemistry, took leave of absence from teaching physics and chemistry and dedicated myself, body and soul, to journalism. Science and journalism are two professions that, although apparently quite different, are similar: both are absolute, wonderful, absorbing, vital to society and totally thankless. Both, moreover, theoretically pursue the same thing: to seek the truth and to make it public. And both need the same skill: great curiosity.

Curiosity was a mortal sin, in the Catholic Church. The Bible recalls that it was Eve's curiosity that made her try a forbidden apple from the 'tree of knowledge', and that condemned mankind. It is no secret how the Bible has harmed the West. The Roman Empire's embrace of Christianity was one of the greatest symptoms of its decline and led to a time of darkness, when the great Greek legacy was lost. This legacy was not political or artistic; its greatest influence was philosophical, especially mathematical and scientific. Once curiosity ceased to be a sin (in about the seventeenth century), the two cultural institutions that defined it began to take root in the West: modern science and journalism, and by extension a media culture.

If they share the same goal and need the same skills (especially curiosity and a degree of personal ego), how can they be such different professions, I wondered. As a journalist, I went through all the phases: I was a scholarship holder and collaborator with local media and, finally, I secured a position with Agencia EFE, the most important in Spanish-language media, where I reported on political and local news.

Next came a time when I missed science—political journalism seemed insubstantial to me—and the newspaper *El Mundo* (the second largest Spanish-language newspaper) gave me the opportunity to cover science news for its 'society' section. Professionally, that was a fascinating stage: I was able to see the difference between science journalism and the other specialties that I had covered. Above all, it gave me the opportunity to come to know at first hand some great researchers and important scientific facilities. I also found out how scientific journals and press offices work.

However, the world of journalism is far tougher than that of the university. With a doctorate now under my belt, I applied for a position as a tenured professor of journalism. During my thesis I had tiptoed over, but now I seriously entered a strange world: the life of a social sciences department (Journalism and Film Studies), so far removed from the inorganic chemistry department in which I had started. It was a fantastic opportunity, because it is not easy for a natural scientist to work in a social science department and describe the differences between the two forms of first-hand knowledge. The university allowed me to do two research stays of one year each at the LSE and at Harvard. In 2016, I switched journalism department from the Universidad Carlos III de Madrid to the Department of Logic, History and Philosophy of Science at the Spanish National University of Distance Education (UNED). My time in Harvard's History of Science Department weighed heavily in this decision.

I must point out that, in both my department of journalism & film studies and in my own university, Carlos III in Madrid, I have sometimes felt rejected because I have a degree in both science and media studies. Instead of being in everyone's camp, I have had the impression of belonging to no one's camp. It is true that at Harvard the History of Science Department was a vibrant community of professors from both disciplines: from its director (during the time that I spent there), Janet Browne, who was a zoologist, to Peter Galison, Gerald Holton, Benjamin Wilson or Ahmed Ragab, who were physicists, David Jones (a doctor) and my faculty sponsor, Everett Mendelshon, who was a biologist. But there were also people who came from history or literature: from Katharine Park (history), Rebecca Lemov (English literature), Sarah Richardson (philosophy and literature), Matthew Hersch (law and political science) and Allan M. Brandt (American history), among others. Although I wanted to investigate whether the approach to science depends on one's earlier training during one's degree, what is clear is that, apart from departments of history of science, the trenches are dug and the departments remain separate.

I would have felt the same—or perhaps worse—if I had joined a chemistry department and contributed research in social science methodology, journalistic writing or trash TV. Obviously, the members of my media studies department would not have understood that, in addition to publications in my field, I want to be valued for my work in the chemical synthesis of manganese (III) fluorides with nitrogen organic bases. They don't think that's the right culture. I had attempted to study phenomena such as the Jahn-Teller effect on manganese orbitals; that is, to analyse aspects of quantum mechanics, and above all to understand how matter behaves. And is there anything more important to human culture than understanding how matter behaves? Has anyone shown that we are anything other than matter?

Although my academic career has been in a Spanish university, and I must admit that Spain is a borderline case in terms of the separation of science and literature, I believe that this widespread isolation is a general problem for the whole West. It is a relevant issue that I will deal with in detail later on.

I have always had great difficulty in being evaluated in two fields—chemistry and social sciences—and the 'evaluators' have been reluctant to add up, as they arrive at less than what I do. I found it hard to believe that there has been no one else in my circumstances. But the two fields are far apart and, in my experience, they basically want to stay that way. This may be one of the causes of both the mismatch between science and journalism and, moreover, the decline of science.

It is true that, just as I felt that I identified with the Department of History of Science at Harvard, with its coexisting academics of science and literature in the same environment, I found peace in my current department of Logic, History and Philosophy of Science at the Spanish National University of Distance Education (UNED), where I finally wrote this book. But, I repeat, they are very small spaces. Such valuable enclaves tend to disappear in the large academic departments of education sciences, sociology, economics, communication or law. History of science departments are the only places in the West in which the model of Greek teaching still survives, in which the study of nature and mathematics (together with rhetoric, history or ethics) is relevant to understanding human beings.

Aristotle in Science and the Arts

My memories of being a university student always bring back to me how the only author whom I studied for a degree in both science and the arts, such as journalism, was Aristotle. In chemistry, I remember, he was at the forefront of many topics: he considered that nature was composed of four elements; he made the first classification of animals and plants; and he came up with a theory about where the Earth was positioned in the universe. His ideas were wrong: matter is not made up of four elements, as he suggested; the Sun does not revolve around the Earth, as he thought, but the other way around. And animals were not created independently, but evolved from each other. However, unlike Plato and his 'world of ideas', Aristotle applied his senses and, especially, used the full potential of logic and rational thought to obtain results from empirical observation. Aristotle did not design experiments; rather, he used only logic. Therefore, while his scientific ideas are erroneous (the first modern scientist was Galileo), his ideas on rhetoric, oratory, politics and poetry are still valid.

Here, I am interested in highlighting the advanced thinking of this Greek philosopher, who was born around 384 BC, in the light of the current Western intellectual hierarchy in the twenty-first century—especially in Latin countries' academies. Aristotle considered that it was necessary to understand botany, zoology and the physics of movement in order to enter into questions such as metaphysics, logic, ethics and politics. This conception—of considering that to be a good political scientist you have to know about botany or physics—is miles away from the current premise, which produces super-specialized intellectuals whom the Spanish philosopher Ortega y Gasset has already called 'ignorant sages'. They know only a portion of knowledge, but they dare to express their opinion on everything with the petulance of an expert, as if that portion of their knowledge were a guarantee for the rest. If politicians, lawyers, journalists and economists were to study botany at university, the planet would not be on its way to environmental disaster, I am sure.

When I started journalism, the professor of public relations told us: 'Everything you need to know about journalism is in the *Rhetoric* of Aristotle. Read

Aristotle's *Rhetoric*, take it up, chew it and you don't need to learn anything more about this profession in college.' And, obviously, I read it.

Rhetoric is the 'art of persuasion', something highly useful for a journalist or public relations officer. Indeed, it is all in the *Rhetoric* of Aristotle; it is difficult to write anything much better. The topics range from typologies of fear to criteria for the length of speeches, the grammatical relationships that we should include and to how to use a metaphor; from oratorical genres and types of speeches to how to conduct a rebuttal. That gave me an idea of the differences between science and literature. One of the problems at the beginning of modern science was having the courage to refute Aristotle.

In physics or biology, we are not going anywhere with Aristotle's knowledge, but in the humanities it is of the essence; the rest is only variations or interpretations of that essence. The great schism between the culture of the arts and the sciences in the West began when philosophers began to glimpse that a modern thinker such as Galileo was better at explaining the physical universe than Aristotle. The Renaissance humanists could remain in the vanguard with Aristotle, but the natural philosophers had to discard him. The Hispanic universities resisted this, and scholars were caught up in not disavowing Aristotle, and failed.

As part of my journalism course, I was taught several topics involving message analysis, or discourse analysis and argumentation. There was almost always a preliminary study of Greek rhetoric, and sometimes its evolution into the work of the great Roman orator, Cicero. I remember a top-flight paper that I had to present on the differences between Aristotle's and Cicero's rhetoric, at the same time relating them to the differences between Greece and Rome. The aim of the professor who set the work was to make us aware that the study of rhetoric exercised by politicians, philosophers and intellectuals in general in each era defined that era better than the historical events themselves. In Greece—basically Athens—the whole populace had to be convinced, because there was a democracy. In Rome, there was a republic and an empire, therefore it was the Senate—an elite—that had to be convinced. Hence, there are some differences between the rhetoric of Aristotle and Cicero. In our age of social networking, Aristotle's rhetoric and Plato's precautions against populism are more vivid than ever.

The curious thing is that this work was requested of me on a course on visual media. Fascinating! Something as modern as television could not be interpreted without Aristotle. From this perspective, the rhetoric used by today's television could explain better what our society is like than the greatest sociological or statistical study. Rhetoric and oratory are such powerful weapons of persuasion that many educational authorities—especially in dictatorial countries or

those with a recent dictatorial past—have removed them from academic programmes. In English-speaking countries, public speaking and debate leagues
are very important at elite universities. In Spain, they are rarely taught outside Jesuit universities, in which there are even internal debate leagues. Above
all, it is shocking that rhetoric and oratory have been abolished from science
degrees—even in English-speaking countries, although this trend is now being
reversed. It has meant leaving the techniques of persuasion of public opinion
in the hands of graduates in communication and other social sciences. Why
doesn't a physicist need to have the tools of persuasion, the same as a political
scientist?

Aristotle's master, Plato, always rejected rhetoric. He thought that there was
only one truth and, if it was obtained, there was no need to persuade anyone
since the truth was itself capable of persuasion. Rhetoric was an art that the
sophists practised and gave one the tools to defend a position. It did not win the
real position, but a position that was defended with better rhetorical elements
and had managed to be persuaded better. This is what lawyers, politicians and
journalists do nowadays: it is not the one who is right who wins, but the one
who better persuades the court, the electorate or the reader.

One of the strangest and most contradictory experiences that I have had
as a journalism professor was preparing a group of students to participate in
the first national university championship of debate and oratory (the Pasarela
Tournament). As captain of my university team, I had to train my student
debaters to defend both a position and the opposite for each of the issues to be
discussed. 'Does business concentration favour society as a whole?' was one of
the issues that we prepared. Although there were two weeks to investigate the
issues, our position, for or against, was drawn immediately before the debate.
This is usual in all debate competitions around the world.

Reason and truth were irrelevant. The important thing was to exercise ourselves in techniques of persuasion so that we could convince ourselves of the
position that we had taken up, which, obviously, did not have to be the one
that we believed in. We won debates in which we defended a position and
then others in which we held the opposite position. The jury was made up
of headhunters from the most important companies and law firms in Spain.
They were looking for students who could convince—regardless of what they
believed—a jury, clients, readers and spectators: students who knew how to
convince themselves of the truth for the contracting company, not of the actual
truth.

The group that I was leading came eighth out of the 24 participants. Was it a
good result? I still don't know whether participating in these tournaments and
training students in rhetoric and oratory is ethical or not. What I do maintain is

that this type of persuasion—the backbone of communication studies—gives much power to those who know how to exercise it in a media society such as that in which we live, in which form and packaging have more value than substance, a media culture in which emotion takes precedence over reason.

At the end of the debate tournament, I remembered an anecdote that describes the effect that direct contact with Greek thought had on the Romans. It is a simple anecdote, but one that demonstrates the great intellectual superiority of the Greeks and, above all, the danger of persuasion techniques in the hands of spurious interests or in which rational thought is not present.

It was 155 BC, and Rome had already conquered much of the known world, including Greece. The Romans had more military and engineering power than the Greeks, but continued to be fascinated by the intellectual potential, especially of the Athenians. They invited to Rome an embassy formed by the heads of various philosophy schools, to instruct young Romans. One day, one of them, the sceptic Carneades, defended justice in political activity. By the end, all the listeners were fascinated by his oratory and rhetoric and, of course, were convinced that, indeed, being fair is the most important thing in being a good politician.

The next day, the young Romans came to the second talk to consolidate their ideas on justice, as they believed. But now Carneades defended the opposite: the importance of injustice in politics. He refuted his argument of the previous day and was able to convince the audience of the excellence of moral evil—injustice—and of the convenience for a government in applying it. The Roman censor, Cato the Elder, was so disgusted by that attitude that he asked that the Greek philosophers should be expelled from Rome immediately so that they would not corrupt Roman youth. He defined the Greeks as 'the most evil and disorderly race'.

Carneades had shown that, with rhetoric and oratory, he could convince a mass of people of what he wanted—even of moral perversion. In fact, he had been sent to Rome to exhibit his great gifts of eloquence. He showed that a good rhetorician with the skills of a speaker could addle the brains and beliefs of anyone, even the educated elite, as was the case with the young Romans in the auditorium. Any manipulative goal, then, could be achieved with an uneducated mass through rhetoric and oratory. It alerted them—without seeming—to the danger, if that mass of people had any kind of power. But all this, including the trap that the audience had fallen into as a rhetorical tool to prove a point, was too sophisticated for the simple Romans, who now drove him from their city.

Carneades was also a defender of a doctrine, scepticism, that affirmed that truth does not exist and that, therefore, nothing can be known with certainty.

According to this belief, everything is in the same category—nothing can be more true than something else. For example, fair and unjust are at the same time synonymous with useful and harmful. Science and magic, for example, would be two forms of truth.

This current of establishing that truth does not exist, that all approaches are equally valid, or that it sustains both the pros and cons of the same thing, has been taken up by postmodern philosophers to attack science. Many social scientists have also defended it as a way of compensating for the serious structural and methodological deficiencies in disciplines such as economics, psychology, sociology, communication and education studies, among others. These are areas that, although they have acquired social prestige thanks to the good rhetoric of their defenders, in terms of their potentialities cannot be compared with physics, chemistry, geology and biology. As we will see later on, this scepticism is tremendously important in the training of journalists and filmmakers at university.

In the social sciences, it is a good plan to defend both one position and the other. In 2013, Eugene Fama received the Nobel Prize in economics for demonstrating the rationality of the markets. He shared it with Robert Shiller, who was awarded for proving the opposite: the irrationality of markets and investors. No one was shocked. However, James Chadwick was awarded a Nobel Prize in 1935 for demonstrating the existence of the neutron. It would be inconceivable for him to have to share the Nobel Prize with someone who could prove that the neutron does not exist. That's why science works and why social studies (I don't regard the field as having the status of a science), such as sociology, economics and education studies, may produce results and also the opposite: no results. Is it worth studying a discipline in which all points of view are valid? If we study it, does it lead us anywhere? Is our mind different if we study only disciplines that do not lead to clear conclusions? This may seem obvious, but it is not so in Western academia, where if a faculty of economics or sociology has more students than a faculty of physics, it has more power. It will use physics or chemistry to support sociology (or economics): since both departments—physics and sociology—are Harvard departments, for example, both are accorded the same credibility.

This has happened with the Nobel Prize in economics. There is no Nobel Prize for economics: there is the 'Bank of Sweden's Prize in Economics in Memory of Alfred Nobel', and it was created in 1969. It was in the bank's interest to give the same status to economics as to physics, chemistry and literature, therefore it is awarded on the same day, and the media simplify the name to 'the Nobel Prize in economics'. Mathematicians do not have the inferiority complex that the economists have, so their Field's Medal is awarded on another day, and it is not called the 'Nobel Prize in mathematics'.

The problem in the West that is many academic economists (or sociologists or media studies or education specialists) believe that, as they earn more than physicists, what they do is on the same level as physics. However, to avoid ruining the reputation of the Nobel Prize (much questioned in non-scientific areas, such as in literature or peace; in fact, these areas do not count towards university rankings), during the last few years a significant number of Nobel laureates in economics have actually been mathematicians or physicists who have preferred to devote their talents to areas that are financially more rewarding than scrutinizing nature.

For the Greeks, the opposite of scepticism was dogmatism. By this I mean that the opposite of a sceptic is a dogmatist. As a philosophical doctrine, dogmatism considers that there is only one truth and a most important thing: that human reason can access knowledge of this truth by following method and order in research, based on evident principles. The fundamental step of dogmatism involves testing the method, always making a thorough critique of appearance. Thus, for example, the rationalist dogmatism of the eighteenth century led to great confidence in reason, obviously after having subjected it to a rigorous examination.

Through a series of historical and philosophical derivations that go beyond the content of this book, the Catholic Church made a very free translation from the Greek of the term dogma. For the Church, dogma identifies those postulates that should be believed only because it is so ordered by ecclesiastical authority. Therefore, they should never be subjected to debate, examination, empirical demonstration or any other form of rational thinking. Dogma, for the Church, is an absolute truth that is revealed by God and that is not open to discussion.

From here on, the adjective 'dogmatic' acquired connotations—at least for a population with little knowledge of philosophy—relating to superstition, intransigence, religion, irrationality, authoritarianism and even pedantry. The antonyms of dogmatic or dogmatism are flexibility, scepticism, simplicity, and even rational and scientific. In other words, dogmatic means both rational and irrational; science, religion and superstition.

This semantic confusion, which occurs in many Western languages such as Spanish and English, among others, has been skilfully used by many critics of the natural sciences to label scientists as dogmatic and, by extension, as intransigent, authoritarian, unresponsive to other beliefs—such as religious authorities—or pedantic. These critics point out that science is just another form of religion, quite exclusive of other alternative interpretations, and that it has no value other than that which its high priests, the scientists, want to give it. We will see all this in greater depth in the following chapters.

Plato and Aristotle, whose intelligences were far superior to those of the sceptical Carneades, were, of course, dogmatic—and to great honour, some would say. Both held that truth exists and that man is able to find it. Aristotle thought, like Plato, that there was only one scientific truth—they called it philosophical truth—rather than several truths. However, the two greatest thinkers of Western history differed in that, while Plato believed that this truth came from abstract reflection, Aristotle saw a fundamental element in observation. That is why there is a gulf between Plato's thought and that of Aristotle, and between Aristotle and Galileo, who added experimentation to observation.

For the purpose of this book, I am interested in highlighting another great difference between the two philosophers. Aristotle, unlike Plato, was well aware that no one would take into account the scientific truth obtained if it were not disseminated with elements of persuasion. While the sophists, creators of rhetoric, defended that the best discourse of action is the one that triumphs in debate or through persuasion and Plato considered that rhetoric was therefore one of the 'evils' of civilization because it gave the possibility to a good rhetorician to convince of a falsehood, Aristotle built bridges. He considered that truth is not obtained from the one who best persuades, but from one with elements of logic, observation and, in short, science. Once this has been obtained, elements of persuasion should be taken into account in order to disseminate it. That is to say, to persuade of a falsehood (something not scientifically proven) is an ethical evil. This is highly topical at the moment, with our social networks that broadcast fake news and alternative facts.

2018 marks the fiftieth anniversary of May 1968. Many say that there's little of that spirit left—and maybe not, in politics or economics. It had interesting elements, such as the whole civil rights struggle, but it heavily infiltrated Western academic ideas. The sophists came out reinforced. All was good, and science was the new enemy because it does not allow several versions of the same fact to come together.

The concepts of 'truth' and 'rational thought', the foundations of the Enlightenment, have suffered an enormous intellectual setback since the second half of the twentieth century in certain Western universities that have sponsored postmodern philosophy. This is since Feyerabend and his idea that there is no difference between science and a fairy tale, summed up by his motto 'anything goes',[13] since Lyotard has maintained that science is nothing

---

[13]P. Feyerabend. (1970). *Against Method: Outline of an Anarchist Theory of Knowledge.* University of Minnesota.

more than a grand narrative (Lyotard 1979)[14]—and since Derrida himself, whose work focuses on criticizing what he considers the totalizing ideology of 'logocentrism'; that is, thinking that is based on logic and reason (Derrida 1967).[15]

In Western universities, a source of distaste for truth and fact has been established, and has led to the emergence of star students such as Sean Spicer, White House Press Secretary (with Donald Trump). In his first press briefing, he accused the media of 'deliberately underestimating the size of the crowd for President Trump's inaugural ceremony'. Many media were alarmed by the small audience, and compared aerial photographs of the ceremony with those of the election crowds for his predecessor, Barack Obama, in 2013. The difference in numbers was significant: with Obama, the Capitol building was full; with Trump, there were many gaps. Spicer stated without embarrassment that the ceremony had drawn the 'largest audience to ever witness an inauguration—period—both in person and around the globe'.[16] Spicer claimed that 420,000 people rode the DC Metro on inauguration day 2017, compared to 317,000 in 2013. He did not offer a source for his claim or clarify the time periods being compared. The data were completely false. The reporters reproached him, but Spicer did not admit any questions (actual ridership figures between midnight and 11 am were 317,000 in 2013 and just 193,000 in 2017. Full-day ridership was 782,000 in 2013 and just 570,557 in 2017).[17]

Spicer's briefing completely contradicted the facts, and the very next day would be a memorable episode in the history of the West. Trump's campaign strategist and counsellor, Kellyanne Conway, defended Spicer's statements in a 'meet the press' interview. 'Why put him out there for the very first time, in front of that podium, to utter a provable falsehood?' Chuck Todd asked Kellyanne Conway. 'It's a small thing, but the first time he confronts the public, it's a falsehood?'

After some tense back-and-forth exchanges, Conway offered this: 'Don't be so overly dramatic about it, Chuck. You're saying it's a falsehood, and they're giving—our press secretary, Sean Spicer, gave alternative facts to that. But the point really is...' At this point, a visibly exasperated Todd cut in: 'Wait a minute. Alternative facts? Alternative facts? Four of the five facts he uttered... were just not true. Alternative facts are not facts; they're falsehoods.'

---

[14]J. F. Lyotard. (1979). *La Condition postmodern: Rapport sur le savoir.* Paris: Éditions de Minuit (Spanish edition by Cátedra).

[15]J. Derrida. (1967). *La escritura y la diferencia.* Paris: Éditions du Seuil. (trans. into Spanish by Antropos, 1989).

[16]Chris Cillizz. (2017). 'Sean Spicer held a press conference. He didn't take questions. Or tell the whole truth.' *Washington Post,* 21 October 2017.

[17]'Alt-fact: Trump's White House threatens war on media over "unfair attacks".' Haaretz: Reuters, 22 January 2017. Retrieved 17 February 2018.

'Fake news' is so yesterday. 'Alternative facts' is where it's at now.[18] In her answer, Conway argued that crowd numbers in general could not be assessed with certainty, and objected to what she described as Todd's trying to make her look ridiculous. She rebuked the journalist who sought the truth and facts, based on Feyerabend's postmodern 'everything goes' philosophy. Conway considered 'alternative facts' to be different versions of reality: 'there were fewer people than ever before at the inauguration of Trump' vs 'there were more people than ever before at the inauguration of Trump'. The followers of each side believed the version of the 'alternative facts' that gave their brains the most pleasure, and that they reproduced it in their respective echo chambers.

This theoretical elaboration of 'alternative facts' (Cooke 2017)[19] in the post-truth era (Peters 2017)[20] has become legendary, and describes the current era in defining the relations between power, public opinion and the media. It also shows the contempt for science and the rational method of approaching truth.

Aristotle reigned not only in the subject of rhetoric but also in fiction, according to the audiovisual narrative course in my journalism studies: in this case, his *Poetics*. In this book, the Greek philosopher laid the foundations for the development of a fiction script. The statements in Aristotle's *Poetics* point out, for example, that a narrative focuses on a major event and that unnecessary and implausible parts must be eliminated. And that dramas don't need to start or end at random. Sometimes he gave curious details, such as that the characters in a tragedy must be illustrious or noble and those in comedy must be ordinary.

Aristotle's *Poetics* is, without doubt, the most important treatise of dramaturgy, defined as the set of narrative principles that seek to establish growing interest in the spectator. For this reason, every professor of audiovisual storytelling will recommend reading Aristotle's *Poetics*, because it is the origin of the so-called 'classical script construction', which has an enormous influence on American cinema and, in general, on commercial cinema. One of my film studies professors once said: 'There are filmmakers who think they are smarter than Aristotle. Obviously, you have to be stupid to think you're better than Aristotle. Therefore, the work of these filmmakers—he was referring to those of the *nouvelle vague*—is usually intellectually stupid. Don't spend money or time watching it.' (He was referring to the phrase attributed to the French filmmaker Jean Luc Godard, that stories (and films) can have an introduction, a climax and a denouement (as Aristotle proposes), but not necessarily in that order).

[18]Aaron Blake. (22 January 2017). 'Kellyanne Conway says Donald Trump's team has "alternative facts". Which pretty much says it all.' *Washington Post*.

[19]N. A. Cooke. (2017). 'Posttruth, truthiness, and alternative facts: Information behavior and critical information consumption for a new age.' *Library Quarterly*, 87, 3, 211–221.

[20]M. A. Peters. (2017). 'Post-truth and fake news.' *Educational Philosophy and Theory*, 49, 6, 567–567.

It is curious that in chemistry my professors never mentioned Aristotle's *Rhetoric* or *Poetics* of Aristotle and that in journalism they did not allude to Aristotle's *Logic* or his studies in zoology or physics. And there is the disagreement: many scientists think, like Plato, that truth does not need to be divulged because it is obvious; and journalists or political scientists believe that truth is not obtained from logic and scientific methodology but from the one who best convinces them, as the sophists argued. On the other hand, filmmakers consider that what is important is that the narrative structure maintains the interest of the viewer and, therefore, that truth or science are irrelevant. Western culture has split Aristotle in two (or three) and, unfortunately, it is necessary to study two degrees—one in science and the other in the arts—in order to have a complete, updated, knowledge of Aristotle.

This book aims to build a bridge between the two Aristotles so that scientists can better persuade and fight against new ideas of irrationality and, above all, so that journalists and filmmakers can understand the scientific method as a way to reach the truth that they must divulge. I will try to put myself in the shoes of them both, because fortunately I have been in many of them. It is in this way—and there will be many others—that I propose to analyse why there is a loss of prestige in science in the West and at the same time a growing interest in magic and irrationality.

3

Is There a Decline in Science in the Western World?

What is the state of science in the West? It's a complex question. It is true that more and more scientific papers are published every day, but it is not clear that there is a large number of relevant discoveries, such as those made in the nineteenth and mid-twentieth centuries. The last really important finding, in the sense of a milestone in the history of science, was the discovery of the structure of DNA in 1953. But a year earlier, in 1952, a memorable discovery was made that demonstrates what science is like: the Hershey and Chase experiment showed that a chemical molecule like DNA (a polynucleotide) is everything we are as people and the heritage we have received from our ancestors. There is the information to make the cells (and not in the proteins as was believed until then before the Hershey and Chase experiment). If the information to build cells is in the DNA, it means that the information to produce living beings is in the DNA. Despite what philosophers such as Kuhn, Popper and many others say, the paradigm that information to build cells resides in DNA remains and will remain. Science, as opposed to other areas of human knowledge, is that definitive. That is their greatness, but also their problem: it is becoming increasingly difficult to obtain new answers. We do not know whether scientific discoveries are reaching their limits (for example, there is a serious problem because it is becoming more and more difficult to obtain new antibiotics) because of the idiosyncrasies of science or because of the way in which science is organized. What is very clear is that without scientists there will be no science. Without young people interested in it, science will have no future. Worse still, what has been discovered so far can be forgotten.

And this other factor is easy to quantify: the vocations of young people to the STEM disciplines (Science, Technology, Engineering and Mathematics). And not only because this training is relevant to becoming scientists or engineers,

© Springer Nature Switzerland AG 2019
C. Elías, *Science on the Ropes*, https://doi.org/10.1007/978-3-030-12978-1_3

but something even more important and which I defend in this book: this training at university level is very pertinent to understanding the world of the 21st century and to gaining access to positions not only as scientific researchers, but also in sociology, history, philosophy, politics, economics, journalism or cinema, among others. That is to say, it is very important that people with studies in science and engineering are in relevant positions in the judiciary, the media, politics or economics. In this book I want to take up the idea of former US President Barack Obama when in 2015 declared: '[Science] is more than a school subject, or the periodic table, or the properties of waves. It is an approach to the world, a critical way to understand and explore and engage with the world, and then have the capacity to change that world…'.[1]

It is a question of having a scientific mentality in the face of the magical, religious, mystical or mythological beliefs that can give rise to other areas, and to irrational decision-making. In fact, as Max Weber reminds us, 'scientific progress can be described as a process of *de-magnification* that has been going on for millennia in Western culture'.[2] Obviously, you have to know about literature, sociology, history, communication and economics, and this is not incompatible with a curriculum in which there is also a strong programme of physics, chemistry, mathematics, biology, geology and computer science.

What is the situation in STEM vocations in the West? We will start with the leading country in science and technology at the moment: the United States. The statistics reflect a complex picture. As Quagliata (2005) points out, the United States is facing a crisis in its education system and the loss of its status as a leader on the international scene because, let us not forget, the world cannot be led without being a leader in science and technology.[3] Since 1985, the American Association for the Advancement of Science (AAAS) has been promoting the 2061 Science for all Americans Project (AAAS 1989)[4] to encourage the general public to take an interest in science.

However, statistics from international organizations certify that American youth is less and less interested in science and technology, and that this has been happening since 1970 (National Science Board 2003).[5] For the time being, the solution is to incorporate scientists and engineers from China, India and low- and middle-income countries. In 2006, the US Congress, under

---

[1] 'Science, technology, engineering and math: Education for global leadership.' U.S. Department of Education, https://www.ed.gov/stem (accessed July 2018).

[2] M. Weber (1959). *Politik als Beruf Wissenschaft als Beruf.* Berlin: Verlag Duncker.

[3] A. B. Quaglatia. (2015). 'University festival promotes STEM education'. *Journal of STEM Education,* 16(3), 20–23.

[4] American Association for the Advancement of Science. (1989). *Science for All Americans: A Project 2061 Report on Literacy Goals Science, Mathematics, and Technology.* Washington DC: AAAS.

[5] National Science Board (2003). *Broadening Participation in Science and Engineering Research and Education.* Virginia: NSF.

George W. Bush, announced its 'American Competitiveness Initiative Act' to encourage the immigration of STEM professionals to the United States. Bush himself stated in 2001: 'The role of government is not to create wealth; the role of our government is to create an environment in which the entrepreneur can flourish, in which minds can expand, in which technologies can reach new frontiers.[6] 'However, with the election of President Donald Trump, this initiative, which was supported by Obama, is in jeopardy.

In January 2017, Silicon Valley's top executives described Donald Trump's anti-immigration measures as shameful, unfair and anti-American. Tim Cook, an engineer and CEO of Apple, in an internal statement said that Apple 'would not exist without immigration'.[7] Mark Zuckerberg, a Harvard mathematics and engineering major and founder of Facebook, wrote on his account: 'My great grandparents came from Germany, Austria and Poland. His in-laws were refugees from China and Vietnam. The United States is a nation of immigrants, and we should be proud of that. Like many of you, I'm concerned about the impact of the recent executive orders signed by President Trump' (Facebook post 27 January 2017). And Sergey Brin, a Russian engineer and mathematician, and a co-founder of Google, criticized the measure: 'I'm here because I'm a refugee.'

In fact, it was not so much a solidarity issue as a business issue: Trump's anti-immigration measure affected a significant percentage of the employees, managers and founders of those companies. According to the Myvisajobs portal (January 2012–January 2017), Google hired almost 13,000 foreign employees using H-1B visas, 28% of its workforce in the United States; at Facebook, it is more than 3,600 of the total 12,700 employees, also 28; 11% of Apple's workforce in the United States is foreign, and 43% at Twitter (one of the highest percentages). On average, 20% of the employees of the first eight technology companies based in Silicon Valley are foreigners. For other firms such as Microsoft, despite being a long way from San Francisco hiring qualified immigrants is just as vital: 31% of its workforce in the United States comes from another country.

A more revealing picture is revealed by statements in a Spanish digital newspaper, *El Confidencial,* by Bernardo Hernández, a Spanish emigrant to the United States, former director of Google and Flickr (Yahoo) in San Francisco and 'general partner' in the investment firm eVentures. In his opinion, the hiring of qualified foreigners through the H-1B visas that Trump intended to

---

[6]https://georgewbush-whitehouse.archives.gov/infocus/aci/text/tech1.html (retrieved in June 2017).
[7]https://www.techtimes.com/articles/194996/20170129/apple-would-not-exist-without-immigration-says-cook-on-trump-administration-immigration-order.htm (retrieved May 2018).

review was the key that explained why Silicon Valley had become the world's largest software and internet innovation centre:

> When I arrived in the United States in 1993, there were no restrictions. Now, the maximum number of H-1Bs awarded per year is 65,000 and you only have three months a year to apply. It's a lottery. Restrictions on these visas have increased in recent years, first with Bush and then with Obama, but now Trump wants to accelerate that process. And because he is so narcissistic, egomaniacal and a bad communicator, he has not been able to explain his plans. The result is chaos and confusion. Four out of five startup founders in the US are immigrants or children of immigrants.[8]

This immigration is necessary because of the decline in STEM vocations among native American youth. Estimates show that the number of STEM occupations will continue to grow: by 2020, the United States will need to fill 20 million new jobs with STEM profiles (Brookings Institution 2014).[9] Data from the National Science Foundation (2014)[10] are similar. The eight areas where more staff will be recruited in the coming years are: (1) Data extraction and analysis (mathematics training); (2) Mental health counselling and therapy (biology and chemistry training); (3) Computer engineering and technology (physics and mathematics training); (4) Research in engineering, mathematics, chemistry, biology and biotechnology; (5) Veterinary (biology and chemistry training); (6) Human health sciences (training in biology and chemistry); (7) Environmental sciences (training in geology, physics, mathematics, chemistry and biology); and (8) Financial and investment markets (mathematics training).

The question is why American-born youngsters prefer to study humanities or social sciences rather than mathematics, physics, chemistry, biology, geology and engineering, if that's where the jobs are. It is not that young people do not study or do not study science, but that the percentage of young people interested in science, engineering and mathematics is low compared to that in the social sciences and humanities.

In addition, there is a racial, cultural and gender factor. Whites and Asians are the most interested in science and engineering studies; however, demographic forecasts indicate that these two groups will decline in the future, as according

---

[8]'Trump incendia Silicon Valley: su plan antiinmigración afecta al 20% de empleados' ('Trump burns down Silicon Valley: His anti-immigration plan affects 20% of employees'). *El Confidencial*, 31 January 2017, https://www.elconfidencial.com/tecnologia/2017-01-31/anti-inmigracion-trump-silicon-valley-google_1323728/.

[9]Brookings Institution (2014). *Still Searching: Job Vacancies and STEM Skills*. Washington DC: Brookings Institution.

[10]National Science Foundation (2014). *Science and Engineering Indicator*. Virginia: NSF.

to statistics the number of African Americans and Latinos, both ethnic groups with a traditionally lower interest in science, will increase.

There is another limiting factor: the science and engineering professions fail to attract women. And because these STEM professions are better paid, women, Latinos and African Americans are paid lower wages. The data collected from different sources show the following table for the United States:

there's a rising demand for STEM job candidates 17% growth in demand for STEM jobs. Between 2014 and 2024, the number of STEM jobs will grow 17%, as compared with 12% for non-STEM jobs. The average median hourly wage for STEM jobs is $37.44. Compared to the median for all other types of jobs in the US, $18.68, STEM-related jobs pay exceptionally well. But the U.S. education system produces a shortage of qualified candidates. Compared to the rest of the world, the U.S. ranks 27th in math.[11]

These data also indicate that 'the U.S. has fallen behind the rest of the world at an alarming rate. U.S. students recently finished 27th in math and twentieth in science in the ranking of 34 countries. Only 36% of all high school grads are ready to take a college-level science course.'[12] This is worrying, because our current digital society is based on the construction of algorithms designed by mathematicians, physicists and computer engineers: 'in 2020, there will be an estimated 1.4 million computer specialist job openings. Unfortunately, US universities are expected produce only enough qualified graduates to fill 29% of these jobs', states the US Department of Labor. According to data from the US Department of Education:

women are vastly underrepresented in STEM education and jobs. When choosing a college major, 0.4% of high school girls select computer science. Contrast this to the fact that in middle school, 74% of girls express interest in STEM courses. Women earned only 17% of computer and information sciences bachelor's degrees in 2014. Compare this to 1985, when 37% of computer science

---

[11] Data collected from the following sources: Boardroom Insiders, 2015 ('Five Facts About Fortune 500 Female CIOs'); College Board AP Program Summary Report, 2015 (Calculus AB and BC, Computer Science A); CRA Taulbee Survey 2014; Department of Labor Bureau of Labor Statistics, Labor Force Characteristics by Race and Ethnicity 2015; Department of Labor Bureau of Labor, Employment by Detailed Occupation 2014 (Occupational Category, 15-1100), Projection includes new and replacement jobs and assumes current undergraduate degree (CIP 11) production levels persist; Department of Labor Bureau of Labor Statistics, Employed Persons by Detailed Occupation, Sex, Race, and Hispanic or Latino Ethnicity 2015; Higher Education Research Institute (HERI), 'The American Freshman: National Norms 2015'; Intel ISEF finalist breakdown by gender, 2015 (unpublished); National Center for Education Statistics (NCES), 2014.

[12] National Math and Science Initiative.

bachelors were awarded to women. This is especially concerning because women have made incredible gains in other areas.[13]

But it is also worrying for two other reasons: the first is that we live in a world where big data, algorithms and artificial intelligence will become relevant to most professions; and the second is that it is not understood why, in an environment in which girls have had technology since childhood, they are less interested in it now than in 1985.

According to data from the NSF:

in 2014, 57% of bachelor's degree recipients were women. 25% of professional computing occupations in the U.S. workforce were held by women in 2015. Compare that to women's share of the overall workforce, 47%. Many minorities are also underrepresented: underrepresented minorities earn only 11% of all engineering degrees. Underrepresented minorities (African Americans, Hispanics, Native Americans) make up 35% of the college-age population, but only 11% of engineering degrees.[14]

These disturbing data, however, have not generated a thorough investigation of the causes among sociologists, anthropologists or cultural studies experts in the United States. They don't dare approach it. The bravest was the former president, Barack Obama, who declared: 'Improvements in STEM education will happen only if Hispanics, African–Americans, and other underrepresented groups in the STEM fields—including women, people with disabilities, and first-generation Americans—robustly engage and are supported in learning and teaching in these areas' (Figs. 3.1 and 3.2).[15]

STEM in Europe

According to European Union statistics, 'despite the economic crisis, employment of physical, mathematical and engineering science professionals and associate professionals is around 12% higher in the European Union (EU) in 2013 than it was in 2000, and this trend looks set to continue'.[16] However, vocational education and training (VET), including that provided at upper-secondary level, is traditionally an important supply line for STEM skills, 'but

---

[13]National Science Board. Science and Engineering Indicators 2016.

[14]National Science Foundation (2014). Science and Engineering Indicator. Virginia: NSF.

[15]https://obamawhitehouse.archives.gov/blog/2016/02/11/stem-all.

[16]http://www.cedefop.europa.eu/en/publications-and-resources/statistics-and-indicators/statistics-and-graphs/rising-stems4.

Underrepresented minorities in the United States earning engineering degrees/certificates

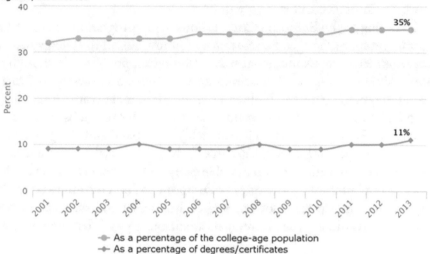

Fig. 3.1 Engineering degrees in the United States, by underrepresented minorities 2009–2013. *Source* US Department of Education

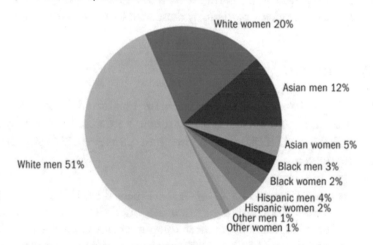

Fig. 3.2 White and Asian workers make up 88 per cent of all science and engineering jobs. Figure 3.1 shows the US science and engineering occupations as at 2013. *Source* US National Science Foundation

there are concerns that the supply of STEM skills may be insufficient and constrain Europe's economic growth.'[17]

According to the European Commission's Work programme 2015, the EU aims to 'promote the agenda, research and innovation in the energy field and to strengthen Europe's industrial base including support for job creation and employability measures'.[18] Disciplines that require a knowledge of mathematics, physics, chemistry, biology or programming are considered not only of special educational relevance but also strategic by the 'Jobs, Growth and Investment Package' (infrastructure, notably broadband and energy networks, as well as transport infrastructure, particularly in industrial centres; education, research and innovation; and renewable energy and energy efficiency).

In its resolution on 'How can the European Union contribute to creating a hospitable environment for enterprises, businesses and start-ups to create jobs?' of 15 April 2014, the European Parliament stressed that member states:

> should invest more in human capital, be more responsive to labour market needs, create variable transition schemes from education to the labour market and foster a partnership approach with a particular view to STEM subjects, retraining and further training of workers, particularly those with low or obsolete skills.[19]

Rising Demand

The same thing is happening in Europe as in the United States: 'demand for STEM professionals and associate professionals is expected to grow by around 8% between 2015 and 2025, much higher than the average 3% growth forecast for all occupations':

> Employment in STEM-related sectors is also expected to rise by around 6.5% between 2015 and 2025, although this masks big differences between different sectors. For example employment in computing and professional services is expected to rise by some 8% and 15% respectively, while the pharmaceuticals sector is expected to see zero employment growth. (Caprile et al. 2015)[20]

---

[17]BusinessEurope. (2011). 'Plugging the skills gap – The clock is ticking (science, technology and maths)', Available at: https://www.businesseurope.eu/sites/buseur/files/media/imported/2011-00855-E.pdf.

[18]http://ec.europa.eu/atwork/pdf/cwp_2015_en.pdf (cited Caprile et al. 2015).

[19]http://www.europarl.europa.eu/sides/getDoc.do?pubRef=//EP//TEXT+REPORT+A7-2014_0101+0+DOC+XML+V0//EN.

[20]M. Caprile, R. Palmén, P. Sanz and G. Dente. (2015). *Economic and Scientific Policy Encouraging Stem Studies Labour Market Situation and Comparison of Practices Targeted at Young People in Different Member States.* Special Study for Directorate General for Internal Policies. European Parliament.

The European Commission points out, however, that having STEM skills is no longer sufficient on its own: 'Graduates at all levels, including those from upper-secondary VET, need personal and behavioural attributes as well as STEM-related skills. Creativity, team working, communication and problem solving are needed as scientific knowledge and innovation is increasingly produced by teams that often combine different nationalities as well as different organisations and enterprises.'[21]

But understanding how technological change affects politics, sociology, economics or the humanities is fundamental to a science- and technology-based society, such as in the twenty-first century. For the European Commission, 'in many cases it is not enough that something works well. It should also be well designed, stylish and desirable for more than just practical features.'[22]

It is vital that STEM undergraduate and graduate degrees are not only technical. It is not a question of training technicians but intellectuals with a great scientific background who, in addition, contextualize this background with historical, political and economic events. Above all, they should know how to communicate pro-scientific discourse well, since those who communicate the excellence of magic, superstition or pseudoscience are not only highly active in social networks but among creators (filmmakers and novelists), and even among university academics in the social sciences and humanities.

In Europe (as well as in the United States), there is an added problem: in the 1960s and 1970s STEM degrees were well regarded, but many of their graduates are now about to retire and have no replacements. Two diabolical circumstances are created that add up to two adverse effects: more and more young people with STEM skills (mathematics, physics, biology, geology, chemistry or engineering) are needed every day for the new emerging professions; and more and more people who have taken these studies are retiring without replacement. According to a European Parliament study, 'between 2015 and 2025, around two-thirds of the anticipated job openings in STEM-related professions will be to replace people working in these areas but who will retire' (Caprile et al. 2015):

> Currently, around 48% of STEM-related occupations require medium (upper-secondary) level qualifications, many of which are acquired through initial upper-secondary level VET. Demand forecasts are difficult to make for highly competitive science- and technology-driven industries. STEM-related sectors such as pharmaceuticals, motor vehicles, engineering and other types of man-

[21] European Commission. (2012). *Assessment of impacts of NMP technologies and changing industrial patterns on skills and human resources.*

[22] European Commission. (2012). *Assessment of impacts of NMP technologies and changing industrial patterns on skills and human resources.*

ufacturing are particularly exposed to the boom-and-bust of the economic cycle. Such sectors are also more prone to restructuring and outsourcing. That demand for STEM-related skills is likely to be highest in professional services reflects how the work has changed. In engineering, for example, work tends to be linked to projects for which external contract engineers are brought in as appropriate. Long-term employment with a single firm has been replaced by temporary assignments that can quickly end when a project ends or the market shifts. These factors affect short- and long-term demand for STEM workers and the skills they need. (Caprile et al. 2015)

Falling Supply?

In 2014 the Royal Academy of Engineering reported that 'the UK needs 100,000 new STEM university graduates every year until 2020 just to meet demand'.[23] The other major technological country in the European Union, Germany, suffers from the same problem: 'Germany's Deutsche Bank points to a shortage of about 210,000 workers in what they refer to as MINT disciplines—mathematics, computer science, natural sciences and technology' (Bräuninger 2012)[24]:

> This is already indicated by the demographically related increasing scarcity of supply. Over the next few years, for example, about 40,000 engineers (excluding IT specialists) are set to retire each year. The number of university graduates will therefore be less and less able to keep pace with this. When the baby boomers leave the workforce in the coming decade Germany's domestic supply of labour will slump further. On top of this, if Germany wants to continue playing in the global economic premier league that is increasingly shaped by aspiring emerging markets, it will have to rely more than ever on innovations to be successful. There will be correspondingly high demand for scientists, engineers, technicians and other well-qualified personnel. It is also clear that in an ageing society the demand for doctors and nursing personnel will rise. (Bräuninger 2012)

It is a problem for the rich countries: according to the European Commission, the phenomenon also occurs in Austria and in Scandinavian countries, such as Sweden. Even in Eastern European countries, such as Hungary, which has based its current economic development on the science and technology education foundation of its communist past, there is a growing shortage of

---

[23] www.theengineer.co.uk, cited in Caprile et al. (2015).

[24] Dieter Bräuninger. (2012). *Talking Point. New stimulus for Germany's skilled workers market.* Deutsche Bank Research. See also: Dieter Bräuninger and Christine Majowski (2011), *Labour Mobility in the Euro Area.* Deutsche Bank Research, EU Monitor 85.

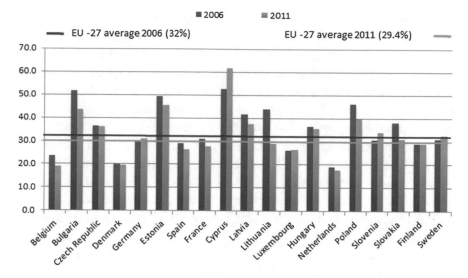

Fig. 3.3 Cedefop (The European Centre for the Development of Vocational Training) calculations based on Eurostat, UOE data collection. EU averages are weighted averages of available country data. Calculations exclude the category of architecture and building. *Source* Caprile et al. (2015)

young people interested in science and engineering. Interestingly, this is not the case in poorer nations:

> At upper-secondary level education, numbers of VET graduates in STEM-related related subjects vary significantly across countries. They account for more than 40% of upper-secondary VET graduates in Bulgaria, Estonia and Cyprus, compared to less than 20% in Belgium, Denmark and the Netherlands (see Fig. 3.3) (Caprile et al. 2015).
>
> The direction and scale of trends are not uniform, but for most countries for which data are available there appears to be a slight decline in the proportion of upper-secondary VET graduates in STEM-related subjects, with the estimated EU average dropping from 32% in 2006 to 29.4% in 2011. But it is difficult to say how significant this decline is. There are also concerns about the quality of maths and science learning at earlier stages of schooling. The 2012 PISA results showed that around 16.5% of pupils aged 15 had low science levels and 22% were low achievers in maths in 27 EU Member States. (Caprile et al. 2015)

The low quality of STEM teaching staff is a difficult problem to solve: bright students who graduate in STEM have much more interesting job opportunities than being a primary or secondary school teacher. On the other hand, as students know less about science and mathematics, the education faculties where teachers are trained have chosen to teach more education studies, psychology and history of education than mathematics, physics, chemistry and programming. This means that science subjects are taught by teachers who have not studied them enough, who do not like them or, what is even worse, are demotivated because, having studied science or engineering, they have not found the job that they wanted and have had to take on Plan B, being a secondary school teacher. This is not the case in the humanities, where students assume from the beginning that their main niche will be in teaching.

Primary and secondary school students perceive a lack of teacher motivation at all levels of education, as well as a lack of teacher interest in teaching science and technology (Strayhom 2010). Other studies (Appleton 2013; Coll and Earnes 2008)[25] state that the lack of scientific training of primary school teachers hinders the exposure to and adequate teaching of advances in science and scientific.

However, the most concerning data for science education are from studies carried out among university students who are preparing to become teachers or primary school teachers: 'Primary school teachers did not have sufficient training in scientific content, and they also had negative attitudes towards science (Appleton 2013, 119). Coll and Earnes (2008) also point out that students in education revealed that they did not like science either, 'because it is very abstract and therefore complicates their understanding' (Coll and Earnes 2008, 98).

Maintaining a Healthy Supply of STEM-Related Skills

Whether or not the supply of STEM skills meets demand depends on more than simple comparisons between the forecasted employment opportunities and the anticipated numbers of graduates. According to the European Parliament Study (Caprile et al. 2015), maintaining a healthy supply of STEM-related skills rests on several factors:

> While young upper-secondary level graduates in STEM-related subjects are needed, it is also important to provide opportunities for adult workers to acquire and update STEM skills throughout working life. Learning opportunities are essential in areas of rapid technological change. Beyond learning, STEM-related jobs and careers need to be attractive both to draw students to STEM-related subjects and to ensure that qualified people do not chose careers in other areas. (Caprile et al. 2015)

---

[25] R. K. Coll and C. Earnes. (2008). 'Developing and understanding of higher education science and enginnering learning communities.' *Reseach in Science and Technological Education*, 26(3), 245–257.

The European Commission notes with concern that Europe cannot compete with countries such as the United States, Australia or Canada in attracting talent, first of all because there are many bureaucratic obstacles, and above all because, after Brexit, only in two European countries (Ireland and Cyprus, one of which is irrelevant) is English spoken, the global language in which science and technology communicate.[26] I know of several English-speaking scientists who have moved to Austria or Germany without having lived there before. Although the universities may teach in English, departmental meetings and everyday business are in German and, at an average age of around 40 years (when they are worthy competitors for a relevant position), learning German (or French, Spanish or Italian) is complicated. In the end, they always return to English-speaking countries:

> Europe is not the preferred destination for many skilled third-country nationals, including STEM professionals. This is due to various things, such as work permit regulations, access to citizenship, language, quality of employment and the extent of ethnic diversity. It needs to be remembered that Europe is not just competing in goods and services but also for talent. (OECD 2012)[27]

However, countries such as Britain have a positive migration balance of scientists and engineers (more in than out). This is also the case in Germany. But the countries of the South—Spain, Portugal and Italy, among others—in addition to the few graduates in these areas, have the additional problem that a large number of them (the brightest, who are awarded competitive scholarships) leave the country. And the migrants that they receive are not graduates, because the jobs in their economies are in tourism, construction and agriculture, not in science and technology. The universities of the poor countries of southern Europe (relatively cheap for students, because most of them are financed by the state, i.e. by taxes) train scientists and engineers who will then go to work in the technical North. This further widens the North–South divide in the European Union. It is a clear fact: where there are more mathematical scientists and engineers per thousand inhabitants, there is greater economic potential. Let's bear in mind that, in this equation, more scientists and engineers means better economics, and a strong economy will result in relevant cultural growth, including in the humanities. Entrepreneurs from countries in the South (including South America) are usually lawyers or economists, while those from very rich areas, such as Silicon Valley, are usually engineers or scientists.

---

[26]European Commission. (2011). *Satisfying Labour Demand through Migration*. European Migration Network.

[27]OECD. (2012). *Connecting with Emigrants, A Global Profile of Diasporas*. OECD.

The State of Science Education

But this does not mean that young people in rich countries study STEM subjects. I repeat, it is in rich countries that science has the worst image among young people. Rich countries obtain their STEM workers from prepared immigration from poor countries. It's a totally perverse system. In poor areas, students study STEM because they know that they will be able to migrate to better conditions. Training is paid for by the states of these poor countries, but the performance of their brightest is in and for the rich areas of the world. The ROSE Project, a highly illustrative study of student attitudes to science, was published in 2005. More than twenty countries participated. One of its conclusions is that 'students' response to the statement "I like school science better than other subjects" is increasingly negative, the more developed the country' (Sjøberg and Schreiner 2005) (Fig. 3.4).[28]

Indeed, there is a 0.92 negative correlation between the responses to this question and the UN Index of Human Development: 'In short, the more advanced a country is, the less its young people are interested in the study of science.'[29] How is it possible that people do not understand that their country's development is thanks to the fact that their ancestors were greatly interested in science and technology? This is confirmed by further studies.

The gap between natural and social and human sciences needs to be bridged. Abandon the message of the gap between rich and poor, because in middle- and low-income countries those who enter university with few resources focus their studies on science and technology, while those with capital opt for financial–legal qualifications (UNESCO 2014).[30] But a country that has graduates only in financial–legal studies will not take long to fall behind, because finance and law do not produce any kind of goods with which to establish a competitive economy. If rich countries take away the scientific and technological talent of the poor countries, the latter will never climb out of their poverty.

This is extremely relevant for several reasons: firstly, because it does not make much sense, a priori, that the more developed countries, which produce science and technology (they are developed because they use science and technology), have less interest in science. It can only be explained by the fact that all these countries have a Western media culture (cinema, television, press, etc.) that is very anti-scientific, as well as teachers of humanities and social sciences who

---

[28] S. Sjøberg and C. Schreiner. (2005). 'How do learners in different cultures relate to science and technology? Results and perspectives from the project ROSE.' *Asia Pacific Forum on Science Learning and Teaching*, 6, 1–16. Available at: https://www.uv.uio.no/ils/english/research/projects/rose/publications/how-to-learn.pdf.

[29] OECD. (2006). *Evolution of Student Interest in Science and Technology Studies Policy Report*. Paris: OECD.

[30] UNESCO (2014). *UNESCO Science for Peace and Sustainable Development*. Paris: UNESCO.

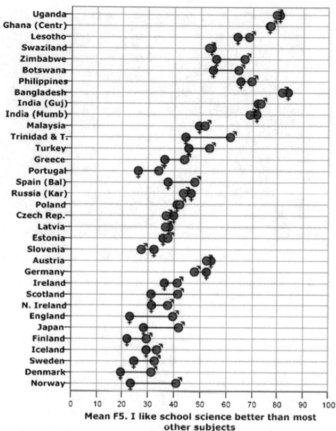

I like school science better than most other subjects

Fig. 3.4   Data from the ROSE study (Sjøberg and Schreiner (2005, 2010): The Rose project: An overview and key findings. University of Oslo) showing students' responses to the question 'I like school science better than most other school subjects' (1—strongly disagree, 4—strongly agree; red = female, blue = male). *Source* Svein Sjøberg and Camilla Schreiner (2010) The ROSE project: An overview and key findings. University of Oslo

have been trained in faculties of arts that are hostile to science, as we shall see in other chapters. European or Westernized Asian countries (such as Japan) have little interest in science, and Asian developing countries (such as India or Bangladesh) and African countries are the most interested.

Another devastating factor is barely addressed by gender studies: why is it that in African countries or, for example, in India, where equal rights for men and women are still not implemented, there is practically no difference between boys and girls in terms of their taste for science? However, throughout the West, especially in countries that are highly developed (England, Norway, Denmark, and so on), not only technologically but culturally and socially in terms of gender equality, the gender gap between boys and girls in terms of

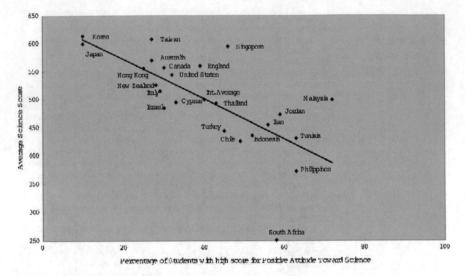

Fig. 3.5 Relationship between student achievement and student attitudes to science for TIMSS data. *Source* Ogura (2006)

liking science is at its greatest. Girls always like it far less than boys. And, I repeat, it is notable that this is not the case in India, Ghana or Uganda. Something is being done wrong in the West: 'Likewise, an analysis of the data from the Third International Mathematics and Science Study (TIMSS), conducted in 1999, and which measured both student attainment and student attitude towards science, shows that the higher the average student achievement, the less positive is their attitude towards science (Fig. 3.5).'[31]

One interpretation of these datasets is:

> that this is a phenomenon that is deeply cultural, and that the problem lies beyond science education itself. Given that learning science is demanding, that it requires application, discipline and delayed gratification – all values which contemporary culture might be said to neglect – there may be some substance to this view. In addition, the immediate relevance of the subject may not be evident to students. (Osborne and Dillon 2008)[32]

It is an obvious fact that science and engineering subjects are much more difficult, and require not only more effort but also vital prior preparation.

---

[31] Y. Ogura. (2006). Graph of Student Attitude v Student Attainment. Based on data from M. Martin et al. (2000), *TIMSS 1999 International Science Report: Findings from IEA's Repeat of the Third International Mathematics and Science Study at the eighth grade.* Chestnut Hill, MA: Boston College/Tokyo: National Institute for Educational Research.

[32] Jonathan Osborne and J. Dillon (2008). *Science Education in Europe: Critical Reflections. A Report to the Nuffield Foundation.* King's College London.

Besides, science, mathematics and engineering have their own language; if you don't use it, you lose it. Humanities or social science subjects use literary language, and students can join in at any age. Thus, almost all graduates with a degree in science and another in the arts have studied STEM first. For example, at the moment there is great demand for journalists who know how to use big data and computer programming: it is easier for a computer engineer to study journalism later than for a journalist or philologist to take up the study of computer science or mathematics.

An analysis entitled 'Science Education and Youth's Identity Construction' (Schreiner and Sjøberg 2007), drawing on contemporary notions of identity, was found to be particularly insightful: 'Their contention, which fits with other research, is not that there is a lack of interest or respect for science and technology but rather that the perceived values associated with science and technology do not match the values of contemporary youth.'[33]

A highly relevant fact of late-modern society has been the 'transformation of the perception of science as a source of solutions to a perception of science as a source of threat'. This would have much to do with what these youngsters are taught, not by science teachers but by teachers of humanities subjects and social studies who have studied in schools that are highly critical of science. The problem of the lack of appreciation of science is not only a question of training science teachers better but, above all, of training humanities and social studies teachers better in science.

Another interesting study, entitled 'Youth, Technology, and Media Culture',[34] found that:

in the European cultural context, as communications and access to travel have improved, there has been a dissolution in the role of traditional structures in establishing individuals' identities. Increasingly, we live in a society where people have more autonomy and choice; routinely question standard practices; and where individuals actively construct their self-identity rather than accept what, in earlier times, was a pre-destined route determined by their social and economic context. (Sefton-Green 2007)

It adds:

Youth, likewise, share in this sense of freedom to choose their social groupings (which are now widely shared through Internet websites such as MySpace, Facebook and Twitter), their lifestyle, religion and values. In addition, contemporary

---

[33]C. Schreiner and S. Sjøberg. (2007). 'Science Education and Youth's Indentity Construction – Two incompatible projects?'. In D. Corrigan, J. Dillon and R. Gunstone (eds), *The Re-emergence of Values in the Science Curriculum*, pp. 231–247. Rotterdam: Sense Publishers.

[34]J. Sefton-Green. (2007). 'Youth, technology, and media culture.' *Review of Research in Education*, 30, 279–306.

societies value creativity and innovation more highly than might have been the case in the past. In the context of school, young people define and communicate their identities through aspects such as their choice of clothing, subject preferences and behaviour. (Sefton-Green 2007)

Adolescence is a complex time during which young people need to build their personality. For this reason, they are enormously impressionable and, also for this reason, the references that they pick up from the environment and the media or social networks are hugely relevant. The need to build one's personality creates many insecurities:[35]

> In some senses, this angst is not new. All young people have had to undertake this process. What is new is that the range of choices presented to contemporary youth is now much greater. For instance, rather than a simple choice between studying the arts or sciences at school, students are now offered an ever-widening range of subjects which can be mixed in different combinations. The result is that subject choice has changed from an issue of being 'What do you want to do when you grow up? to one of 'Who do you want to be when you grow up?' (Fielding 1998)[36]

This is one of the greatest transformations of the educational process in the West: who I want to be, instead of what I want to do. And that means that celebrity culture and the media are enormously important:

> Education in such a context becomes a means of self-actualization and finding personal meaning – a value reflected in the contemporary obsession with celebrity. In such a context, personal interest becomes the dominant factor in subject choice not the possibility of any future career it affords. Hence, whilst science might be perceived as quite interesting, it is seen as 'not for me' (Jenkins and Nelson 2005)[37] by many young people as it is identified with becoming a scientist or engineer – careers which are strongly associated with the advancement of technology rather than aiding people, and not as a means of self-realisation. (Fielding 1998)[38]

Another case of failure is that, after so many policies and projects over the past thirty years to engage girls in physical sciences and engineering,

---

[35] J. Head. (1985). *The Personal Response to Science.* Cambridge: Cambridge University Press.

[36] H. Fielding. (1998). 'The undesirable choices?' Undergraduate dissertation, School of Education, King's College London, cited in Osborne and Dillon, 2008.

[37] E. Jenkins and N. W. Nelson. (2005). 'Important but not for me: students' attitudes toward secondary school science in England.' *Research in Science & Technological Education*, 23, 41–57.

[38] H. Fielding. (1998). 'The undesirable choices?' Undergraduate dissertation, School of Education, King's College London, cited in Osborne and Dillon, 2008.

'girls still remain in a minority. Percentages of female maths, science and technology graduates vary from 19.5% in the Netherlands to a maximum of 42% in Bulgaria, with an average of 31% across Europe.' Why it is that 42% of girls in Bulgaria graduate in science and only 19.5% in the Netherlands, where there is an established tradition of gender equality, is a question that Europe urgently needs to answer. From my point of view, it has much to do with the media (according to which there are hardly any scientists who are women and whose character is interesting) and the fact that, as there was a previous gender selection in careers, in secondary school the teachers of science, mathematics and technology are usually men, and those of history, literature, art or philosophy are usually women:

> Whilst there is still some debate about whether such differences are innate or cultural, there is a high level of concern that both girls and science are losing out. Girls, because their lack of engagement with the further study of science forecloses a number of career options, and science because it is failing to attract a large number of students who potentially have a very significant contribution to make. (Osborne and Dillon 2008)

If we analyse in depth some details of the ROSE study for a country with a great scientific tradition such as England, we can see the magnitude of the problem. The ROSE questionnaire presents 108 topics that students might like to learn and asks its respondents to rate them on a scale of 1 ('not at all') to 4 ('very interested'). Between English boys and girls there were 80 statistically significant differences. The top five items for English boys and girls are shown in Table 3.1 (Jenkins and Nelson 2005).

Table 3.1   The top five items that English boys would like to learn about in science and the top five for English girls

English boys	English girls
Explosive chemicals	Why we dream when we are sleeping and what the dreams might mean
How it feels to be weightless in space	
How the atom bomb functions	Cancer—what we know and how we can treat it
Biological and chemical weapons and what they do to the human body	How to perform first aid and use basic medical equipment
Black holes, supernovae and other spectacular objects in outer space	How to exercise the body to keep fit and strong
	Sexually transmitted diseases and how to be protected against them

Source Jenkins and Nelson (2005)

The British Case: When Science Faculties Close Down

How did this situation come about? How is it possible that England, the country that has done the most for modern science—from Francis Bacon to Newton, Dalton, Darwin or Francis Crick, among many others—is the country that is the least interested in it and where the gender gap is widest? The hallmark of England should not be football but physics, chemistry or biology, yet it is perfectly possible for someone to graduate in England with a degree in humanities or social sciences without having had any science training at university. My idea is that English universities should use a badge system for courses to indicate that their graduating sociologists, journalists and historians have studied some natural science topics during their degree.

The problem was detected a few years ago. On Sunday, 12 March 2006, the British newspaper *The Observer* reported grave news for the history of science: the University of Sussex had decided to close its chemistry department due to the absence of students interested in this discipline. But the Sussex chemistry department was not just any chemistry department: it had produced three Nobel Prize winners in chemistry: for example, more than the whole of Latin America combined, and more than Spain throughout its entire history—a country that has not generated any Nobel Prize winner in the basic sciences, either physics nor chemistry. In Spain, France, Italy and Germany, scientific departments cannot be closed because, at present, many of their professors have security of tenure. No young teaching professors with bright ideas are able to get in, but the older ones are able to wait peacefully for retirement. A slower, more agonizing death is occurring in these countries.

The death of the Sussex chemistry department was not the first in Britain. Before it, and in the previous five years, the same had happened at the universities of Exeter, King's College London, Queen Mary's London, Dundee and Surrey. This was a real bloodletting that alarmed the Royal Society. The University of Sussex communiqué stated that it was closing its department of chemistry because it wanted to concentrate on those studies 'that students were demanding'. And, among those studies, it expressly cited English philology, history and, above all, communication sciences. It left the mathematics department as a scientific department, because it was cheap to maintain and, in addition, there were people interested in studying economics, and this includes mathematics.

The decision was highly controversial. 'Can a university be considered as such if it does not have a powerful and prestigious chemistry department?' Sir Harry Kroto, the Nobel Prize winner in chemistry, threatened to return an honorary award from the University of Sussex. The British parliament appointed a commission. Meanwhile, the Head of the Sussex chemistry department, Gerry Lawless, mobilized the media and students across the country to protest and march to the Executive Director of the British Council for Higher Education, Steve Egan. However, the most anticipated statements were those of Professor Alasdair Smith, the university's vice-chancellor responsible for the closure decision. In a very empirical British style, he said: 'It closes because it's too expensive to maintain and almost nobody registers.' That's how honest he was with the press.

These statements—a fact that confirms the decline—are supported by worrying empirical evidence. According to the Higher Education Statistics Agency, there was an 11% increase in undergraduate enrolment between 1995–96 and 2001–02. In this period, there was a 22% increase in enrolment in the biological sciences in Great Britain. However, there was an 8% decrease in those studying engineering and an alarming 20% decrease in physical sciences. The fall in the number of those enrolled in chemical undergraduate programmes was described by the Royal Society as 'terrifying': 31%.

This decrease, given that the number of students in British universities increased overall during this period, made the problem 'worrying'. Sir Alistair MacFarlane, Head of the Royal Society's Education Committee, told *The Guardian* (17 December 2003):

> These downward trends will inflict huge damage on the UK if not reversed, adversely affecting prosperity and the quality of life in the UK. Labour market projections show a growth in the demand for employees with training in science, engineering and technology over the next few decades. We live in an increasingly technological world, and we need, as a nation to have a workforce that includes highly skilled scientists, engineers, doctors and technicians, and a citizenry able to understand, appreciate and act upon the consequences of advances in scientific knowledge.[39]

And where have the British students gone? To the same place as the rest of Europe: to media degrees. Thus, the number of British students enrolled in media studies increased from 6,888 in 1995–96 to 15,905 in 2001–02. That's a 43% increase. The most important aspect is that the numbers then increased by a further 67.8% in just three years, since in the 2004–05 academic year the enrolment had reached 26,685.

The 2004–05 academic year was a landmark in Great Britain, as it was the first in its long and fruitful academic history in which those enrolled in

---

[39]'The crisis in science education.' *The Guardian* (17 December 2003). https://www.theguardian.com/education/2003/dec/17/highereducation.uk2.

media studies exceeded the numbers on any of its science degree programmes. Thus, compared to 26,685 in media studies, 26,290 were enrolled in biology, 18,520 in chemistry and 14,610 in physics. There were 7,570 students enrolled in journalism alone. To understand the magnitude of this problem in Great Britain (especially the decline of vocations in subjects such as physics) we can point out that, compared to those 14,610 students enrolled in physics in 2004–05, almost forty years earlier, in 1966, more than double that number had been enrolled. Specifically, according to statistical data from educational records, in 1966 there were 31,831 physics students in British universities....[40] But in 1966 the 'space race' was still in full swing. Readers should note that those who studied physics as students in 1966 are now around 70 years old, and are either retired or on the verge of retirement.

Another interesting fact that illustrates this change in trend in Great Britain is that in 1966, according to the source mentioned, there were only 28,149 students enrolled in social sciences in all British universities. That is to say, there were more students in physics than in all the social sciences combined (including economics, sociology, media studies, psychology, education studies and many others). However, forty years later, in media studies alone, there would be 26,685, to which must be added the 7,570 of journalism alone. That is to say, at least in terms of vocations, it seems that the social sciences have triumphed and the natural sciences are in decline.

It is illustrative of a trend to note that in 1966 the 'social sciences' were not called that as such, at least not in the statistics consulted, but simply 'social studies'. In my opinion, this is a wise way to avoid conflating what is studied in them with pure science. However, as they have gained more power in the university, social scholars have changed their name to 'social scientists', a circumstance that is insufficiently criticized by natural scientists (and sometimes even endorsed), and this is another symptom of the decline of science.

But trends can change, and that change may be due exclusively to the media and the stereotype of scientists in fiction series. In November 2011, *The Guardian* published a very interesting news item entitled 'Big Bang Theory fuels Physics Boom'. The article described how:

A cult US sitcom has emerged as the latest factor behind a remarkable resurgence of physics among A level and university students. *The Big Bang Theory*, a California-based comedy that follows two young physicists, is being credited with consolidating the growing appetite among teenagers for the once unfash-

---

[40] Statistics of Education, 1966. In the Spanish journal *Las Ciencias. Revista de la Asociación Española para el Progreso de las Ciencias*, 34, 72.

ionable subject of physics. Documentaries by Brian Cox have previously been mentioned as galvanising interest in the subject.

In the report by journalist Mark Townsend, one pupil, Tom Whitmore, aged 15 from Brighton, acknowledged that *The Big Bang Theory* had contributed to his decision, with a number of classmates, to consider physics at A level, and in causing the subject to be regarded as 'cool': '*The Big Bang Theory* is a great show, and it's definitely made physics more popular. And disputes between classmates now have a new way of being settled: with a game of rock, paper, scissors, lizard, Spock,' he said.

According to the Higher Education Funding Council for England (HEFCE), there was a 10% increase in the number of students accepted to read physics by the university admissions services between 2008 and 09, when *The Big Bang Theory* was first broadcast in the United Kingdom, and 2010–11. Numbers in 2011 stood at 3,672. Applications for physics courses at university were also up more than 17% on 2010. Philip Walker, an HEFCE spokesman, told *The Guardian* that the recent spate of popular television shows had been influential yet was hard to quantify.

The number studying A level physics had been on the rise for five years (from 2011), up 20% in that time to around 32,860. Physics was among the top 10 most popular A level topics for the first time since 2002—and the government's target of 35,000 students taking the physics A level by 2014 seemed likely to be achieved a little ahead of schedule. It is a far cry from 2005, when physics was officially classified as a 'vulnerable' subject.

The number of those entered for AS level had also increased by 27.8% from 2009, up from 41,955 to 58,190. The number of girls studying physics AS level had risen by a quarter to 13,540, and of boys by 28.6% to 44,650.

Experts at the Institute of Physics (IoP) also believe that this television series is playing a role in increasing the number of physics students. Its spokesman, Joe Winters, told *The Guardian*: 'The rise in popularity of physics appears to be due to a range of factors, including Brian's (Cox) public success, the might of the Large Hadron Collider and, we're sure, the popularity of shows like *The Big Bang Theory*.' Alex Cheung, editor of physics.org, said: 'There's no doubt that TV has also played a role. *The Big Bang Theory* seems to have had a positive effect and the viewing figures for Brian Cox's series suggest that millions of people in the UK are happy to welcome a physics professor, with a tutorial plan in hand, into their sitting room on a Sunday evening.'[41]

---

[41] https://www.theguardian.com/education/2011/nov/06/big-bang-theory-physics-boom.

High School Students Don't Want Science

What was the state of the natural sciences in British secondary school before the *Big Bang Theory*? The IoP, the largest international organization dedicated to the study and teaching of physics and with more than 37,000 associates, produced a report on the evolution of secondary science. Data for England, Wales and Northern Ireland show that between 1991 and 2003 the number of A level students rose (the university preparatory course increased by 7.4%). Students enrolled in biology increased by 11.0%. However, in those 12 years there was an 18.7% drop in chemistry enrolment, a 25.4% drop in mathematics and a scandalous 29.6% drop in physics. In 1993, the three sciences—biology, chemistry and physics—together with mathematics accounted for 30% of all A level students in Great Britain; in 2003, it was a modest 23.2%.

The Confederation of British Industry (CBI) had older figures, and it claimed that the number of students choosing physics at A level had dropped by an alarming 56% in the last 20 years (1983–2003). The fall in chemistry during the same period was 37%. The CBI's member employers estimated in 2003 that by 2014 the United Kingdom would need 2.4 million scientifically skilled workers.

However, the situation appears to have been partly reversed. According to data from HESA (experts in UK higher education data and analysis) and the report 'Patterns and Trends in UK Higher Education 2015', using HESA data:

> changes in student choice at the level of broad subject groups between 2004–05 and 2013–14 also demonstrate how the student body is altering over time, and how universities are adapting to a changing population and economy. The three subjects with the largest increases in absolute numbers were biological sciences (up by 61,945 or 42.6%), business and administrative studies (up by 46,145 or 15.9%), and engineering (up by 26,985 or 20.4%). In the same period two of what had been the largest subject areas, subjects allied to medicine and education, fell by 23,635 (8.0%) and 25,105 (12.7%) respectively, and computer science fell even more, by 36,795 (28.7%). There has been considerable growth in the number of students entering all science, technology, engineering and mathematics (STEM) degrees except computing.[42]

The report gives a clear warning: 'However, it is worth noting that a large proportion of these students are international, so these subjects remain vulnerable to any volatility in the international student market.' In other words, these are students from developing countries who go to England to study sci-

---

[42]"Patterns and Trends in UK Higher Education 2015" https://issuu.com/universitiesuk/docs/patternsandtrends2015_23aa97d68a4d62/5?e=15132110/31968656.

ence and engineering because, in their countries of origin, these subjects have a good image. In addition, there is another factor: since science and engineering use a universal (mathematical) language, foreign students without native English language proficiency may opt for studies where literary language is not so necessary. After all, a good mathematician, like a good musician or a good chemist, can practise the profession in any country without knowing the local language. Between the 2004–05 and 2013–14 academic years, social sciences increased from 189,425 students to 210,580 and mass communication from 45,720 to 49,525, despite the crisis in the communication sector.

However, media reports continue to warn of the lack of STEM graduates in Britain and how this will affect its economy. One of the most prominent to warn of this problem is Sir James Dyson, one of the United Kingdom's leading entrepreneurs, and *The Telegraph* in 2013 ran the article: 'Shortage of engineers is hurting Britain'.[43] He insisted that Britain 'produced 12,000 engineering graduates a year—and there are currently 54,000 vacancies. It's predicted that in two years' time there will be 200,000 vacancies. India produces 1.2 m engineering graduates a year. The Philippines produces more than us, so does Iran, so does Mexico. It's not a sustainable situation.' He said that he planned to hire a further 650 engineers this year (2013), 300 at Dyson's UK base in Malmesbury—where 2,000 people are currently employed—and 350 employed between its two plants in Asia: 'We'll get all the workers we need in Singapore and Malaysia,' Sir James told the *Daily Telegraph*, 'but we have to be realistic in Britain. If we can get 300 we'll be doing well. We would recruit 2,000 if we could. We have got the technology and the ideas. We just need the people.'

This type of statement, over time, becomes insistent. The Director-General of the CBI, Richard Lambert, told the BBC in August 2006 that British (entrepreneur) 'employers are increasingly worried about the long-term decline in numbers studying A level physics, chemistry and maths, and the knock-on effect on these subjects, and engineering, at university.'[44] He added: 'They see, at first hand, the young people who leave school and university looking for a job, and compare them to what they need—and increasingly are looking overseas for graduates. China, India, Brazil and Eastern European countries were producing hundreds of thousands of scientists and engineers every year.'

Lambert believed that the problem did not lie in the youth, but in the Western system that his parents and grandparents had formed, which had led

---

[43] 'Shortage of engineers is hurting Britain', *Telegraph* (5 September 2013). https://www.telegraph.co.uk/finance/newsbysector/industry/engineering/10287555/Shortage-of-engineers-is-hurting-Britain-says-James-Dyson.html.

[44] BBC (13 August 2006). 'Schools "letting down UK science".' http://news.bbc.co.uk/2/hi/uk_news/education/4780017.stm.

to an educational system that does not encourage science to media that give more coverage to a footballer than a physicist:

> This is not a criticism of young people – they work hard to achieve the best possible grades in the system provided. But it is clear we need more specialised teachers to share their enthusiasm for science and fire the imaginations of pupils, and to persuade them to study the core individual disciplines to high levels. We must smash the stereotypes that surround science and rebrand it as desirable and exciting; a gateway to some fantastic career opportunities. (Lambert, at BBC in 2006)[45]

In 2006, no one in the United Kingdom thought that there would be a Brexit referendum and that it would leave the European Union. Everything was more favourable to immigration. However, Lambert warned that if British youth still did not study physics, chemistry, biology, mathematics or engineering, 'the United Kingdom risks being knocked off its perch as a world leader in science, engineering and technology. We cannot afford for this to happen.'

In 2005, Robert J. Samuelson, an economic analyst at the *Washington Post*, warned of the situation in an interesting article entitled 'It's Not a Science Gap (Yet)'[46]:

> A nation's economic power could once be judged by tons of steel or megawatts of electricity. But we have moved beyond these simple indicators or even updated versions, such as computer chips. All advanced societies now depend so completely on technology that their economic might is often measured by their number of scientists and engineers. By that indicator, America's economic power is waning. We're producing a shrinking share of the world's technological talent. China and India are only the newest competitors to erode our position. We need to consider the implications, because they're more complicated than they seem. (Samuelson, *Washington Post*, 2005)

If that were so, the economic power of the United States would have gone into decline, as its production of technological talent is decreasing. At the moment, it is China and India that are leading the way in scientific and, above all, scientific production. The United States can survive because it imports a great deal of highly skilled labour from these countries. According to a 2002 study by Stanford University, Chinese and Indian skilled emigration is, in fact, what maintains the highly competitive level of the American technolog-

---

[45]http://news.bbc.co.uk/2/hi/uk_news/education/4780017.stm.
[46]Robert Samuelson (2005). 'It's not a science gap (yet).' *Washington Post* (10 August 2005).

ical heart, Silicon Valley.[47] However, as we have seen, this situation may be reversed since the election in 2016 of US President Donald Trump and his anti-immigration laws. And the same can happen in Britain.

Other Western countries, as I have already mentioned in the case of STEM vocations, do not succeed in attracting them. One of the most desperate is the European locomotive: Germany. A study by the ZEW (Research Centre for European Economics) found that 73,000 jobs for scientists and engineers were vacant in Germany in 2006.[48] The German economy lost 0.8% of its GDP that year; that is, 18 billion euros, due to the lack of such professionals. The West, meanwhile, is full of sociologists, economists, lawyers and, above all, experts in top-flight interpretations of the influence of German cinema on American movies.

And the problem worsens every year. In 2017, a study by the German Institute for Economic Research[49] warned that the German economy was suffering from 'an alarming deficit of 237,500 professionals in the so-called STEM [MINT, in German] careers', and warned that the 'hole' in the German economy of scientists, engineers and mathematicians had grown by 38.6% since 2016. In May 2017, it was at its highest level since 2011, when these six-monthly studies began: 'This bottleneck will become even greater if foreign employees do not start to work in this field at a higher rate than they are present in society,'[50] said Thomas Sattelberger, president of the German private-sector initiative 'Achieving the Future MINT' (German for 'STEM').

In 2013, Germany began an aggressive policy to attract scientists (chemists, physicists and biologists), engineers and mathematicians to the country and it has paid off: from 2013 to 2017, the number of foreign professionals with STEM training increased by 43%. However, the study stressed that this strategy of attracting scientific and technological talent must be intensified to close this gap.

For years, the German government and the country's powerful industrial sector have been warning about this growing problem, which is caused by the progressive ageing of the German population, and also that young Germans prefer to study other types of careers, influenced by environmental movements and by teachers of the arts and social sciences in schools that are highly critical of science and technology. On the other hand, Germany's aggressive campaign in

---

[47]R. Dossani. (2002). 'Chinese and Indian engineers and their networks in Silicon Valley.' (APARC). Stanford University Working Paper.

[48]Carmen Vela. (2007). 'El "made in Germany" peligra por falta de ingenieros e informáticos.' *Expansión* (10 December 2007, 40).

[49]Founded in 1925, DIW Berlin (German Institute for Economic Research) is one of the leading economic research institutes in Germany. It analyses the economic and social aspects of topical issues.

[50]News article (10 May 2017) published by the EFE Agency in Berlin under the title 'Germany suffers a record deficit of scientists, engineers and mathematicians', and in the Spanish press.

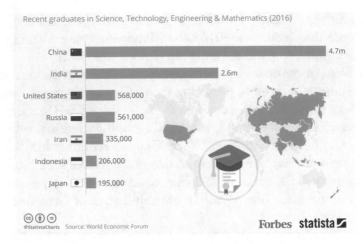

Fig. 3.6 Forbes report on countries with the most STEM graduates. *Source* World Economic Forum from Forbes Report

southern European countries, where low levels of scientific and technological development mean that these professionals may be unemployed or on low wages, has not been as successful as expected. One of the possible reasons is the language, as most scientists and engineers speak English, not German. The United States and the United Kingdom have a huge competitive advantage in attracting scientific talent: they speak English.

A report[51] published by Richard Freeman indicates that in 1975 the United States generated more Ph.Ds in science and technology than all of Europe and, above all, that the number of Ph.Ds in science and technology in the United States was three times higher than that of Asia. However, by 2005 the outlook had reversed, and Asia was now ahead of the United States. According to Freeman, in 2005 the number of such Ph.Ds awarded in a single Asian country, China, was now almost half that of the United States, and the forecasts were that it would overtake it in 2010, and this has indeed happened. The case of the increase in Chinese engineers is spectacular: according to a report by the National Science Foundation, 220,000 engineers graduated in China in 2001, compared to 60,000 in the United States. In 2017, *Forbes* published the data: by 2016, 4.7 million young people graduated in STEM disciplines in China. It was followed by India, with 2.6 million and the United States in third place with only 568,000. The United States was followed by Russia (561,000), Iran (335,000) and Indonesia (206,000) (Fig. 3.6).

---

[51]Richard Freeman. (2005). 'Does globalization of scientific/engineering workforce threaten U.S Economic leadership?'; Working paper 11.457, part of the NBER Science and Engineering Workforce project.

The *Forbes* report by data analyst journalist Niall McCarthy emphasized that:

> since the turn of the century, China has experienced a revolution in third-level education. It has overtaken both the United States and Europe in terms of the number of graduates and, as of 2016, was building the equivalent of almost one university per week. That progress has caused a massive change in the world's graduate population, a population that the United States used to dominate. Last year, India had the largest number of graduates from any country in the world with 78.0 million, while China followed closely behind with 77.7 million. The United States now ranks third with 67.4 million graduates, and the gap behind the two major countries is widening. Some estimates indicate that the number of Chinese graduates between the ages of 25 and 34 will increase by 300 percent by 2030, compared to only 30 percent in the United States and Europe. According to the World Economic Forum, STEM (Science, Technology, Engineering and Mathematics) has become a major problem in China's thriving universities. In 2013, 40 percent of Chinese graduates completed a Bachelor of Science degree in STEM, more than double the proportion in third-level U.S. institutions. (*Forbes*, 2017)[52]

The Golden Age of Science

The professor who managed my time at the LSE, Martin W. Bauer, constantly reminded me that, if I am talking about decline, I have to start early by establishing a 'golden age'. 'If there is a decline, it is that at one time it was better,' he said. What conditions existed at that time—if any—that are not present now? What sum of factors triggered the decline?

When speaking of decline, as we said at the beginning, the reader can argue that a decline in interest or in vocations is not quite such, since science now has more resources and publishes more results than ever before. This is true, but the whole of this current system is actually moving forward with the impetus given to it in the 1950s and 1960s, and is being maintained and increased by its own momentum. The scientists who are now in power—and who are concerned about this phenomenon—were trained in those years. But science needs new blood. It needs a society that is excited about science, not disenchanted.

Economic resources are important, but even more so is talent and creativity. Talent is the lifeblood, the oxygen that nourishes science. If there are factors cutting off the supply of this fresh blood, science dies, regardless of the resources

[52]https://www.forbes.com/sites/niallmccarthy/2017/02/02/the-countries-with-the-most-stem-graduates-infographic/#48264999268a.

that we commit to it. Science can only advance if it is attractive to the brightest young people.

Some may argue that the golden age of science was the seventeenth century, with Newton, or the early years of the twentieth century when the theory of relativity was published. However, establishing its date is not our objective, as it would be a debate on the history of science. In my opinion, these were glorious moments in science, but scientists as a body of influence and social prestige have never been stronger than they were during the Cold War. At that time, the rulers needed scientists in the same way as the kings of the Middle Ages required knights. Public opinion saw scientists as twentieth-century knights. Industry, economics, and war were advancing with the scientists.

From the mid-1950s to the 1960s, economic prosperity in the West meant that, for the first time in history, higher education could be a goal for the children of the working class: a goal of social and economic advancement. But, at the same time, there was a concurrent idea that the scientist was a modern knight, so it was a source of pride that a brilliant son or daughter defended the country as a scientist, just as it would have been a source of pride in the past if a son had been a warrior. Science, moreover, was considered to be something difficult that required a much effort and personal sacrifice, precisely the values that prevailed after World War II in the West. Literature, journalism and cinema were seen as dilettante activities. Science was seen as something practical that could help society, which had suffered so much from the war, to live more comfortably.

World War II did not end in global peace but with two victors who were fierce enemies. Political power in both blocs saw science as a way to perpetuate the system. To beat the enemy. That is why the speeches and initiatives in support of science at that time were so important: because they were actually formulated and delivered. They did not flourish among politically correct yet empty ideas, as they do now. Science may again be needed to perpetuate itself in power, but politicians and, above all, society, are not aware of it in the same way as at that time.

All the factors of that time—the spirit of sacrifice, the massive arrival of young people at university, political and social support for science as a way of survival—are very favourable to science as we know it. Money and the progressive increase in the university population, together with a policy of improving social welfare, also meant that the working conditions of all those scientists were less competitive than they are now. Therefore, it can be said that this was a golden age, in the sense of maximum capacity for recruiting talent, good working conditions and social prestige for scientists, of great public interest and, above all, financial and political support.

The Cold War therefore required a major political offensive to popularize science in the United States and Europe, among other reasons because there was a clear rivalry with the Soviet Union in space technology. Americans needed a great deal of funding—and support from public opinion—for their expensive projects. And they also needed a great many scientists.

The technological race and, by extension, science, became the hallmark of the American people—and of the Soviets—and considerable funds were diverted to these programmes of space-race rivalry, which were, in fact, ideological propaganda. The starting point of this deliberate strategy to popularize science began a little earlier, in 1945, with the 'Report to the President: Science, the Endless Frontier' by Vannevar Bush, which sought to alleviate the negative effects of the public's opinion of science caused by the bombing of Hiroshima and Nagasaki. Bear in mind that these bombings and the nuclear threat, among other reasons, also gave rise to the genesis of a new, unscientific, philosophical thinking.

The thought of the time equated the West with freedom won after World War II. Scientific proselytism began, not only by governments but by other institutions, relating to 'science with freedom and democracy'. Indeed, at the height of the Cold War in 1960, Warren Weaver, an official with the Rockefeller Foundation and the AAAS, went so far as to say in his now famous text 'A Great Age for Science', 'It is imperative that the citizens of our democracy have a better understanding of what science is, how it works and the circumstances that make it thrive'.

But for me, the high point is something else. I discovered it in an interesting seminar on oratory from which I never expected to learn so much: the investiture speech of US President John F. Kennedy in 1961. Kennedy, who was previously registered to be a student at the LSE, gave a speech that is considered a masterpiece of political rhetoric. Its end is memorable and well known across the English-speaking world: 'And so, my fellow Americans: ask not what your country can do for you—ask what you can do for your country. My fellow citizens of the world: ask not what America will do for you, but what together we can do for the freedom of man.' A few paragraphs earlier, Kennedy had spoken about science and, from my point of view, explained very well how non-scientists felt about science: fear and fascination. It seemed to be absolute power:

> Finally, to those nations who would make themselves our adversary, we offer not
> a pledge but a request: that both sides begin anew the quest for peace, before the
> dark powers of destruction unleashed by science engulf all humanity in planned
> or accidental self-destruction. (…). Let both sides explore what problems unite
> us instead of belaboring those problems which divide us. Let both sides, for

the first time, formulate serious and precise proposals for the inspection and control of arms—and bring the absolute power to destroy other nations under the absolute control of all nations. Let both sides seek to invoke the wonders of science instead of its terrors. Together let us explore the stars, conquer the deserts, eradicate disease, tap the ocean depths and encourage the arts and commerce. (President Kennedy's Inaugural Address, 20 January 1961)

Just a few months later, on 25 May 1961, Kennedy pleaded with Congress for additional appropriations for NASA: 'I believe,' said Kennedy, 'this nation should commit itself to achieving the goal, before this decade is out, of landing a man on the Moon and returning him safely to the Earth'.

In 1962, in another Kennedy speech (at Rice University), science was fully present:

Despite the striking fact that most of the scientists that the world has ever known are alive and working today, despite the fact that this nation[1]s own scientific man-power is doubling every 12 years in a rate of growth more than three times that of our population as a whole, despite that, the vast stretches of the unknown and the unanswered and the unfinished still far outstrip our collective comprehension. No man can fully grasp how far and how fast we have come, but condense, if you will, the 50,000 years of man[1]s recorded history in a time span of but a half-century. Stated in these terms, we know very little about the first years, except at the end of them advanced man had learned to use the skins of animals to cover them. Then about 10 years ago, under this standard, man emerged from his caves to construct other kinds of shelter. Only five years ago man learned to write and use a cart with wheels. Christianity began less than two years ago. The printing press came this year, and then less than two months ago, during this whole 50-year span of human history, the steam engine provided a new source of power. Newton explored the meaning of gravity. Last month electric lights and telephones and automobiles and airplanes became available. Only last week did we develop penicillin and television and nuclear power, and now if America's new spacecraft succeeds in reaching Venus, we will have literally reached the stars before midnight tonight. (…) I am delighted that this university is playing a part in putting a man on the Moon as part of a great national effort of the United States of America. Many years ago the great British explorer George Mallory, who was to die on Mount Everest, was asked why did he want to climb it. He said, 'Because it is there.' Well, space is there, and we're going to climb it, and the Moon and the planets are there, and new hopes for knowledge and peace are there. And, therefore, as we set sail we ask God's blessing on the most hazardous and dangerous and greatest adventure on which man has ever embarked. (President John Kennedy's Rice Stadium Moon Speech, 12 September 1962)

In Great Britain, the political climate for science was also highly enthusiastic. In October 1963, the leader of the Labour Party, Harold Wilson, even stated at the annual conference of his political development: 'We are redefining and refounding our socialism, but now in terms of a scientific revolution'. The following year, Wilson won the election and became prime minister.

The Decline of Science Begins

In 1971 Shirley Williams, then a Member of Parliament and later (1976–1979) Secretary of State for Education and Science, spelled out an unmistakable warning:

> For the scientists, the party is over… Until fairly recently no one has asked any awkward questions… Yet there is a growing suspicion about scientists and their discoveries… It is within this drastically altered climate that the dramatic decline in expenditure on scientific research in Britain is taking place.[53]

Basically, the policy of direct support for the popularization of science had ended in 1969 with the arrival of the first man on the Moon. The space race had been won by the Americans. But, curiously enough, Kennedy marked another turning point: the beginning of the media age. Television became widespread and it was at that time that its power was first measured. In fact, Kennedy beat Richard Nixon in the 1960 election because he knew how to handle media language better. Nixon, who was the favourite, despised the power of debates on television, and Kennedy realized that in them and in offering hope for a film script—his Camelot court, according to Noam Chomsky[54]—was the future of politics.

The predominance of television cannot be seen only from the point of view of television as a medium or as a technology. The intellectual formation of those who work in it is highly relevant; besides the fact that those who are producing television's messages (screenwriters, journalists, etc.) are basically not from a physics or telecommunication engineering background (the very disciplines that we have to thank for television and telecommunication). They are the ones who are most critical of science: they come from departments of literature, literary theory, cultural studies, media studies, journalism, and so on.

From the second half of the twentieth century onwards, these faculties of the arts and social sciences began to criticize political power, which extended

---

[53] In T. Theocharis and M. Psimopoulos. (1987). 'Where science has gone wrong.' *Nature,* 32 (15), 95.
[54] Noam Chomsky. (1993). *Rethinking Camelot. JFK, the Vietnam War, and U.S. Political Culture.* Boston, MA: South End Press.

to science itself and to the scientific community: the humanities and arts intellectuals accused the scientists of being 'an oppressive elite allied with the political and economic elites'[55] in their classes and books. There are several reasons for this twist. Reason is blamed for totalitarianism and its consequences: from the Nazi concentration camps to the Soviet gulags, to the atomic bomb and napalm (and the iconic photo of the girl victim of this explosive in the Vietnam War, a Pulitzer Prize photo). On the other hand, in the middle of the twentieth century another decolonization process began and science was accused of both making Western colonialism possible and justifying it, as if Europe blamed Greek mathematics for Roman technology and its colonization.

Under the guise of objectivity and progress, these arts intellectuals conceal an ethnocentrism that privileges the West over other cultures. There is a whole series of philosophers (mainly French, from the French Theory of Deconstructivism, and also others such as the Austrian Feyerabend) who flood departments of humanities and communication with theories such as that science is not a universal knowledge but a European or Western ethno-theory and, in the words of Feyerabend, no more valuable than another interpretation of the world, such as magic, witchcraft or the beliefs of primitive peoples. Moreover, science is a dangerous theory, because of its imperialist and destructive character, and all this is reflected in the cinema and in many humanities and social science teachers (theory of literature, sociology, philosophy) who, while learning in their arts faculties, pass on this interpretation to pupils in primary and secondary schools, in which they teach not only mathematics and science but literature and social studies.

Television creates other values, from individualism to the search for quick success. Effort is replaced by the culture of consumerism and easy living, as it appears in the movies. It also spreads unscientific ideas. These ideas, as we will see, came earlier and appeared in the cinema, but with television the cinema becomes massive; let us not forget that cinema is the master of television. All experts conclude that, from the 1970s onwards, cinema has been clearly unscientific, and this must be emphasized. So, that is where the decline of science begins.

The dominant philosophy in the West, in the early 1960s, was still the logical positivism of the early 1960s, which stated that we can only know that something is true if it can be demonstrated in a logical or empirical way: rational thought. Positivists regarded mathematics and science as the supreme sources of truth.

---

[55] Andrés Carmona. (2017). 'Cientificismo y política', in *Elogio al cientificismo* (ed. Gabriel Andrade). Pamplona: Laetoli.

Science, therefore, continued to rise sharply against other disciplines, and the philosophers Popper, Kuhn, Lakatos and Feyerabend and the French Theory, each in their own way and for different reasons, tried to confront any attitude that was flattering towards science. With the potential that physics and chemistry showed in the 1950s and 1960s, many thought that no one brilliant would want to study any other discipline. The arts faculties would be emptied as there would be interest only in physics. However, some defenders of the philosophers listed above argued that they acted in good faith towards science, and that they believed that by sowing doubts in the scientific method they would strengthen science and the advancement of human knowledge.

The detractors of these philosophers (basically the scientists) argued that their writings sought simply to bring scientists down from their intellectual and social altars, because they saw that philosophy, and the arts in general, were losing the battle for the interpretation of the world. The truth is that in this twenty-first century the writings of these philosophers serve as an argument for the anticipation and advancement of irrational thinking. Thomas Kuhn's book, *The Structure of Scientific Revolutions*, was published in 1962 and has probably been the most influential treatise on how science advances—or does not advance. For many, the decline began with that book. To me, he is just another agent strangling rational thinking.

The book was read religiously in the faculties of humanities and social sciences, and was therefore incorporated into their academic programmes. By the end of the 1960s, it had some effect, and anti-scientific journalists, as well as those intellectuals who were critical of science, gained more presence in the press. Much money had been spent on going to the Moon and, in addition, science intellectuals had gained considerable power—economic and political—in universities and in society. The defenders of arts-based studies, the anti-scientists, attacked as they did because our media culture is wholly supervised and created by people from the arts and social sciences.

By the beginning of the 1970s, several circumstances began to come about that had been in incubation since the 1960s, extending their influence into the 1980s and exploding in the 1990s, generating the current phenomenon of declining vocations.

In 1959, Snow gave his famous and controversial lecture (published in 1960), 'The Two Cultures'[56]:

> Young scientists now feel that they are part of a culture on the rise while the other is in retreat. It is also, to be brutal, that the young scientists know that with an

---

[56]C. P. Snow. (1959). *The Two Cultures.* The citation is from the 1961 edition, published by Cambridge University Press. New York.

indifferent degree they'll get a comfortable job, while their contemporaries and counterparts in English or History will be lucky to earn 60 per cent as much. No young scientist of any talent would feel that he isn't wanted or that his work is ridiculous. (Snow 1960, 19)

I am not sure that a scientist could deliver a speech with such an optimistic approach to science and scientists in this twenty-first century. In fact, the pro-Cold War spirit of science,[57] which lasted until the mid-1980s, is responsible for the fact that there are still large numbers of active scientists aged between 40 and 70 who must now compete for fewer resources because—despite them saying otherwise—science is not really on the political agenda as it was in the 1950s and 1960s. This means that science has also become competitive, creating precarious working conditions that are also aired by the media. All these factors, as we shall see, add up to an image of science as a field without opportunities. They serve to explain, in my opinion, the decline of scientific vocations in countries with a Western media culture.

Golden Age and Scientific Decline in the Media

This interpretation of the rise and fall of science that we have described is obviously based on historical data. But is it documented in the media's image of science? It was a difficult but necessary challenge to answer this question and, further, this one: 'Can a golden age of science be established from the media point of view?' A report that was most helpful to me in answering[58] was published by Bauer and others in 1995, in which they quantified all the scientific news published from 1946 to 1990 in the British newspaper, the *Daily Telegraph*.

One of the variables that were evaluated was whether the approach (which is known as the frame, in the theory of communication, as we will see later on) to the scientific news was favourable or unfavourable towards science. Adding the favourable approaches of each year and subtracting the unfavourable ones, they obtained a relative numerical parameter. If positive, it implied that there was more news with a beneficial approach than with a harmful one. That is to say, whenever the result exceeded zero, it meant that, in the final computation,

---

[57]There are many dates to establish the end of the Cold War period. One of them is the fall of the Berlin Wall (1989) and another is the disintegration of the USSR (1991). But one of the dates that may serve our purpose is the signing of the first arms reduction treaty between the US and the USSR in 1987.

[58]M. Bauer, J. Durant, A. Ragnarsdottir and A. Rudolfdottir. (1995). 'Science and technology in the British Press 1946–1990.' *London Science Museum Technical Reports*, 1–4. Cited in Bauer, M.W. and Gregory, J (2007) 'From journalism to corporate communication in post-war Britain' in *Journalism, Science and Society* (Bauer & Bucchi editors). London: Routledge (p. 33–51).

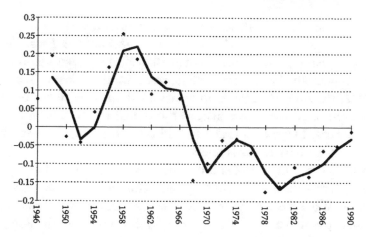

Fig. 3.7 Average tone of science news stories rated as relatively favourable/unfavourable towards science' in the Daily Telegraph. *Source* Bauer et al. 1995 cited in Bauer and Gregory (2007). M. Bauer, J. Durant, A. Ragnarsdottir and A. Rudolfdottir. (1995). 'Science and technology in the British Press 1946–1990.' London Science Musseum Technical Reports, 1–4. Cited in Bauer, M. W. and Gregory, J (2007) 'From journalism to corporate communication in post-war Britain' in Journalism, Science and Society (Bauer and Bucchi editors). London: Routledge (pp. 33–51)

the approach in favour of science won. If the sum was equal to zero, it indicated that the number of news items with positive approaches equalled the number of negative ones. If the result was less than zero, it implied that the science was portrayed as negative (Fig. 3.7).

Viewing the data for the whole period, they observed that there were two clearly distinguishable stages: one from 1956 to 1966, in which the media spoke well of science and the summation, therefore, was always positive. The peak in the media's enthusiasm for science came in 1961/62 (above +0.20), after which it began to decline. It is curious that the maximum coincided with Kennedy's speech that we mentioned. The next phase started in 1967, when the curve reached zero and continued with a sharp decline that had its lowest point in 1970, below −0.1. It began to recover slowly, with a small rebound in 1974 that reached −0.025, but quickly dropped to the historical low of the series being studied. This minimum was reached in 1980, when the value for the attitude towards science fell below −0.15. From the 1980s onwards, it recovered again, but during the years studied the sum of the approaches to scientific news never become positive again. That is to say, at least in the *Daily Telegraph*, a negative approach to science has prevailed.

It can be argued that this is just data for Britain and cannot be extrapolated. But, just as I arrived in London, together with some Bulgarian colleagues

Bauer was about to publish a comparison with Bulgaria[59] using a similar methodology and applied to the newspaper *Rabotnichesko Delo*. The curves converged at many points, even though Bulgaria had been behind the Iron Curtain, so it might be felt that the representation of science would differ. Since 1972, the sum of negative information about science has been greater than that with a positive approach, so the overall sum has been below zero. This in Bulgaria. In general, it can be said that, from this time on, the trend has been for negativity towards science.

Obviously, more newspapers and more countries—including Spain, and above all China—need to be investigated before we can draw more general conclusions. What we can do is to rewrite the history of the rise and fall of science that appeared a few paragraphs back, this time from the point of view of the power of the media.

Basically, from these data on the media's approach to science, what we can summarize is that there was indeed very positive treatment from just after World War II until the late 1960s. Jane Gregory, Professor of Public Science Communication at the University College of London, explains that this was because in the post-war period public opinion recognized the crucial role that science had played in the Allies' victory. This recognition coincided with the abolition of the laws of secrecy about the scientific advance that dominated the war period (1939–1945). All this meant that the media no longer had any problem with disseminating not only the achievements of science in winning the war but the promise that scientific knowledge represented for social and economic progress.[60]

I am interested in highlighting the phenomenon of unfulfilled curiosity during the laws of scientific secrecy in force during World War II, which led to a spectacular interest in science among the population. That is to say, in this strange field that is the public communication of science there may be a paradox whereby non-disclosure can enhance interest, and over-information disinterest.

In any case, the curves in Bauer's media approach also show a decline in the late 1960s, which, despite some recovery since the mid-1980s, has not regained the level of scientific optimism of the early 1960s. One of the BBC's most famous science fiction television series writers, Gerry Davis (1930–1991)—creator, among others, of the *Dr. Who* series—describes very well what happened at that time:

---

[59]M. W. Bauer, K. Petkova, P. Boyadjieva and G. Gornev. (2006). 'Long-term trends in the public representation of science across the "Iron Curtain": 1946–1995.' *Social Studies of Sciences*, 36(1), 99–131.
[60]J. Gregory. (1998). 'Fred Hoyle and the popularisation of cosmology.' Doctoral thesis, University of London.

By the end of the 1960s, earlier optimism in media coverage had given way to political scepticism over science's ability to contribute to the economy, and to public criticism of science and technology that centred on its military connections and its adverse environmental impact. While many of the new science journalists were primarily enthusiasts for science, the developing climate of social criticism in the 1960s also gave space to journalists who challenged science. This critical tone was echoed in science-based drama: in 1970, one television screenwriter noted that 'the days when you and I marvelled at miracles of science… are over. We've grown up now – and we are frightened…. The honeymoon of science is over. (Davies, cited in Gregory and Miller 1998, 44)[61]

These statements are extraordinary, because they take us back in time to the moment when the media began to offer a negative image of science. They reproduce what was discussed in those newsrooms and television studios in the late 1960s, which were not accessible to natural scientists. I believe that this type of statement by the media leaders who lived through this period explains the state of the issue more than the numerical results of the content analysis that we mentioned above.

These results, which are extremely interesting, must always be qualified, because the meaning of 'negative image' needs to be studied further. For example, if the negative approach refers to criticism of human behaviour by scientists, it does not necessarily imply a decline; on the contrary, even. If those criticisms were true, they show only a symptom of good, critical journalism. Scientists, like any other human group, are likely to commit inbreeding, embezzle resources, provoke harassment at work, submit to political dictates to improve a mediocre scientific career, pollute the environment and collaborate with the military for unethical purposes. Obviously, all this must be denounced in the media and, furthermore, it is beneficial not only for journalism but for science itself, purging it of scientists who, like any human beings, may be corrupt.

However, a different parameter is the criticism of science itself as a method of obtaining truth, or contempt for a scientist's work against that of other collectives. That would be a most dangerous thing, because it represents a gateway to irrationality and esotericism and, in short, a return to mediaeval magical thought. Interestingly, in recent years cinema and especially television have shown an unusual interest in telling stories in a mediaeval setting, which allows the author to mix reality and magic in a credible way.

In addition to these statements by BBC scriptwriter Gerry Davis, which illustrate the climate of social criticism that prevailed in the late 1960s and its influence on the media's approach to science, this period coincided with

---

[61] Davis' statements quoted in J. Gregory and S. Miller (1998). *Science in Public: Communication, Culture and Credibility*, (p. 44). New York: Plenum.

an important phenomenon in the history of journalism: the replacement of scientists who collaborated with newspapers by professional journalists.[62]

Until the mid-1960s, a significant percentage of science news coverage was in the hands of active scientists or high school science teachers looking for a spot on their agenda to inform the public about scientific fact. In the late 1960s, newspapers had to compete with television, now becoming increasingly powerful, and one of their survival strategies was an increase in the number of pages. This increased the coverage of all types of news, including scientific news. The *Daily Telegraph* went from 2,000 science news items a year in 1946 to 6,000 in 1990. This is most interesting, because it confirms the fact that publishing more about science is not necessarily positive in itself; it can even be harmful if the approaches are negative, inaccurate, sensational or unscientific.

The increase in the number of pages and the number of news items meant that the collaborators had to go full time. If a scientist did not want to give up science or science education, he or she was replaced by a professional journalist. The problem was that most of these professionals had a background in the humanities, social studies, journalism, media studies or cultural studies, and these groups were fascinated with French Theory.

Journalism, in general, has always paid worse than science, especially science journalism. In other words, the professionalization of science journalism coincides with the increase in the negative image of science, at least in terms of dates. Throughout this book we will delve into this last idea, because it constitutes one of the central axes of the decline of science in the West. We will show that this is basically the result of the intellectual and university training of professional journalists, many of whom are involved in reporting on science.

The Public Understanding of Science (PUS) Movement

If we look at the scientific decline from the point of view of vocations or interest through surveys, as we have done, we see that the trend is always negative and, furthermore, that this trend worsened in the 1980s, 1990s and especially at the beginning of the twenty-first century. This is contrary to the Bauer curve that we are talking about, because according to that study, at least in the *Daily Telegraph*, from the 1980s onwards there has been an upturn in the image

---

[62] M. W. Bauer and J. Gregory. (2007). 'From journalism to corporate communication in post-war Britain.' In M. W. Bauer and M. Bucchi (eds), *Journalism, Science and Society. Science Communication Between News and Public Relations,* pp. 32–51. New York: Routledge.

of science. However, it must be repeated, the sum of articles with a positive attitude to science news has never since exceeded that with negative attitudes.

Although more experiments designed not to push forward the boundaries of science but to attempt to replicate earlier results, to verify them, are essential, this phenomenon can be explained in several ways. First of all, vocations are decided at an early age, and young people in the 1980s, 1990s and, especially, at the beginning of this century did not read the press. For them, cinema and television were more important, and that is why we will dedicate space to such channels in this book. The image of science and the scientist offered by film and television does have a significant influence on vocations, as has been seen in the case of *The Big Bang Theory*.

On the other hand, studies to replicate earlier experiments are a necessity. While I greatly appreciate the effort and hard work of Bauer and his collaborators in analysing the focus of the scientific news that appeared from 1946 to 1990 in the *Daily Telegraph*, I recognize that this is a single newspaper; it probably only has a single science journalist. It could be enough for this journalist to have changed for the approach to become different.

If someone were to carry out, for example, an analysis of the content of the approach that the newspaper *El Mundo* gave to the Spanish National Research Council (CSIC) or to university inbreeding in Spain between 2000 and 2006—more than 90% of the professors of a university have read their thesis there and had no contract elsewhere, as has been criticized in *Nature*, no less—they would observe that there are two clearly differentiated periods. One is from 2000 to the end of 2002 and the other is from that date to 2006. The first period is critical, both to the CSIC and its political dependence and to Spanish university inbreeding. Is there a sociological, media or philosophical justification for this change in trend?

The explanation is biographical. In the first period, I was responsible for the scientific information of the newspaper. I had previously researched for my doctoral thesis the use and abuse of public relations by politicians and journalists of scientific officials.[63] I had many scientific friends who were suffering from the perverse effects of Spanish university inbreeding, so I found these science policy issues interesting and proposed them to my bosses. Pablo Jáuregui, who was in charge of science reporting from the end of 2002, had not had these experiences, thus his approaches were different. I mention this because, when journalistic information is in the hands of a few journalists—as in the case of science—the biography of the editor can be fundamental to explaining the trends.

---

[63] C. Elías. (2000). 'Flujos de información entre científicos y prensa.' Doctoral thesis, Universidad de La Laguna. Servicio de Publicaciones.

However, I am willing to admit that the upturn in the image of science in the 1980s—as described by Bauer for the *Daily Telegraph*—may be generalizable to the entire British press. It may make sense, yet in my opinion this does not discredit my hypothesis of scientific decline in the West because of media culture so much as how we interpret this upturn, in any case. This is what Bauer himself does as a consequence of the influence of the so-called PUS movement, which could be termed the public understanding of science.

The PUS movement was promoted by the Royal Society in 1985, alarmed by the negative media image of science. The strategy was defined in its now famous report, *Public Understanding of Science*, and was based on two very well-designed lines of action: to make scientists aware that they should not only communicate with the public but that they should 'consider it an inescapable professional duty'; and, at the same time, to inspire a proactive policy of 'infiltration' of science graduates into the fields of communication. In a doctoral thesis on this movement, it is pointed out:

> Science students are being encouraged to take part in this infiltration: the institutionalisation of science communication as an accredited component of formal science education in the UK began in the late 1980s. The courses were hosted by scientific institutions rather than journalistic institutions, and admission criteria were scientific qualifications. (Davis 2000)[64]

One of the side effects of the PUS movement was that, in programming numerous degrees and postgraduate degrees in scientific communication, the public relations (journalists working for press offices) proliferated in scientific institutions. The doctoral thesis that we mentioned about the PUS movement points out:

> Science communication remains a growth industry in UK universities. Employment rates among graduates from these programmes are high and their destinations varied: many join the growing ranks of PR agencies devoted to promoting scientific and commercial interests. In contrast to the days of the isolation of the academic 'ivory tower', the biggest growth areas for public relations in the UK have been the universities and the health and biomedical sciences. (Davis 2000)

Today, there is no scientific institution or university that does not have its own science journalist to produce science news for the web. The NASA website has millions of views; however, this information does not translate into a huge quantity of young people interested in studying for STEM careers. There are

---

[64]A. Davis. (2000). 'Public relations, political communication, and national news production in Britain 1979–1999.' Doctoral thesis, Goldsmith College, London (in M. W. Bauer and J. Gregory, 2007, 40).

more in China or India, where there is less disclosure. It is an important problem: by making scientific information so understandable, people may not want to strive to learn mathematics or physics to understand what the Big Bang theory actually is. But, without observations, equations and theorems, the Big Bang theory, explained only with literary language vocabulary, is just another story, and perhaps of worse quality than Genesis. This may be another of the errors of Western science: if you think that it can be understood without equations or formulae, why study it, considering how difficult it is?

This is the situation that we are in at the time of writing this book. In principle, it might seem that we are on the right track. However, science communication is such a complex and fascinating area that the paradoxical fact is that the more people who are engaged in it (especially in public relations in press offices), the more vocations are being lost and the more disinterested the population becomes. We will describe in detail how this may be due to the fact that, since there is greater competition for 'selling' the message to the media from scientific sources, the caricature of science, news events or news stories are more important.

4

Reason on the Ropes

Scientific American is the oldest continuously published monthly in the United States (it became monthly only in 1921), with 171 years of history. It includes among its authors the most important American scientists (including Albert Einstein). The October 2017 issue included a monograph with a provocative title, 'Reason on the Ropes'. The issue analysed in just 10 pages how Western science was going into decline and, by contrast, how China was becoming stronger and stronger. The periodical highlighted two political tragedies—the election of Donald Trump and Brexit—that threaten science in the two most important research countries of the West: the United States and the United Kingdom.

In the United States, a growing anti-science movement has arisen, led by the far right, which believes that scientific facts are opinion, as if they were views on the history of film. It is in this sense that the debate on climate change is situated: as if one had to choose between two opinions or both were equally valid, as is the case in the humanities. In literary criticism or in philosophy, one can affirm both one thing and the other—and even in the social sciences. On the other hand, Brexit will mean that European science will cease collaboration or, in any event, there will be even more bureaucratic obstacles. That is always a negative for the production of knowledge.

The Left, on the other hand, also accuses science of multiple evils. This was not always the case. The beginning of the concept of Left and Right was born in the ephemeral National Assembly of the French Revolution (it lasted only for June and July in 1789). To the right of the President of the Assembly were the defenders of religion and the divine right of the king. To the left were the representatives who were in favour of equal rights, of the revolutionary idea

© Springer Nature Switzerland AG 2019
C. Elías, *Science on the Ropes*, https://doi.org/10.1007/978-3-030-12978-1_4

of 'freedom, equality and fraternity' and, above all, of the Enlightenment; and who, remember, were the product of the scientific revolution.

With these beginnings, it is logical to expect the Right side to act against reason and science. The Left was for scientists, for encyclopaedists. The best thing about the French *Encyclopaedia* was the contribution by scientists and the idea of making magical–religious thought irrelevant. Give Newton more room than St. Augustine or St. Paul, more of Francis Bacon than of St. Teresa or Jesus. More space for technology and machines than for the lives of kings. The Left was to be the descendants of the encyclopaedists: the mathematician and inventor Jean-Baptiste de La Chapelle; the pioneer of the defence of atheism, Paul-Henri Thiry, Baron d'Holbach; the engineer and naturalist, Jean-Baptiste Le Romain; the chemist and doctor Gabriel Venel; and the physicist, mathematician and engineer, Jean-Baptiste D'Alembert.

If there was a difference between Right and Left, it was that the Right was religious; therefore, it was prone to magical-esoteric thinking and, in many cases, opposed to science, especially science that questioned magical-mystical-religious thinking. And another aspect of the Right was its belief in the divine origin of life, which had repercussions on medical and biomedical issues: abortion, human embryo research, genome variations… The Left was the opposite: rational thinking; empiricism versus mysticism; engineering and machinery to avoid the arduous work of the lower classes; science versus belief.

However, from the middle of the twentieth century onwards, with the proliferation of postmodern philosophers and especially the influence of cultural studies (which value any kind of cultural belief) on communication studies, and the enormous influence of these professionals on Western culture, the Left began to be characterized by an attitude against the Enlightenment and by follow-up theories opposed to scientific explanation. The journalist Mauricio Schwarz, in his book *The Left Feng-Shui: When science and reason ceased to be progress*,[1] argues that in Western societies the Left embraced the American anti-intellectual tradition of 'my ignorance is as good as your knowledge', counter to Asimov's criticism of this attitude, and that its view is now shared by social justice movements and the extreme Right. The chemist and intellectual Isaac Asimov had complained that the idea that 'ignorance is as good as your knowledge' was valid in the United States, and was maintained by the American extreme Right, and that was what Asimov criticized. What is said in Schwarz's book is that this way of thinking is also being adopted by the Left, which was previously rational and enlightened.

---

[1] Mauricio Schwarz. (2017). The left Feng-Shui. When science and reason ceased to be progress. Barcelona: Ariel.

The title of Schwarz's book is revealing, because it mentions how the West has misinterpreted the *feng shui* of ancient China as its mystical beliefs. While the West, through its cultural studies, lays emphasis on Chinese anti-scientific mysticism, the Asian country has renounced it and, paradoxically, is the greatest current promoter of Western science and technology. Faced with the esotericism of *feng shui*, which believes in mysterious energies that no one can measure and that do not allow reproducible experiments, China has taken on the only thought that has made the West great: science. It is not its history, philosophy or art, but physics, chemistry and biology. It's the world upside down. And this universality of science is something that bothers arts intellectuals. The ideas of the greatest expert on Kant, Shakespeare or Velázquez are irrelevant in Asia, not to mention the ideas of the postmodern philosophers. However, chemistry, mechanics, physics and biology are studied with commitment, and this is humiliating for intellectuals in the arts and social sciences.

The West (the United States or the European Union) carries on, using the momentum of scientific thought from the Enlightenment. But there are developing countries that, instead of embracing science, have embraced *feng shui* thought in the West and, obviously, have done very poorly, as did China until it embraced physics, chemistry, biology and engineering as some of the greatest cultural products of mankind.

Curiously, as we will see, cultural studies do not deal with science unless to discredit it. They prefer magic to science. That's why their students, who will later become writers, filmmakers, journalists or cultural critics, prefer magical thinking to scientific thinking, and why there are more series—from *Charmed* to *Game of Thrones*—in which magicians appear than those featuring scientists.

Latin America is a clear example. From the nineteenth and mid-twentieth centuries, it was in a better position than China in terms of educational level, let alone wealth (especially in terms of raw materials), yet it had a disastrous scientific policy for two reasons. The first would be cultural: having been conquered by Spain in the midst of the Counter-Reformation, Iberian mysticism (hence the magical realism of García Márquez there) prevailed in the inherited culture, instead of English empiricism. Mysticism (*feng shui* thinking, Spanish style) does not only imply political action or unreasonable ideologies but an even worse problem. This is that the ideological professions—law, communication, sociology—are valued more than the material sciences: chemistry, physics, mechanical engineering and computer science. It is not easy to find scientists or engineers among South American politicians or the elite. Another consequence is that social studies and the humanities not only despise science but study bizarre theories about it, and this does much damage to these countries.

A good example is that of the Argentine Óscar Varsavsky (1920–1976) who, although he studied chemistry (at the University of Buenos Aires), had an attack of mysticism (and was influenced by Thomas Kuhn and his erroneous conception of the scientific paradigm) and wrote a delirious pamphlet, 'Science, Politics and Scientism'[2] (1969). Varsavsky's thesis is that all science is ideology: from discovery to consolidation and application. He does not practise it, but he writes it (all that is not natural science: everything is valid, at least for publication and study). He believes that if a scientist does not commit to producing science that can change the political system, that science is not worth it. Worse still, it may lead to the installation of an imperialist ideology.

According to Varsavsky, who despises everything from theoretical physics to topology, the aim of science must be to support and sustain a revolution aimed at the establishment of a 'national and socialist' political movement. Although Varsavsky is Left wing or, for some, Far Left, these ideas may seem inspired by Nazi ideologue Philipp von Lenard (1862–1947). They had the same education, since Lenard, like Varsavsky, studied and earned his doctorate in science—in physics—before 'philosophizing' on science policy.

Lenard was a remarkable scientist. He discovered many properties of cathode rays, earning him the Nobel Prize in physics in 1905. Varsavsky is not known to have made any great scientific achievements worthy of a Nobel Prize. But Lenard, like Varsavsky, suffered a mystical-ideological outburst when he proclaimed that science was also ideology and that it was 'national'. He argued that it should not be ideologically neutral and, just as Varsavsky had proclaimed a Latin American science, Lenard advocated an 'Aryan physics' that would be useful to Germany.

Lenard fought with all his might against 'Jewish physics', especially Einstein's theory of relativity. His stance, backed up by the Nazi politicians, was to prevent much teaching of Jewish-produced physics—as if physics were an ideology, such as law, economics or philosophy. Interestingly, this worked against Germany: many Jewish physicists emigrated to the United States, and it was there—thanks to them, among others—that the atomic bomb was developed, which the Germans had thought could be done only with 'Aryan physics'. The atomic bomb put an end to fascism, but it may have been one of the beginnings of scientific decline (atomic physicists started being referred to as nuclear physicists as 'atom' and 'atomic' acquired a negative image).

To determine that science is political ideology is like saying that if Newton, instead of being English, had been Spanish, perhaps the force that a body

---

[2]Oscar Varsavsky. (1969). *Ciencia, política y cientificismo*. [Science, politics and scientism]. Buenos Aires: Centro Editor de América Latina. A complementary view can also be read in Varsavsky, Óscar (1972). *Towards a national scientific policy*. Buenos Aires: Centro Editor de América Latina.

experiences would not be its mass multiplied by its acceleration (f = m.a), or that the law of universal gravitation would be different if Newton had been French or German. Without the results of science we can make absolutely no progress. In Venezuela or Argentina, no matter how much they study Varsavsky, if they want to put a communications satellite into orbit they have to know Newton's science. And the same goes in America or China. By contrast, in any of these countries they can do without books by Varsavsky, Kuhn, Derrida, Foucault and Feyerabend. That is the greatness of science compared to other, non-scientific disciplines.

Argentine physicist Gustavo Romero argues:

> the idea of Aryan science disappeared with the Third Reich. The idea of Latin American science, on the other hand, still has many defenders. Varsavsky's thought transcended the borders of Argentina and had a great influence on those sectors of the Latin American intelligentsia concerned with breaking the patterns of cultural and economic dependence. (Romero 2017, 98)[3]

One of the countries where Varsavsky's ideas were most influential is Venezuela, where he lived for several years. Hugo Chávez (1954–2013) frequently quoted him in his speeches, and some, such as the neo-Marxist economist Alfredo Serrano, considered him to be one of Chávez's leading ideologues,[4] describing him as 'the example of a scientist committed to the development of a national science'. And, as the Argentine physicist Gustavo Romero recalls: 'In reality, what is national are the policies of scientific development, which in Venezuela have been non-existent since 1999' (Romero 2017, 99).

For Varsavsky, the scientist in an underdeveloped nation 'is a perpetually frustrated man':

> To be accepted into the upper echelons of science he must devote himself to more or less fashionable subjects, but as fashions are implanted in the North he always starts out at a disadvantage of time. If you add to this the least logistical support (money, labs, helpers, organization) it's easy to see that you're in a race you can't win. His only hope is to maintain close ties with his Alma Mater – the scientific team where he did his thesis or apprenticeship –, to make frequent trips, to make do with complementary or filling-in jobs he does there, and in general to reach a total cultural dependency. (Varsavsky 1969)

---

[3]Gustavo Romero. (2017). 'Confesiones de un maldito cientificista.' In Gabriel Andrade (ed.), *Mario Bunge. Elogio del cientificismo*. Pamplona: Laetoli.

[4]Alfredo Serrano. (2014). *El pensamiento económico de Hugo Chávez*. Mataró: Intervención Cultural.

When Varsavsky, who has had much influence on the organization of science in Latin America, especially in countries such as Venezuela or Argentina, wrote this pamphlet in 1969 these countries had a scientific, cultural or economic position far superior to China. Nothing was preventing them from collaborating, as in the European Union, or adopting China's science or technology policies.

The Chinese never believed these theories. They preferred Newton, Darwin and Einstein to Varsavsky and the postmodern philosophers, from Kuhn to Popper, Feyerabend and all those who espoused the French Theory, as did Foucault, Derrida, Lacan and Deleuze, among others. And so, as we shall see, by preferring Euclid to Aristophanes and Newton to Foucault, China will soon be the world's leading scientific power.

Being the first scientific power means being the first technological power and, obviously, the first economic and cultural power. There are only three ways for a country to generate wealth: the conquest of other territories and the appropriation of their wealth (but the entire planet is already conquered); natural resources (which always give rise to extractive elites that appropriate those resources); and science and technology, which need human talent and, therefore, must enhance access to education for the entire population, if only for the economic interest of someone who starts up a great business development. And this has been understood in Asia, but not in Africa or South America.

One of the turning points of Western science could be seen at a specific moment: on 2 June 2017. On that day, US President Donald Trump revealed his most radical beliefs and decided to break with the Paris Accord against climate change, which he called 'debilitating, disadvantageous and unfair'. The withdrawal of the pact, signed by 195 countries, marked a new era. This is not only because the president of the most powerful nation in the world turned his back on science—proven scientific data—and deepened the fracture with Europe, but also because it was China that took his place. China, the world's most polluting nation, did want to support the agreement. In other words, China was in favour of science, making political and economic decisions based on scientific data (physics, chemistry, biology and geology).

There were two men there, Donald Trump and Chinese President Xi Jinping, the two leaders of the most powerful nations, with two biographies that explained their relationship to science and, by extension, their position on climate change. While Trump had studied business administration (an intellectually poor science background), Xi had graduated in chemistry and chemical engineering from the prestigious Qinghua University in Beijing. This is a centre that was founded in 1911 with the idea that its students continue their studies in American universities. Xi himself studied in the United States and his only daughter was a discreet and highly studious Harvard student who graduated in 2015.

Those two men who met in Paris, the cradle of the Enlightenment and where reason triumphed over irrationality, were like two icons of their culture in the whole past of humanity—the West and the East—and also the whole future and how to face one of the greatest challenges. America had led the world for the last 200 years because it had had politicians who believed in science: from Benjamin Franklin to Kennedy. But now Trump was declaring, without a blush, that climate change is an invention of the Chinese, that vaccines cause autism and that immigrants spread disease.

Trump's vice-president, Mike Pence, did not study science either. He received his Juris doctorate in Indiana. However, despite his lack of scientific training—or precisely because of it—he questions Darwin's theory of evolution in the media and is a recalcitrant defender of the theory of intelligent design. Trump and Pence, in addition to their scientific illiteracy, have something in common: both have become famous (before they became politicians) for talking rubbish in the American media: Pence on the radio, on the *Mike Pence Show*, which has been broadcast since 1994 on almost twenty radio stations; and Trump on reality television shows (*NBC* and *Prime Time*) or the Miss Universe beauty pageant. Both represent Western intellectual degradation, although both are university trained (Trump studied business and Pence law) but not scientists or engineers. Western universities can graduate an individual who is absolutely illiterate in science, and this is the clearest symptom of their decline.

In the West, in general presidents have studied either law or economics, but China prefers engineers. Xi Jinping, a chemical engineer, replaced Hu Jintao, a hydraulic engineer, who in turn replaced Jiang Zemin, an electrical engineer. Since 1993, when Jiang came to power—that is, for the last 25 years—China has been run by engineers. Previous politicians came from the army. The fact that a country's political elite is made up of engineers is the clearest indicator of the role that society accords to science and technology.

There is another interesting approach within the United States: the Asian racial minority (mostly from China, but also from India) has great academic success, despite coming from middle-income migrant families. These Asian families have a great appreciation of science and technology, which they pass on to their children. A few weeks after the Paris meeting, there was another important event, this time in Thailand: on 15 July 2017, the United States made history at the 49th International Chemistry Olympiad, a secondary school event, by winning four gold medals for the first time, taking first place.

It is well known that American secondary education does not have a high level of achievement (this is the preserve of the elite universities), so how did they manage this unprecedented success? It was not explained well, although it was interesting: the four Americans who had won, and who appeared smiling in the photo in *American Chemical Society*,[5] were of Asian origin: Steven Liu, Joshua Park, Harrison Wang and Brendan Yap. The runners up (with three gold medals and one silver medal) were, of course, an Asian team: Vietnam.

Why it is that Asians, including those living in the United States and in Europe, are not carried away by the anti-scientific Western culture is a very interesting phenomenon that the West has hardly paid any attention to. When I was at Harvard, I was able to see the huge number of Asian-Americans in the Science Center library. I asked about it, but it seemed taboo. Nobody wanted to talk. However, there was some discomfort in the Asian community, because it felt that its members were being discriminated against for being better at their studies (they did not express it so clearly, but perhaps they thought so) and, in some conversations, some considered that a discriminatory treatment had come to light: there were greater demands on them, just because they were Asian.

Some compared it to the pre-World War II discrimination against intellectually brilliant Jews in American and European universities. Asians' culture of effort has before been shared, in the West, especially in the Calvinist and Protestant countries in which work was a way of honouring God. This was not the case in Catholic culture, in which work was a divine 'curse'. However, the influence of celebrity culture and the media, with films in which heroes mock science nerds, has undermined the culture of Europe and Protestant America that prevails among Asians, who are impervious to the culture of Hollywood, Almodóvar or the French *Nouvelle Vague*.

I discovered this phenomenon during my year at Harvard (2013–14), but it was nothing more than simple conversations in the Science Center cafeteria. It exploded in the media in August 2017 in an article in the *New York Times* entitled 'Justice Dept. to Take On Affirmative Action in College Admissions.'[6] The report brought to light a thorny problem: Asian-Americans may be required to score higher to enter the Ivy League campuses, based on affirmative action laws that must promote (on equal merit) the acceptance of students from racial minorities. Asian-Americans are a racial minority in the United States; however, these regulations serve to require more from them than from others (even the White majority) to qualify.

---

[5]Linda Wang. (2017). 'U.S. team makes history at International Chemistry Olympiad'. *Chemical & Engineering News* (24 July 2017), 95(30), Available at: https://cen.acs.org/content/cen/articles/95/i30/US-team-makes-history-International.html.

[6]https://www.nytimes.com/2017/08/01/us/politics/trump-affirmative-action-universities.html?_r=0.

The matter was picked up a few days later by other means. The *Atlantic*, under the headline 'The Thorny Relationship Between Asians and Affirmative Action',[7] suggested that Asians may be disadvantaged but that this was difficult to substantiate and, in any case, failing to introduce such racial parameters could significantly increase the number of Asians on campuses, despite them being a minority:

In 2005, a Princeton study found that Asian-Americans must score 140 points more than white students of otherwise comparable caliber on the 1,600-point SAT in order to be considered equals in the application process; it also found that they'd have to score 270 points higher than Latino students and 450 points higher than black students. Other studies[8] have found that eliminating race-conscious admissions policies would see a drastic increase in the number of Asian students admitted. Just look at the California Institute of Technology, which bases admission strictly on academics: Asian enrollment at the school grew from 25 percent in 1992 to 43 percent in 2013. Similarly, Asian-Americans account for roughly a third of students at the University of California yet make up just 15 percent of the population in California, which prohibits race-conscious admissions. In other words, based on these examples, Asian students appear to be extremely overrepresented in relation to the general population when their institution doesn't practice race-conscious admissions. (*Atlantic*, 2017)

The *New Yorker* was also talking about the issue around the same time in an article, 'The Uncomfortable Truth About Affirmative Action and Asian-Americans',[9] by Jeannie Suk Gersen. This stated: 'In a federal lawsuit filed in Massachusetts in 2014, a group representing Asian-Americans is claiming that Harvard University's undergraduate-admissions practices unlawfully discriminate against Asians.' Gerson added:

At selective colleges, Asians are demographically overrepresented minorities, but they are underrepresented relative to the applicant pool. Since the nineteen-nineties, the share of Asians in Harvard's freshman class has remained stable, at between sixteen and nineteen per cent, while the percentage of Asians in the U.S. population more than doubled. A 2009 Princeton study showed that Asians had to score a hundred and forty points higher on the S.A.T. than whites to have the same chance of admission to top universities. The discrimination suit survived

[7] https://www.theatlantic.com/education/archive/2017/08/asians-affirmative-action/535812/.

[8] Lauren Robinson-Brown. (2005). 'Ending affirmative action would devastate most minority college enrollment.' Princeton University. Available at: https://www.princeton.edu/news/2005/06/06/ending-affirmative-action-would-devastate-most-minority-college-enrollment?section=newsreleases.

[9] https://www.newyorker.com/news/news-desk/the-uncomfortable-truth-about-affirmative-action-and-asian-americans.

Harvard's motion to dismiss last month and is currently pending. (*New Yorker*, 2017)

By the end of November 2017, the case had been fully prosecuted. According to the Associated Press (21 November 2017), the US Department of Justice threatened (in November 2017) to sue Harvard University in order to obtain records as part of an investigation into the school's admissions practices, following a lawsuit by a group of Asian-American students.[10]

On 17 November 2017, a letter from the Department gave Harvard until 1 December to turn over a variety of records that Justice officials had requested in September, including applications for admission and evaluations of students. The department said that Harvard had pursued a 'strategy of delay', and threatened to sue Harvard if it did not meet the department's deadline: 'We sincerely hope that Harvard will quickly correct its noncompliance and return to a collaborative approach,' the letter said, adding that 'Harvard has not yet produced a single document.' The Associated Press remembered that the inquiry was related to a federal lawsuit filed by a group of students in 2014, alleging that Harvard limits the number of Asian-Americans that it admits each year. A similar complaint was made to the Justice Department. A statement from Harvard on November 2017 said that it will 'certainly comply with its obligations' yet also needed to protect the confidentiality of records related to students and applicants. The university said that it has been 'seeking to engage the Department of Justice in the best means of doing so.' This move indicates that the situation was felt by Harvard to represent a threat. A letter dated 7 November from a Harvard attorney said that it was 'highly unusual' for the department to open an investigation into a complaint more than two years after it had been filed and while it was still being decided in court.

On 30 August 2018, the Justice Department lent its support to students who were suing Harvard University over affirmative action policies that they claimed discriminated against Asian-American applicants. This case, according to the *New York Times* (30 August 2018), could have far-reaching consequences in college admissions:

'Harvard has failed to carry its demanding burden to show that its use of race does not inflict unlawful racial discrimination on Asian-Americans,' the Justice Department said in its filing. The filing said that Harvard 'uses a vague "personal rating" that harms Asian-American applicants' chances for admission and may be infected with racial bias; engages in unlawful racial balancing; and has never

---

[10]http://www.foxnews.com/us/2017/11/21/feds-threaten-to-sue-harvard-to-obtain-admissions-records.html.

seriously considered race-neutral alternatives in its more than 45 years of using race to make admissions decisions.'[11]

Obviously, minority like the Asian-American must be defended. But this can also have a dark side: the Trump administration want to use this case to eliminate affirmative action laws whose objective is that other minorities also access elite universities and study, among others, STEM disciplines. It would be terrible if this demand of the Asian-American were used to eliminate the possibilities of other minorities.

China Versus the Western World

There is a huge difference between China and the Western world: the Chinese do not have a feeling that empires are crumbling, because the Chinese Empire, which was born at the same time as the Roman Empire, remains intact. In Europe, we have asked and rethought why empires are falling. Since Herodotus, and especially in the last few centuries, we have been fascinated by how the Roman Empire fell, from the greatness and decay of the Romans described by Montesquieu to Edward Gibbon's history of the decadence and fall of the Roman Empire, among many others. This way of understanding how the Roman Empire collapsed has extended to why it disappeared, from the Carolingian to the Spanish, the Napoleonic and the British empires... And, at the moment, the whole of the West believes that the end of the American (and, by extension, European) Empire is coming.

The first Americans were fascinated by the Roman Empire—hence the existence of a Capitol and a Senate in Washington—and they established a series of rules to prevent a president of their newborn republic from becoming an emperor: it is the origin of a democracy with many counterweights and stands between Trump and absolute power. Indeed, he has lost many of his proposals in the Senate. And the same thing happened with Obama. There are interesting debates about why the Roman Empire succumbed: many say that in its final days it accepted many foreigners and its rules were too soft on granting citizenship, and that this led to the loss of classical Roman values. Others argue against this, citing the case of Sparta, which disappeared because it was so demanding of those to whom it granted citizenship.

We must also analyse this decline from the point of view of the relationship between empire and technology. Technology gives power, because it makes we mammals with few natural abilities (unable to fly by ourselves, run fast

---

[11] https://www.nytimes.com/2018/08/30/us/politics/asian-students-affirmative-action-harvard.html.

or swim well…) highly effective. Some may dispute that it is intelligence, not technology, that makes *Homo sapiens* powerful. But it's not true. The intelligence of one group can overcome that of another group if it is able to produce technology that makes it stronger or more influential. Intelligence produces technology, and this gives that group more power than those who do not have it.

Roman engineering (from roads to buildings to aqueducts) still fascinates us. And Rome was the one that favoured empire (not religion, literature or laws). Latin disappeared as a European language when Roman roads became impassable and were left unrepaired. Since there was no contact, Latin evolved into several languages that are now incomprehensible to the speakers of another. However, it was not a problem of language but, above all, of the fact that the technology that kept Roman Europe connected now succumbed. When the Roman Empire fell, nothing interesting was cooking there, from a scientific or technological point of view.

Technology also led to the Spanish conquest of America and, let us not forget, it was the basis of the British Empire (its industrial revolution) and the American Empire after World War II. One of the problems of the fall of the Spanish Empire was precisely the influence of the Counter-Reformation among the Spanish elites, who closed themselves off from the scientific and technological currents right in those years of the rise of modern science. If the Protestant Reformation and the Catholic Counter-Reformation had not coincided with the beginnings of modern science, perhaps the world would be very different now. But the fact that Protestants saw science as a tool to fight Rome and that Catholics saw it as an entry point for Protestantism has been relevant to the world's subsequent economy. And this may be happening again.

China discovered the three technological artifacts with which the West sustained its power in the modern age: the compass, with which it was able to navigate across planet Earth; paper (and moveable type printing); and gunpowder. Why didn't China conquer the world? It's an enigma. Maybe it's because it wasn't interested in expanding. But there is another, more complex, question: Why wasn't the scientific method born in China? Many believe that it has to do with the fact that there was no Galileo or Newton in an Asian country, but why did Galileo or Newton appear in Europe and not in China?

One answer may be that, in the Renaissance, Greek rationality was taken up again. Archimedes, Euclid and Democritus were re-read (Galileo was an enthusiast of Greek physicists and mathematicians). But those authors are so ancient that they were probably somehow already known in China or India also. Alexander the Great (bear in mind that he had Aristotle as his teacher) reached India in 326 BC.

Another approach that China did not take with the scientific method might be its Mandarin system: the way to select Chinese officials was a very difficult examination. This system lasted 1,300 years, from 605 to 1,905, and changed over time. The difficulty was always great, but little by little less importance was given to mathematics, technology and knowledge of nature, and other, more protocol, normative and empire history approaches were added. This is what happens in many Western diplomatic or political science schools: more importance is given to the history of treaties than to physics or biology.

The Chinese imperial system of examinations to elect its officials theoretically promoted social advancement, although in practice only the well-off classes could afford to have their children not work but study for many years. However, it was a meritocratic system that, for example, was not imposed in bourgeois Europe until the nineteenth century and that many British-heritage countries question. They do not understand whether the acquisition of knowledge that has been created by others should be rewarded as much as the creation of original knowledge; that is, whether it is preferable to select a teacher who knows all the authors and children's stories, or one who is capable of creating his own stories. It is as if, in order to be a university professor, you have to pass an examination in a subject that others have researched or else to evaluate what the candidate has produced himself.

The current format of the Mandarin exam is the *gaokao*, considered to be the most difficult university entrance examination in the world. Fewer than half of the candidates who are presented pass, and only the first placed enter prestigious universities such as Qinghua, where Xi, the current president of the republic, studied chemical engineering. Two types of exams are taken: the arts examination (on Chinese, foreign languages, history, geography, political science and mathematics, and easier than the science option); and the science examination on physics, chemistry and biology, and highly complex mathematics (including difficult problems in differential calculus or geometry). The advanced mathematical preparation of Chinese pre-university students is the country's greatest asset, producing complex technology, and it represents the greatest threat to the West.

The great advantage of scientific language is its universality: the problems of trigonometry are the same all over the world and do not depend on the cultural zone. It is no wonder that China leads the rankings of 15-year-old students in science and mathematics tests, as they must prepare for *gaokao*. And the Chinese elite comes from that science test. Physics, chemistry and biology—even much of mathematics—are a product of Western culture. The formulation of chemical molecules is performed in the Latin alphabet, not in Greek or Chinese. And it is done in that alphabet because chemistry was born in Europe, starting with Lavoisier and Dalton, among others.

However, the Chinese do not feel that they are not neglecting their own culture in *gaokao* although, apart from its Chinese element, it is from Western culture. They know that only in this way can China aspire to be a scientific and technological power, synonymous with being an economic and cultural power.

The scientific and industrial revolution that took place in Europe (from which, I reiterate, Spain was left out, and the cause of the loss of its empire) is the one that produced social and material well-being in the West, in the face of the hunger and hardship that still existed in China in the twentieth century (in the Maoist era).

I think that this is the most important lesson that China has learned from Western history and, in that sense, it explains why China is giving science and technology a huge boost. There is hardly any criticism of science in China. The scientific denial that is happening in the West under the cover of 'democracy' and 'freedom of expression' does not occur there. By 1980, the scientist and disseminator Isaac Asimov already feared that scientific anti-intellectualism would take over the elites: 'it's nurtured by the false notion that democracy means that my ignorance is just as good as your knowledge'. This is now law in the media and, above all, on the internet.

The *Scientific American* report highlights that, in the face of the problems of science in the West (from lack of funding, to the difficulty of progress of its graduates who have low salaries or loss of interest of the population towards these studies and professionals), China will spend 1.2 trillion dollars between 2016 and 2020 on scientific research. It will invest 373 million dollars in renewable energies alone. Its 'Made in China 2025' programme will encourage the development of artificial intelligence, cloud computing, robotics, biotechnology and electric vehicles.

China already invests more in science and technology than the European Union and, according to the Organization for Economic Co-operation and Development, is one step ahead of the United States. It will indeed be ahead in 2020. In the decade from 2006 to 2016, China's scientific publication output went up from 13 to 20 % of all such articles published in the world. If they decided to publish in Chinese, the West would not understand any part of the science that was produced. If we consider high-quality items, China is surpassed only by the United States. According to the *Scientific American* report:

> it now lays claim to the world's longest electric high-speed rail network, the largest radio telescope and the top two fastest supercomputers. It is launching

a carbon emissions cap-and-trade market this year that will dwarf the planet's largest one, in the E.U. Also the top producer of rare earth metals for high-tech manufacturing, China leads the globe in solar, wind and hydropower capacity and is or will soon be the strongest market for goods such as electric cars, nuclear reactors, smart devices, industrial robots and 3-D printers.[12]

China's data on science is impressive. For example, the Shenzhen-based BGI (formerly Beijing Genomics Institute) has 5,000 workers in 47 laboratories, with the intention of sequencing the genomes of as many organisms as possible, from ancient hominins to rice to giant pandas. In July 2017, it announced its pilot project to sequence the entire genome of 10,000 plants and microbes. Those data will revolutionize biology. China built in record time the world's largest radio telescope (with a spherical aperture of 500 m), located in Guizhou. This dwarfs the next largest radio telescopes in the world at the moment: the Arecibo radio telescope in Puerto Rico (United States), at 305 m; Green Bank (United States); and Jodrell Bank (United Kingdom). Its mission will be to search for extra-terrestrial life.

The scientific area on which China is focusing (and where it faces the most future restrictions from the West) is biotechnology. There are many fields that exemplify China's biotechnological commitment; one that illustrates its transformation as a country is the production of pig corneas for human transplants. By 2016, more than 200 operations had already been performed. A few decades ago, the main source of organs in China was executed prisoners, yet now it has a huge biomedical research programme for animals to produce organs and tissues for humans. In this sense, China has a huge advantage over the West: it does not profess a religion (or a culture with religious influence) that, at the moment, may restrict the freedom of research in sensitive areas of biology.

In the West, by contrast, even if they are secular, most countries have a Christian (or Muslim or Jewish) cultural past that considers a human being as a child of God. And this condition influences its scientific policies in biomedicine, with bioethical approaches whose foundations, although possibly well intentioned, may restrict the freedom of research the same as, in the past, did the ethics of the Counter-Reformation in Catholic countries with astronomy or atomic physics.

On November 2018, a scientist in China announced that he had created the world's first genetically edited babies, twin girls who were born that month.[13] The researcher, He Jiankui, said that he had altered a gene in the embryos, before having them implanted in the mother's womb, with the goal of making

---

[12]Lee Billings. (2017). 'China's moment.' *Scientific American,* October, p. 73.

[13]He has not published the research in any journal and did not share any evidence or data that definitively proved he had done it at the end of 2018.

the babies resistant to infection with H.I.V.[14] *The New York Times* explained: 'While the United States and many other countries have made it illegal to deliberately alter the genes of human embryos, it is not against the law to do so in China[15]'.

While a priority line is biotechnology (including robotics and human-robot interaction), another arm of China's scientific and technological challenge is its space programme. China plans to visit Mars in 2020 and to study the hidden face of the Moon in 2019. 'The Soviet Union and the United States have descended several times to the Moon, but both have descended from the front (the Earth-facing side), never from the far side,' Wu Weiren, design director of the Chinese lunar programme, told the BBC.

One problem with the history of science is that it is not politically correct. Discoveries of important theories, chemical elements, reactions or laws have been found by Westerners, especially from countries such as Britain, Germany, the United States and France. The Chinese interest in being a pioneer is therefore understandable: in the granite mountain of Daya Bay, in the south of the country, 300 metres underground, they have built the most complete facility to discover more about the most fascinating subatomic particles: neutrinos. And at a shipyard south of Shanghai they are assembling an oceanographic vessel (the largest of its kind in the world, to date, at 100 m long) to explore the ocean floors. It will serve as a platform to launch submarines and investigate the marine depths that have scarcely been explored so far. One of their unmanned submarines reached a depth of 4,000 m, but they want to reach 11,000 m in the Marianas Trench in the Pacific. The first voyage to that place was in 1960 and the second—and last, to date—was by filmmaker James Cameron in 2012, and for television (a documentary for *National Geographic*) more than for scientific interest.

Another interesting aspect is that the Chinese want to be pioneers in space mining. It is a debate that has not yet begun: who will discover what there is in space? Well, whoever has the technology to be the first to discover it and then use it. Sociologists, political scientists and economists have little to say about this.

---

[14]'Chinese Scientist Claims to Use Crispr to Make First Genetically Edited Babies'. *The New York Times*. Nov. 26, 2018. https://www.nytimes.com/2018/11/26/health/gene-editing-babies-china.html.

[15]'Chinese Scientist Claims to Use Crispr to Make First Genetically Edited Babies'. *The New York Times*. Nov. 26, 2018. https://www.nytimes.com/2018/11/26/health/gene-editing-babies-china.html.

5

Science and Economics: A Complex Relationship

Introduction

The relationship between science and bourgeois capitalism is interesting, and there is no consensus on which is the most important for scientific and technological advancement: capitalism or state investment. By 1917, the German sociologist and economist Max Weber stated in his book (from a lecture):

> The large institutes of medicine or natural science are 'state capitalist' enterprises, which cannot be managed without very considerable funds. Here we encounter the same condition that is found wherever capitalist enterprise comes into operation: the 'separation of the worker from his means of production.' The worker, that is, the assistant, is dependent upon the implements that the state puts at his disposal; hence he is just as dependent upon the head of the institute as is the employee in a factory upon the management. For, subjectively and in good faith, the director believes that this institute is 'his,' and he manages its affairs. Thus the assistant's position is often as precarious as is that of any 'quasi-proletarian' existence and just as precarious as the position of the assistant in the American university. (Weber 1919 (1946), 131)[1]

It seems obvious, however, that the best thing is collaboration between the two. As for the state, the clearest case is the space programme (both Soviet and American), which involved considerable public funds. It is true that Americans have since taken more advantage of this research due to their system of

---

[1] *Max Weber: Essays in Sociology*, pp. 129–156, New York: Oxford University Press, 1946. From H. H. Gerth and C. Wright Mills (trans. and ed.). Published as *Wissenschaft als Beruf*, Gesammlte Aufsaetze zur Wissenschaftslehre (Tubingen 1922), pp. 524–555. Originally a speech at Munich University, 1918, published in 1919 by Duncker & Humboldt, Munich.

© Springer Nature Switzerland AG 2019
C. Elías, *Science on the Ropes*, https://doi.org/10.1007/978-3-030-12978-1_5

transferring these results to private companies: the development of the digital society owes much to state research during the space race or large installations. In fact, the internet was born at CERN.

By the beginning of the twentieth century, Weber already feared that the German university would become 'Americanized' and would no longer cover the sciences, just the humanities or social sciences. His vision in this regard is most interesting:

> In very important respects German university life is being Americanized, as is German life in general. This development, I am convinced, will engulf those disciplines in which the craftsman personally owns the tools, essentially the library, as is still the case to a large extent in my own field. This development corresponds entirely to what happened to the artisan of the past and it is now fully under way. (Weber 1919 (1946), 131)

This idea still survives to some extent in the twenty-first century, especially in the humanities. The possibility of being able to research without money (as in my case, for example) until one secures a permanent position in the academy, and, above all, to be able to be the sole author and the absolute owner of the means of production, is a vital aspect for people with talent and an independent spirit, favouring the humanities over the sciences. In the social sciences, it is becoming increasingly important to have funding to use the expensive methodologies, such as the surveys that social science journals love. But there is also a danger: whether it's the state or private enterprise, if they finance you it is so you can say (with a scientific ring of truth) what they want you to say or investigate. But Weber was already warning that the competitive advantages of a capitalist and bureaucratized company are unquestionable.

Collaboration with companies is also important. One of the most interesting phenomena to explain what science and technology are is the process of synthesis of ammonia from atmospheric nitrogen: the so-called Haber-Bosch synthesis. Air is mostly nitrogen (78%), an inert gas. Thanks to the German chemist Fritz Haber (1868–1934), a way was found to convert this nitrogen into ammonia, and from there into nitric acid and nitrites and nitrates, the basis of chemical fertilizers. But the process would not have been anything but an intellectual curiosity if it had not been for another German chemist, Carl Bosch (1874–1940) who, in addition to being a chemist, was a chemical engineer and, above all, an industrial entrepreneur. Bosch patented the method of producing the reaction on a large scale to manufacture fertilizer, one of the world's most important industries. Thanks to fertilizers, the population has been able to increase without starvation, although many soils have since been eroded. One of my chemistry teachers said that the world should erect a monument to Haber and Bosch on every street, as we are alive because of them.

The British economist and demographer Thomas Malthus (1766–1834) had predicted (on the basis of a mathematical model) in his book *An Essay on the Principle of Population* (1798) that, according to the data for population growth during the eighteenth century, extrapolated by his mathematical model, there would be insufficient resources on the planet to feed humans in the nineteenth and twentieth centuries. Karl Marx (1818–1883) replied in a note in *The Capital* that progress in science and technology would allow exponential growth in resources and, therefore, people would not go hungry. We owe this much to Haber and Bosch, yet that same technology was also used for the chemical weapons that were used in World War I in the first example of mass destruction. It became evident that the best soldiers are not those who ride horses but those who know of science and technology.

Capitalism is based on the law of supply and demand, among other procedures, but also on the formula for rewarding talent. There is a certain consensus that one of the reasons why many brilliant young people are not engaged in science and engineering in the West is that these professions are underpaid, especially given the enormous effort involved in studying these subjects.

There is another problem: the STEM professions are the best example of global professions, with all their positive and negative aspects. Precisely because science is true (we have seen how the economist Malthus made a mistake in his mathematical model of population by not taking chemistry into account) and the laws of technology work the same everywhere, emerging countries such as Brazil and, above all, India and China produce more and better and better scientists and engineers every day.

These professionals can achieve the same as their Western counterparts because, despite what academics in the humanities and social scientists who study the natural sciences say—such as that science depends on the cultural environment in which it is generated—it is not true: the laws of science and technology are not affected by the cultural environment. The law of gravity works the same in all environments, although why China did not lead the scientific revolution in the seventeenth century may have a different interpretation for a Chinese historian than for an Englishman: in any case, both are valid.

The functioning of science and technology over and above cultural conditioning is perhaps one of their greatest aspects, something that does not happen in the social sciences or in the humanities. A lawyer, a historian, a journalist or a Chinese sociologist does not have to work in the West and their results do not have to be made relevant to overseas contexts; but a chemist, physicist, engineer or biologist does. The rise of Asian higher education over the past thirty years has led to a dramatic increase in the number of people engaged in science and engineering worldwide. And this has had two effects on Westerners. The first is immigration of Asian scientists and engineers to the West. Lawyers or historians do not usually emigrate, because they are literature-based professions; in other words, the level of knowledge of the language must be higher than that of an average native speaker, because in these professions language is the main tool, something that does not happen in science and technology, which have their own languages. The second is the relocation of Western technology companies to Asia, where they find workers who are, in many cases, better trained than Westerners and who, above all, will work for lower salaries. Some companies, such as chemical and pharmaceutical firms, also benefit from more permissive environmental legislation. There is a great deal of debate about whether the know-how that the West has transferred 'for free' to Asia or has taught altruistically to Asian postdoctoral scholars who have moved to the United States or Europe has been positive for the West. At present, for example, the level of science and technology—especially in information technology and biotechnology—in Asia is in many ways higher than in the West.

If we are talking exclusively about market effects, an oversupply of graduates would imply low salaries. However, the professional market for lawyers, sociologists, political scientists, economists, historians and Western journalists does not suffer from these educational advances in Asia: as I have said before, a Chinese lawyer is no danger to an American or Spanish lawyer. But the market in the West for STEM professions is affected by both the immigration of qualified professionals and the relocation abroad of companies that have hired local STEM workers.

The science economist Paula Stephan, in her book *How Economics Shapes Science,* observes how, while postgraduate law programmes in the main American universities usually include how much their graduates earn after completing their studies, this information rarely appears for chemistry, physics, mathematics or engineering. Maybe this is so as not to discourage future students.[2]

Many reports from the late 1980s (Adam 1990[3] or Bowen and Sosa 1989,[4] among others) warned of a dramatic reduction in the number of scientists and engineers in the United States over the following two decades. Specifically, Bowen and Sosa referred to the period 1987 to 2012. However, their

[2]Paula Stephan. (2012). *How Economics Shapes Science.* Harvard University Press.

[3]James Adams. (1990). 'Fundamental stocks of knowledge and productivity growth.' *Journal of Political Economy,* 98, 673–702.

[4]William Bowen and Julie Sosa. (1989). *Prospects for Faculty in the Arts and Science: A study of factors affecting demand and supply 1987–2012.* New Jersey: Princeton University Press.

predictions have been wrong—as has often been the case with the predictions of social scientists, since Malthus—because they were based on the retirement rate of American scientists and engineers who started working in the 1950s and 1960s. They did not contemplate—because they did not yet know it—the fall of the Berlin Wall and the collapse of Eastern Europe and, above all, the dismantling of the Chinese Maoist regime that isolated Chinese scientists and engineers.

In 2018, not only is there no such decline as Bowen and Sosa predicted in 1989 but the number of postgraduates in science and engineering has not stopped growing in the United States. Immigration and the exchange of talent are always positive, both for science and for the host country. However, growth obviously has economic, employment and scientific effects on the professionals in that country: it increases considerably the competition for both places (so local scientists who are not very bright are left behind, as those who migrate there are often of higher calibre) and publication, which negatively affects local scientists and engineers, as it reduces the scientific production of native speakers.

This was demonstrated in a very interesting paper (Borjas and Doran 2012)[5] on the impact of the collapse of the former Soviet Union in 1992 on American mathematics. Soviet and American mathematicians had been poles apart since 1922, when Stalin had decreed the isolation of Soviet science. Mathematics is a universal—and even eternal—language, yet this lack of contact led to two different schools. After the collapse of the Soviet Union, many mathematicians emigrated to the United States, but no Americans went to the former Soviet Union. This led to an increase in mathematical production in the United States but, as Borjas and Doran show, it also led to increased unemployment among US mathematicians, reduced productivity (they now had to compete with Soviet mathematicians) and changes in the disciplines most widely published by US institutions: after emigration, the approaches and fields of the Soviet school gained importance. The United States as a country may have won from this immigration, but young, unconsolidated mathematicians did not see it in the same way. If, instead of studying mathematics, they had studied law, the collapse of the Soviet Union would have had no impact on their job prospects, whether as lawyers, journalists, sociologists or political scientists.

The same is true of the current immigration of Chinese and Indian scientists and engineers: it is highly positive for American companies (they pay lower salaries because the graduate market is larger and they have more talent at their disposal). However, an American boy or girl may be discouraged by this

[5] George J. Borjas and Kirk B. Doran. (2012). 'The collapse of the Soviet Union and the productivity of American mathematicians.' *Quarterly Journal of Economics*, 127(3), 1143–1203. https://doi.org/10.1093/qje/qjs015.

immigration from studying STEM professions, since in these professions the competitive advantage of being an English-speaking native is almost irrelevant, since Chinese students have been able from high school to solve differential equations to pass their *gaokao*.

In this sense, another interesting study (Borjas 2007) shows that natives, especially White students, stop enrolling in doctoral programmes at elite universities if these programmes have a high rate of foreign students.[6] They understand that competition will be very high and they do not even try. The problem that faces the United States itself is what would happen if the foreigners who study on the best research programmes left the country? There would be a significant loss of talent. However, there is strong competition for places on these programmes—worldwide—as the living conditions of the United States make it a goal for brilliant professionals, as the truly brilliant have no problems in succeeding, encouraging many to apply for permanent residence in the United States.

There is an added problem. While some authors (Osborne and Dillon 2008; Stephan 2012)[7,8] argue that all this—especially the fact that science and engineering are global labour markets—leads to an increase in the number of these professionals, causing unemployment and precarious working conditions (and that this discourages vocations among young people who do not want to compete), others argue (Barth et al. 2017),[9] in recent research funded by the National Science Foundation, that the problem lies in the focus. They believe that studying science and engineering at university provides these students with ways of working and tackling the problems that make them highly effective workers in any environments in which they find themselves. The research shows that scientists and engineers increase the productivity of the places in which they work. The conclusions show that: (1) most scientists and engineers in industry are employed in establishments that produce goods or services, and do not perform research and development (R&D); (2) productivity is higher in manufacturing establishments with higher scientists and engineers proportion (SEP), and rises with increases in SEP; (3) employee earnings are higher

---

[6]G. J. Borjas. (2007). 'Do foreign students crowd out native students from graduate programs?' In *Science and the University*, 134–149, ed. Paula Stephan and Roland Ehrenberg, Madison: University of Wisconsin Press. And also in 'Immigration in High-Skill Labor Markets: The impact of foreign students on the earnings of doctorates', in *Science and the Engineering Careers in the United States: an analysis of markets and employment*, 131–162, ed. Richard Freeman and Daniel Goroff. Chicago: University of Chicago Press.

[7]Jonathan Osborne and Justin Dillon. (2008). *Science Education in Europe: Critical Reflections. A Report to the Nuffield Foundation.* King's College London.

[8]Paula Stephan. (2012). *How Economics Shapes Science.* Cambridge, MA: Harvard University Press.

[9]Erling Barth, James C. Davis, Richard Freeman and Andrew Wang. (2017). 'The effects of scientists and engineers on productivity and earnings at the establishment where they work.' Working papers forthcoming in *U.S. Engineering in a Global Economy*, ed. Richard Freeman and Hal Salzman, Chicago, IL: University of Chicago Press.

in manufacturing establishments with higher SEP, and rise substantially for employees who move to establishments with higher SEP but only modestly for employees in an establishment when its SEP increases. The results suggest that the work of scientists and engineers in goods- and services-producing establishments is an important pathway for increasing productivity and earnings, separate and distinct from the work of scientists and engineers who perform R&D (Barth et al. 2017).

These conclusions are highly relevant, because they point out that the labour market for scientists and engineers cannot be evaluated solely on the basis of the number of research positions available in the academy in these areas. The market is affected by the loss of potential that areas other than science and engineering may suffer if they lose these professionals. In this sense, it would be highly positive for a country if these scientists and engineers were to work in other services, and—as I can well appreciate—it would even be most important for there to be a greater number of professionals with an in-depth background in science and engineering—as is the case in China—in journalism, politics, history, education studies, philosophy, sociology and economics, as already exists in finance.

However, as documented by Paula Stephan, a professor of Economics of Science, academia considers industry or another field outside the university as the 'dark side' and does not train graduates to use their science knowledge in fields other than science (Stephan 2012, 159). It is short-sighted to point out that science and engineering education—which is the best way to search for the truth—only serves professions in that field. A decline in the number of graduates in these areas would be very dangerous, especially if a shortage of these professionals is filled by workers from the humanities or social sciences in which everything is taught, from French Theory or postmodern philosophy, stating that everything is valid, to other programmes with no science or technology content. We will lose enormous opportunities for progress.

But this trend is not easy to reverse in the West, where one of the relevant parameters in education is the search for happiness. The Declaration of Independence of the United States of America of 1776 states as one of its fundamental principles the search for happiness (it seems that was due to the American scientist and journalist, Benjamin Franklin). Are scientists happy?

Harvard economics professor Richard Freeman, who is also an expert on how a worker's happiness increases their productivity, clearly documents that American scientists and engineers are not well paid if one considers their professional qualifications and, above all, in the case of doctorates, the time taken to obtain their degrees.[10] According to Freeman, between 1990 and 2000 the average income of doctors of engineering rose from \$65,000 to \$91,000 (approximately €52,000–€74,000); that is, engineers increased their salary by 41%. Other professional groups' average incomes increased by more than those of engineers. Medical doctors' increased by 58%, from \$99,000 to \$156,000 (from about €80,000 to about €126,000). Lawyers had a 49% increase, from \$77,000 to \$115,000 (approximately €62,000–€93,000). Compared to the 41% of engineers or 59% of doctors, for doctors of natural sciences this increase in average income was only 30%: from \$56,000 to \$73,000 (from €45,000 to €59,000). As their salary is lower than that of other groups in absolute terms and their increase is also lower in percentage terms, the reality is that they are poorly paid. All this suggests that the market does not value doctors of science.

Freeman tells a curious anecdote that demonstrates the mentality of many older and established scientists to young people. During the height of the physics employment crisis in the 1970s, he gave a talk to the physics department at the University of Chicago:

> When I finished the presentation, the chairman shook his head, frowning deeply… 'deeply… got us all wrong,' the chairman said gravely. 'You don't understand what motivates people to study physics. We study for love of knowledge, not for salaries and jobs.' But… I was prepared to give… arguments about market incentives operating on some people on the margin, when the students – facing the worst employment prospects for graduating physicists in decades – answered for me with a resounding chorus of boos and hisses. Case closed.[11]

This way of thinking—that the scientist is like a monk who is interested only in knowledge, not money, already around in the 1970s, harms those talented but not wealthy young Westerners who need to study something to live on. The achievements of rich scientists—from Lavoisier to Humboldt to Darwin—who used their fortunes to produce science are laudable, but there are many others—just as talented—who seek in science not only a passion for knowing the world but a way of life worthy of life. And, in this sense, there is an increasingly standardized practice in the West that is strangling science: the phenomenon of postdocs, many of whom have funding from outside the institution where they work. Moreover, this funding has a specific time interval of from two to five years, both for doctoral studies and for postdoctoral work. Imagine being a company—from industries to schools, hospitals

---

[10]Richard Freeman. (2005). 'Does globalization of scientific/engineering workforce threaten U.S Economic leadership?' Working paper 11.457 as part of the NBER Science and Engineering Workforce Project.

[11]Cited in P. Stephan, 2012, 159.

or civil servants—whose workers have no labour rights and wages are financed by external entities, for that is what happens in the science market, and not with low-skilled workers but possibly the most skilled in the West. In a capitalist system—where the labour market is governed by supply and demand of workers—the only way to address this situation is with a huge shortage of scientists and engineers.

In fact, this drop in wages and working conditions is defended by many as the health of the system. This is approximately the central thesis of Richard Monastersky. In a newspaper report entitled 'Is there a Science Crisis? Maybe Not' in the *Chronicle of Higher Education* (7 July 2004), he observes that several parameters indicate that there has been a very slight increase in the number of science students in the United States since 2000 after the great downturn, the lowest point of which was 1998.

However, some of the interviewees for this report qualify that, perhaps, the decline is because not as many science and engineering graduates are needed as one would expect. Others interviewed by Monastersky point out that the case of the slight increase in the number of students in science baccalaureates and universities since 2000 may not be matched by that of high-quality (brighter) students. These students would opt for degrees where more is earned, and this would also be a symptom of decline.

What everyone agrees on (including Freeman, obviously) is that the best option for the truly intelligent is to do an MBA (Masters in Business Administration) at Harvard University. A few months after completing their Master's degree (which lasts two years), the 891 graduates in 2005 earned on average $100,000 per year. The average age of those who finished the Master's was 27 years.

However, if a doctorate in science is chosen, it will take an average of eight years to achieve a position similar to that secured by an MBA by the second year. In addition, as we have seen, the employment situation is even more precarious than the salary.

The Decline in the Working Conditions of Science

Some may think that a Harvard MBA graduate should earn more because an MBA is so difficult to achieve. However, the data show that studying for a Ph.D. in science is even tougher than an MBA. Research conducted by Freeman, published in the journal *Science* and referring to the most dynamic area of science, biosciences, shows the harshness of the profession.[12] What's even worse, this deteriorates daily. This study is undoubtedly one of the most rigorous and exhaustive on how a scientific career works and, among other results, it shows that there is a wide gap between the expectations of students at the beginning of their doctorate and the actual professional opportunities. It has a rather sinister conclusion: team leaders are not able to talk to their potential student-workers about how difficult the work situation is, because they are afraid of losing those who really do the hard work in science.

In October 2017, *Nature* published a huge survey of more than 5,700 doctoral students worldwide.[13] Under the descriptive title 'A love-hurt relationship', it revealed that 'despite many problems with doctoral programmes, Ph.D. students are as committed as ever to pursuing research careers'. The survey was the latest in a biennial series that aims to explore all aspects of Ph.D. students' lives and career aspirations. The respondents to the 2017 survey came from diverse scientific fields and from most parts of the world. Asia, Europe and North America were all strongly and equally represented.[14]

Respondents indicated high levels of satisfaction with Ph.D. programmes overall, but also revealed significant levels of worry and uncertainty: more than a quarter listed mental health as an area of concern, and 45% of those (or 12% of all respondents) said that they had sought help for anxiety or depression caused by their Ph.D. studies. Many said that they find their work stressful, worry about their futures and wonder whether their efforts will pay off to earn them a satisfying and well-compensated career.

The *Nature* survey also found that students with anxiety don't always have an easy time getting help. Of those who sought assistance, only 35% said that they found helpful resources at their own institution. Nearly 20% said that they tried to find help at their home institution but didn't feel supported. But the big paradox is that in the *Nature* survey, nearly 50% of students who reported seeking help for anxiety or depression said that they were still satisfied or very satisfied with their doctoral programme.

Ph.D. anxiety can have a variety of causes. Among other issues, the survey uncovered widespread concerns about future employment. Only 31% of respondents said that their programme was preparing them well or very well for a satisfying career, but more than three-quarters agreed or strongly agreed that it was preparing them well for a research career, suggesting that many

---

[12]Richard Freeman, Eric Weinstein, Elizabeth Marincola, Janet Rosenbaum and Frank Solomon. (2001). 'Competition and careers in biosciences.' *Science*, 294(5550), 2293–2294.

[13]Ibid.

[14]The survey was advertised through links on nature.com, in Springer *Nature* digital products and through e-mail campaigns. The data (which are available in full at go.nature.com/2kzo89o) were fleshed out by interviews with a small group of respondents who had indicated that they were willing to be contacted.

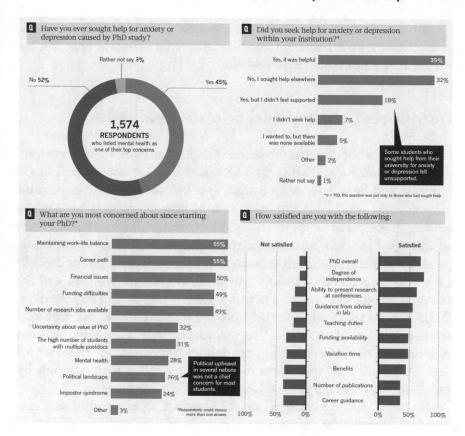

Fig. 1 Concerns and satisfaction with Ph.D. *Source* Chris Woolston. (2017). 'A love-hurt relationship'. Nature, 26 October, 550, 549–552

see a significant distinction between a research career and a 'satisfying' career. Although two-thirds of respondents said that a doctoral degree would 'substantially' or 'dramatically' improve their future job prospects, one-third had a more tepid outlook.

Yet a sizeable proportion of survey respondents are unhappy with the mentoring. More than half of respondents in 2017s survey agreed that their supervisor was open to their pursuing a career outside of academia, which echoes findings from the 2015 survey. The survey responses suggest that many Ph.D. students lack a clear vision of their future. Nearly 75% of respondents said that they would like a job in academia as an option after they graduate, whereas 55% said that they would like to work in industry. That might partly be down to indecision: nearly half of respondents indicated that they were likely or very likely to pursue a career in either sector.

When asked how they arrived at their current career decision, almost two-thirds chalked it up at least in part to their own research on the topic. Just 34% credited advice from their careers advisor (Fig. 1).

Tournament Theory and Science Evaluation

I confess that one of the most suggestive theories that I learned of in my course on the economics of science at the LSE was that of the 'tournament model'. It may be well known to many readers, but I was surprised that something that I thought was highly intuitive, such as human resource management, could be described in terms of an elaborate theory. The tournament theory was described in 1981 by economists Lazear and Rosen. They argued that a worker's performance (both evaluated and the incentive) can be modelled as a tennis tournament. There is only one prize: either you win and have everything or you lose and have nothing; glory is only for the first. And most US/UK experts who analyse the behaviour of science consider that this model fits the profession of a scientist like a glove. Why? Because everyone knows the winners of the Tour de France, but nobody remembers the runners up. In science, it means if anyone comes to a result or a patent second, if only by a few hours, it is of no use to him. Only the first has any glory.

Note to the reader: to come in second place you also have to be very good, and possibly in the case of current science the difference is simply that one of them works harder. In this tournament system, there is a disproportionate incentive to win, to be the first, encouraging excessive competitiveness. To sustain that, people endure harsh conditions, duped into trying for the first prize. Athletes are able to tire themselves nearly to death. In the case of science, researchers are aware that small advantages (having more fellows or working at weekends, or both) can be the difference between success and failure, between glory or the dishonour of having come so close. All that time spent on getting to second place is worthless, in this model, because the publication or patent is for first place only.

In 'hot' areas; that is, scientifically thriving areas, this competition can be brutal. So, according to the scientists surveyed in the work published in *Science* by Freeman, one of the formulas to achieve success is not to be smarter or brighter but to work more hours.

In this sense, data from the US National Science Foundation indicate that 33% of researchers in molecular biology (considered to be a 'hot' area) work an average of 60 h per week, while in other areas this percentage is reduced by 24%. The vast majority of scientific biologists between the ages of 35 and 44 (the most productive age) work about 50 h a week.

These studies on the professional career in the biosciences also show that these long working days have a special influence on those who, in addition to a professional career, want to have a family. In this regard, it is noteworthy that among scientists without children, the women work longer hours than the men. However, if we evaluate the hours worked by scientists with children, the study concludes that women work far fewer hours than men. 'This decrease in the hours worked by mothers irreparably damages their career progression in such a rapidly advancing field of life sciences' (Freeman et al. 2001, 2293). In fact, in this area, the half-life of scientific citations is the shortest of all scientific fields.[15]

This tournament model is much more demanding in science and engineering than in the humanities or social sciences, and this also discourages science vocations. Precisely because of the greatness of science, with identical laws in all countries and cultures, there will be many scientists who are interested in a subject. This does not happen in areas that are closer to their own cultural traditions. An example of my own is that, when I was working as a chemical researcher in an area as specific as the synthesis of manganese fluorides with nitrogen bases, many others around the world—including in China—were also working on it. A few days off could mean that I would slip behind in the game—as I did, in fact—and all my work would be for nothing. However, in my current research on how the media in Spain uses data journalism, I do not have that global pressure to be first in the tournament. The citation rate will be lower, but in my current field—journalism—the rates required for the same job level are lower than those for chemistry, because the statistical average is taken into account.

It should be remembered that the citation rate is one of the most important parameters (if not the most important) for evaluating a scientist. This explains why it is so difficult to return to a high-level trajectory after a break or a planned reduction in work, as it takes many years to achieve a globally competitive citation rate. If that permanence is lost quickly, as happens in powerful areas of science, the competition is brutal, favouring terrible working conditions.

It is necessary to publish a great deal, in competitive areas. Data from the US National Science Foundation demonstrate the close link between the number of hours worked and scientific success. For example, between 1990 and 1995 bioscientists published an average of 6.7 articles each year, compared to 4.7 in other fields.

In biosciences, according to the study published in *Science*, working five hours of additional work per week is associated with one more published article

---

[15] A. Preston. 2000. 'Do differences in rates of skill depreciation across scientific fields impact labor market outcomes?' Haverford College, in Freeman et al. 2001.

per year. Each scientific article implies an increase of 0.9% in a researcher's salary: 'Bioscientists who work more hours publish more and those who publish more earn more', the study says.

This tournament model is a very efficient way for entrepreneurs to develop the competitiveness of their employees. It is also the one that has been chosen by scientists. No one from outside has imposed it on them. Moreover, when politicians or intellectuals try to question it, they are criticized by scientists. The model is madness because, as many scientists have the talent and equipment to make a great discovery, what makes the difference is the terrible working conditions. However, everyone is aware that the grand prize of being first eclipses everything else.

Regardless of the negative effects that this model may have on the personal frustrations of those who have worked hard but do not achieve the goal of being the first, or of those who have to give up a family life, it is also proving detrimental to science itself.

The first negative effect is that the tournament system does not guarantee the best absolute performance—the best science and the best scientist—but the best relative performance: the best of those who compete. Therefore, the recruitment of scientists depends on the competence at each moment and place, thus is susceptible to the election of very poor candidates if there are many positions and few candidates, a circumstance that has begun to occur in some Western countries.

By contrast, having few places and many applicants can also exclude bright people. In this case, which could be beneficial to science, what reigns is a meritocratic system, which normally chooses the one with the most merit, and this does not necessarily coincide with the most brilliant one. Merit is quantified by the number of publications and their impact. Also, it relies on the citation index; that is to say, all aspects of the media, not just scientific ones.

This could discriminate against bright young people who are outside of the system, which is, as the history of science shows, the best source of innovative scientific ideas. Einstein might have been eliminated because his publications on the theory of relativity were not in *Nature* or *Science*. Moreover, the meritocratic system can be subverted in such a way that often, in today's science, a mediocre older man with 100 mediocre publications—which he has done simply because he has had time—is selected ahead of someone young and brilliant with an innovative idea or publication. But what does science need: 100 mediocre publications or one revolutionary one? Maybe both, but you can't overlook the revolutionary one. What is more, that candidate must be sought, at any price, if leadership is to be achieved.

The Spanish novelist Pío Baroja (who was not only a great writer but a graduate in medicine) explained in a dialogue between himself and Dr. Gregorio Marañón that 'the transcendental age' for these revolutionary ideas is between 25 and 30 years old:

> In all great men, philosophers, writers, scientists, the lantern for observation is formed in the midst of youth. Age adds neither more clarity nor more radiance to them. (…) What seems obvious is that one can be a great strategist and politician like Alexander the Great, Hannibal or Napoleon; an extraordinary musician like Mozart; a mathematician like Pascal; a painter like Raphael, in his youth. Philosophy itself, which seems to be a work of old age, is also the product of young people and from Plato to Schopenhauer the most important works of the philosophers were written in youth rather than in old age. Now that the man of talent or genius is recognized by the environment, sooner or later, it does not mean anything for his conditions, which are already developed, not in germ, but matured and formed at that age of twenty-five to thirty years.[16]

However, the current procedure for recruiting scientists means that, just at their most creative age, our researchers are excessively subordinated to their older bosses, without the chance of developing their own ideas, and living in precarious conditions. The system, moreover, rewards the merit of the 100 mediocre publications, not the revolutionary one.

Due to the obvious shortcomings of the meritocratic system, some prestigious US/UK universities try to compensate by introducing a section into the selection of the scientist in which candidates must solve difficult research cases (this is the recruitment formula of large technological multinationals, such as Google). This idea, which in the fields of knowledge management is considered innovative, is only a copy of the old Greek strategy of 'solving the puzzle'. The same enigma—under the same conditions—was put to all the candidates—both princes and subjects—and the winner was the one who solved it properly.

How Google and Other Private Companies Recruit

In 2014, Matthew Carpenter, a former employee of both Google's Resources Department and Human Resources Department, unveiled some strategies to

---

[16]Pio Baroja. (1976). 'La edad transcendental' ('The transcendental age'). *Obras completas* ('Complete Works'). Biblioteca Nueva 1976, 1109.

catch the world's best talent.[17] None of them is applied in universities or research centres. Nor are they valued for funding research projects. Among these strategies are the more or less obvious: cognitive ability; aptitude for the job; technical knowledge; and one called 'Googliness', or the ability to become used to the Google culture. But the most groundbreaking (because it breaks with Western trade union tradition) was to appreciate talent rather than experience. Google has a history of finding the best college graduates and giving them important responsibilities from the start, despite their lack of work experience. According to Carpenter, 'Google believes it is possible to teach knowledge, but difficult to cultivate talent. That's why Google values less the experience the person has when they enter and more the value they can add.' Another very important element is to avoid overvaluing degrees, university prestige and academic qualifications (this is what is valued, for example, in research grants).

In its early days, as Carpenter recalls, Google was known for hiring only people who had studied at the best universities. 'Over time, Google gradually began to downplay the credibility of the academic institution, and to measure the differences in accountability between students at recognized and less recognized universities.' They found that there was little difference in performance, and as a result, he dismissed the policy of giving great weight to the person's academic institution—although the grades achieved are still taken into account for many jobs. Other elements are also taken into account, such as rapid decision-making, a passion for the work that you do and the search for unusual solutions to common problems.

However, the practice of 'solving the enigma', favouring talent over experience or passion, is minimal in today's science. What is usually encouraged is, I repeat, meritocratic competitiveness, which has another perverse effect: that of the need to publish a great deal and quickly, making it impossible to keep up with a moderately active discipline. The result is plenty of scientific publications but not much science.

In fact, according to science journalist John Horgan, since the 1950s, when the structure and function of DNA were discovered, there have been no revolutionary advances in science, yet the number of scientific publications is growing worldwide, making it impossible to keep up with the times.[18] Stephen Hawking[19] gives the following information: in 1900 some 9,000 articles were published, in 1950 some 90,000 and in 2000 some 900,000. In 2009, an esti-

---

[17] http://blogs.evaluar.com/10-cosas-que-hace-google-para-la-seleccion-de-personal.

[18] John Horgan. (1996). *The End of Science. Facing the limits of knowledge in the twilight of the scientific age*. Helix Books.

[19] Stephen Hawking. (2002). *El universo en una cáscara de nuez*, p. 158. Barcelona: Crítica.

mated 2.5 million articles were published each year, and that year the number of articles published since the seventeenth century exceeded 50 million.[20]

At its most basic level, we've seen a substantial increase in the total number of academic journals. In 2014 there were approximately 28,100 active scholarly peer-reviewed journals. Add to this the increasing number of predatory or fake scientific journals, producing high volumes of poor-quality research, and you have a veritable jungle of journals to wade through.

It is true that, as has already been explained, in 2009 the number of scientists in the world is much higher than it was in 1950 or 1900. But it is also true that the number of publications required of the new generations is growing. If one compares the articles that Blas Cabrera (one of the best physicists in the history of Spain) needed when he took out his professorship in electromagnetism at the Central University of Madrid in 1905, one can see that it is very small.[21] It is almost certain that Cabrera would not get past the postdoctoral scholarship phase at present.

However, a day was the same 24 h in 1905 as in 2005. Cabrera's publications are very relevant, but in no way detract from those of the 2017 professors. I very much doubt, however, that any current professor of the Complutense (descendant of the Central University) or of any other Spanish university will give their name to the same number of schools, institutes, books, tributes and congresses in the century to 2117 as Blas Cabrera has in 2017, so is this a symptom of scientific decline?

The meritocratic system, increasingly established in the West, hires because of what a scientist has done in the past (merit), not because of what that scientist can do in the future. This does not happen in any other professional field and, in my view, it may discourage innovative science. Einstein published the theory of relativity (special) in 1905 in only three scientific articles (I repeat, none in *Nature* or *Science*). And yet these three articles are among the most important in all twentieth-century science.

The problem is that, like the mass media and contaminated by them, science is creating a system that values the publication of volumes of expendable results. Even when the research is relevant, it generates what Richard Smith, director of the *British Medical Journal* for 13 years, has called the 'salami effect': in other words, the results are cut up again and again to produce more publications. There are distorted or incomplete approaches that benefit journals'

---

[20] Arif Jinha. (2010). 'Article 50 million: An estimate of the number of scholarly articles in existence.' *Learned Publishing*. https://doi.org/10.1087/20100308.

[21] A compilation of Cabrera's scientific work can be found in Francisco González de Posada. (1995). *Blas Cabrera ante Einstein y la Relatividad. Amigos de la Cultura Científica* ('Blas Cabrera before Einstein and Relativity: Friends of scientific culture'). Madrid: Departamento de Publicaciones de la Escuela Superior de Arquitectura de la Universidad Politécnica de Madrid.

and scientists' assessments yet harm science. This diagnosis of one of the evils of current science is similar to that of the mass media: the excess of information and its enormous fragmentation (to generate more publications) produce disinformation, disorientation and disinterest.

The other terribly negative effect is that to publish a lot and quickly—whether it is mediocre or good but fragmented—demands harsh working conditions. When it comes to competitive areas, where many laboratories work in the same field, the conditions can become semi-slavery.

A principal investigator (or group leader) interviewed in the aforementioned study published in *Science* says it without hesitation: 'If I have three postdoctoral scientists and we all work all the time, I am more likely to be the first to achieve results than if I have two postdoctoral scientists and, on top of that, they rest for the weekend'.

This can also be stated by a builder or a shopkeeper. But in the case of science, readers should note, we are not talking about unskilled employees who work under harsh conditions because they are from developing countries or because they are illegal immigrants without labour rights. In science, we are talking about young and enthusiastic people who are brilliant because they have a career in science and who, in addition, have obtained the highest possible qualification in the educational system around the world: the doctorate.

Why do young scientists agree to such a sacrifice? Veteran scientists who appear in the research cited recognize that 'their students should have more free time for their personal lives' but that they adapt to the competitive incentives of the tournament model. They also acknowledge the economic difficulties of young scientists (including those with doctorates). But, curiously, it is the tournament system that pays them so little so that they are always competing hard for positions in which the prize is unique (i.e. it is competitive, with 15 candidates per place) and that it is to earn a little more or have improved working conditions. If you pay them more at first (or just enough), you stop being so competitive. This is how this model works.

It is curious that many established scientists still question the causes of the decline in vocations in the West: what brilliant young scientist wants to work in these conditions? The study makes it clear that the conditions that earlier applied to low-paid professionals such as waiters, domestic workers, day labourers and even prostitutes now prevail in science. The scheme is that, first, the standards for local workers to secure a job are reduced. In other words, you don't need to achieve such good results in high school and your career to be a scientist, because there aren't enough scientists to go around. And, if there are none, foreign labour is imported—that's what has happened in the United States—to take up the jobs that the natives don't want.

The study on careers in biosciences that was published in *Science* also denounces how it takes more and more years to secure a consolidated position. The data from 1970 to 1990 indicate that this period may be at least three years, but this is optimistic. All of this means that professional scientists—throughout the West—are comprised of the ones who later become contributors. 'Many scientists don't start contributing to a retirement pension until they're 35 years old,' says the *Science* study, using data from US scientists. This is completely irresponsible with regard to what they may receive as a pension when they retire. If a worker starts contributing to retirement at 25 and another at 35, there is 10 years' less contribution. The retirement salary will be lower for the one who obtains their first contract at 35. In science, until that point, they only have scholarship funding.

On the basis of all this information, we conclude that scientists are the 'ideal' workers in the tournament system for unscrupulous employers: they are young, highly trained, intelligent, idealistic, competitive, settle for very low pay, work up to 12 h a day, including weekends, without complaining, and have no employment rights whatsoever. Moreover, under these conditions, as everyone believes that they will win the tournament prize—a permanent place or being first in a scientific result—they postpone their family responsibilities to devote all their time and effort to work.

Since they can't raise money on their own to conduct research, they are completely dependent on their superiors, so they must be totally docile if they want at least a chance. Veteran and established scientists are delighted with this system but, in my opinion, they do not realize that they are killing science.

But now comes the toughest question: What moderately intelligent young scientist wants this life? How can they be reproached for preferring to do a Master's degree in business, for only two years, and earn $100,000 at the age of 27 instead of reaching 35 on a scholarship and working 12 h a day? How can anyone, with even a minimum of moral responsibility, recommend to a brilliant young scientist this kind of professional future dedicated to science? However, the question is not easy to answer. It is true that in law or in business more money can be earned, but the dynamics of precariousness have also entered.

The employers' call to increase the number of STEM graduates may be because they want the option of paying lower wages. But it is also true that even after pointing out that there is an important work niche in science (not just in the humanities, sociology and communication), there is an inexplicable reluctance to study these STEM subjects. Low wages and abusive work practices are found not just in science, but are something that needs to be changed. If it is complicated in science research, it is worse in certain areas of the humanities or social sciences, where there is much less money for research. And yet there is more demand for these careers than scientific ones. Above all, the decline in engineering cannot be understood from an economic-wage point of view: engineers' salaries have always been competitive. Moreover, there are fewer and fewer people in this field.

Let us not forget that this pernicious system is the product of the publishing madness and the enormous store set by publication. And here is the connection to media culture: public communication of scientific results, rather than being a consequence of them, has become the reason why science is being conducted and, what is worse, is degenerating into unsustainable working conditions in developed countries.

Hence, it is Asia that is currently most rapidly increasing its scientific presence. I should point out that in democratic, rich and Westernized Asian countries (with free media), such as Japan or South Korea, the problem of the decline of science is similar to that in the West. Japan has acknowledged that it has had a serious lack of scientific vocations since 1980, and in the 1990s the government launched repeated campaigns to get young people interested in science.[22] In South Korea, the authorities are so concerned about this issue that, if a young man chooses a career in science, he is exempt from compulsory military service. However, it seems that even this is not enough of an incentive.[23]

It is curious that the Chinese model of industrial production has not been able to impose itself on the West. However, the model of scientific production—and modern science itself—that has been designed in the West has been successfully exported to China. Now, what the United States is doing is importing labour—scholarship holders—from that country so as not to relocate the scientific production. This is not the fault of the philosophers, nor the sociologists, filmmakers or journalists; actually, it is exclusively the scientists who are to blame.

Researchers conceal these working conditions from undergraduate students and scholarship applicants. The National Research Council of the United States says in a report that scientists must tell their students the truth about the job outlook.[24] It denounces how, in the scholarships and contracts offered, the salary and working conditions are rarely made known. Neither is it clear what the career path of someone who secures the position or scholarship is, as there

---

[22]K. Makino. (2007). 'Japan: a boom in science news.' In M. W. Bauer and M. Bucchi (eds.) (2007). Op. cit. (pp. 259–262).

[23]These data were mentioned at the World Science Public Awareness Meeting held in Seoul, South Korea, in 2006.

[24]National Research Council. (1998). *Trends in the Early Careers of Life Scientists*. Washington, DC: National Academic Press.

are usually no commitments in this regard from the departments concerned. As Snow asked at his 1959 lecture on the 'Two Cultures': For a scientist with average qualifications, is it a comfortable job?

Is Science Only for the Rich?

The heading above is the suggestive title of an article in *Nature* (September 2016) in its special issue, 'Science and inequality'. It stated: 'In every society on Earth, at least some fraction of the citizens find their talents being sacrificed to poverty, prejudice, poor schooling and lack of opportunity.' And the editorial warned: 'Socio-economic inequality in science is on the rise. Current trends indicate that research is starting to become a preserve of the privileged.'[25]

The article explored a number of countries very different from each other—the United States, China, the United Kingdom, Japan, Brazil, India, Kenya and Russia—and concluded:

> The good news is that science is keeping up with modern trends. The bad news is that trend seems to be towards wider inequality, fewer opportunities for those from more disadvantaged backgrounds and a subsequent smaller pool of people and talent for research to draw on. From the United Kingdom and Japan to the United States and India, the story is alarmingly consistent. In many places, careers in science tend to go to the children of families who belong to the higher socio-economic groups.[26]

The problems for disadvantaged groups start in early, as revealed by *Nature*'s Andrew Campbell, dean of the graduate school at Brown University in Providence, Rhode Island. 'It starts in high school,' he says. 'Government-supported early education is funded mainly at the state and local level,' he notes. And, because the courses that are the most expensive per student are science courses, few schools in relatively poor districts can afford to offer many. Students from these districts therefore end up being less prepared for university-level science than their wealthier peers, many of whom have attended well-appointed private schools.

The same is also true in the United Kingdom: one study showed that only 15% of scientists are from working-class households, which comprise 35% of

---

[25] http://www.nature.com/news/socio-economic-inequality-in-science-is-on-the-rise-1.20654. *Nature* (26 September 2016).
[26] Editorial. *Nature*, 537, 450 (22 September 2016). https://doi.org/10.1038/537450a.

the general population.[27] Another study—Leading People 2016—found that over the past 25 years, 44% of UK-born Nobel Prize-winning scientists had gone to fee-paying schools, which educate just 7% of the UK population.[28] 'There's a class barrier to the professions,' says *Nature*'s Katherine Mathieson, chief executive of the British Science Association, 'but it's more extreme for science.' Another barrier could be that UK students who are interested in a science career often need to abandon other subjects at the age of 16. 'People from lower-income backgrounds who are unaware of the range of possible science careers might see it as a high-risk gamble,'[29] says Mathieson. 'The danger,' she adds, 'is that a failure to represent all backgrounds will not only squander talent but increasingly isolate science from society.'

That disconnect was apparent in the Brexit referendum in June 2016, when more than half of the public voted to leave the European Union, compared with around one in 10 researchers. 'That diverging world view is a real problem,' says Mathieson, 'both for the quality of research and for scientists' place in society.'

This risk is present in all countries where registration fees are expensive. The risk of failing in a science and engineering career is higher than in the humanities or social sciences. Moreover, in those fields research does not depend on outside resources. One can do a thesis on literature or journalism in one's free time while working elsewhere, but in science or engineering you need money and labs to conduct research.

That is why, when *Nature* published its biannual survey of scientists' salaries in 2016, it found that, 'Although the United States and Europe are slowly rebounding from the recent recession, it is a precarious time to be a scientist. Government funding is flat or flagging in many nations, and that, along with other concerns, has many scientists worried about their futures.'[30] *Nature*'s survey, which in 2016 drew responses from 5,757 readers around the world, found plenty of success stories, and nearly two-thirds of the 3,328 who responded to the question said that they were happy with their current jobs. But the survey also uncovered widespread unhappiness about earnings, career options and future prospects. And although the dissatisfaction spanned the globe, it seemed to reach its greatest heights in Europe, a continent that is struggling to find a place in the budget for research.

[27] Daniel Laurison and Sam Friedman. (2016). 'The class pay gap in higher professional and managerial occupations.' *American Sociological Review*, 81(4), 668–695. https://doi.org/10.1177/0003122416653602.

[28] https://www.suttontrust.com/research-paper/leading-people-2016/.

[29] 'Is science only for the rich?', http://www.nature.com/news/is-science-only-for-the-rich-1.20650. *Nature* (26 September 2016).

[30] Chris Woolston. (2016). Salaries: Reality check. *Nature*, 537, 573–576. https://doi.org/10.1038/nj7621-573.

However, I believe that the scientific system is based on the passion that many still have for science. Thus, the aforementioned *Nature* survey on wages concluded that:

> despite widespread misgivings, 61% of all respondents say they would recommend a research career. But the commitment comes with a cost, they say. More than half have sacrificed a good salary for the sake of science, some 42% have sacrificed work–life balance, and close to one-third say that their relationships have suffered. The flip side is that more than 80% say that their work is interesting, and 62% feel sense of accomplishment.[31]

This is the true engine of science: money does not bring happiness. But the rope is getting tighter and tighter.

The Case of China: Low Pay Powers Brain Drain

If this state of affairs reflects the capitalist world, the Chinese model is not much better. It is true that, as stated in the article 'Is Science Only for the Rich?' quoted in *Nature*, in China there is an equal opportunity policy for poor people in the country's interior or those from disadvantaged ethnic minorities to gain access to university. They are awarded extra points in the fearsomely difficult *gaokao*. There are even 12 minority-only universities. But, according to the article, there is a dark side:

> Beneath the surface, however, the reality of Chinese science often falls short of its egalitarian ideals. Children of senior government leaders and private business owners account for a disproportionate share of enrolment in the top universities. And students hesitate to take on the work-intensive career of a scientist when easier, and usually more lucrative, careers await them in business. According to Hepeng Jia, a journalist who writes about science policy issues in China, this is especially true for good students from rich families.

In other words, in this sense, Chinese millennials will no longer emulate their parents by studying more science or engineering, but will opt for the most sought-after career in the Western elite: business. One of the problems of this, however, is that 'business' is not a productive economy but a speculative one.

---

[31] *Nature*'s biennial salary and job-satisfaction survey, which in 2016 drew responses from 5,757 readers.

As a result, says Jia in a *Nature* article, scientists usually come from poorer families, receive less support from home and have a heavier financial burden. The situation is exacerbated by the low salaries, he says. The average across all scientific ranks is just 6,000 yuan per month, or about one-fifth of the salary of a newly hired US faculty member. Things are especially tough for postdoctoral researchers or junior-level researchers, 'who can hardly feed their families if working in bigger cities,' says Jia to *Nature*.

> This leads many scientists to use part of their grants for personal expenses. That forces them to make ends meet by applying for more grants, which requires them to get involved in many different projects and publish numerous papers, which in turn makes it hard to maintain the quality of their work. Many Chinese researchers escape that trap by seeking positions overseas. Thousands of postdoctoral researchers will go abroad in 2016 with funding from the China Scholarship Council, and many more will find sponsors abroad to fund them.[32]

The *Nature* article also deals with other countries such as Brazil, Kenya and Russia, where there are special programmes to increase the presence of young people from disadvantaged classes in science faculties. However, as the article highlights, the scientific productivity of these countries is poor. The case of Russia (a former scientific power in the Soviet era just a few decades ago) stands out. 'There is a national consensus in Russia regarding the value of equal opportunities in education for the modernization of our country,' says *Nature*'s Dmitry Peskov, who directs the young professionals' division of the Moscow-based Agency for Strategic Initiatives, which promotes economic innovation in Russia. But his conclusion is the same as across the West: 'Lucrative jobs in finance, business administration or industry are much more popular among well-trained young Russians than is a risky academic career.' In other words, two things are clear: the wages of the speculative economy must be reduced (perhaps by imposing very high rates on their yields); and those of scientists and engineers must be increased.

On the other hand, there must be greater synergy between the productive economy (which comes from the results of science and technology) and the curricula of these disciplines. In other words, young people should know that the best way to get rich is to study one of the STEM careers, from which they can start new businesses. They must not study only how to run companies.

---

[32]http://www.nature.com/news/is-science-only-for-the-rich-1.20650.

The End of Science?

The science journalist John Horgan, whom we have already mentioned, published his book *The End of Science*[33] in 1996. The book became a bestseller in the United States and Great Britain, but I think its main merit has been to reopen the crude debate in English-speaking academic circles (and others, too, as the book has been translated into 13 languages). Horgan, after interviewing some of the most prominent American scientists, concludes that, precisely because the postulates of science are true (no one, for example, argues that it is the Sun that revolves around the Earth or that its orbit around the Sun is circular), the era of scientific discoveries may be coming to an end. Success in science triggers a stage when there is no need for further research, because you already know almost everything that is important to what the scientific method can be applied to.

Scientists interviewed by Horgan point out that, indeed, since 1953, when the structure of DNA was discovered to be responsible for heredity, there has been no other spectacular discovery. The genome, which has received so much hype from the media, is really nothing more than the development of DNA theory and its technical application. The internet is a technological development, yet it has not involved a revolution in scientific theory. It is already known how the universe was born, how it developed, the structure of matter, how the evolution of species on Earth took place and how inheritance is transmitted.

Theories to contradict this knowledge are not expected to emerge, because everything indicates that it is true. It is possible to refute small details, the product of poorly done measurements, but not big ideas. All this makes science less attractive and more mechanical every day: the cloning of Dolly the sheep (as we will see later on) was more the result of hundreds of repetitions of a technique than of the technique itself.

It is important to note that the production (in number) of scientific articles has increased dramatically, yet this is not necessarily related to the production of science—or, at least, the relationship does not have to be direct. Whoever considers the two things to be the same (article production and science) is thinking like a journalist (or newspaper entrepreneur), not like a real scientist.

Scientists tell Horgan that they are indeed pessimistic, because they need more and more resources to research perfectly dispensable subjects. They add that the research careers of today's scientists admit only small contributions that support the great theories established up to the mid-twentieth century.

---

[33]John Horgan. (1996). *The End of Science: Facing the limits of knowledge in the twilight of the scientific age.* Reading, MA: Helix/Addison-Wesley.

And this would explain the decline in science: every day we have to work harder and earn less in order to obtain results that are not very scientifically relevant. In addition, as we will see later on, authorship has shifted from co-author to contributor, eliminating the epic of solo scientific work.

This lack of prospects for great discoveries could be one of the reasons why so few people want to devote themselves to scientific disciplines anymore. In fact, some researchers interviewed by Horgan compare physics, chemistry or biology with physical geography. By the middle of the twentieth century, there was nothing left to discover on Earth, and it was accepted that the era of geographical discoveries was over.

According to this view, governments may be considering a gradual reduction in budgets for basic scientific research, as it has not been as profitable since the 1980s. This issue deserves much more attention and needs to be deepened. However, I am not sure that there is a consensus among scientists that, because there is nothing more to discover, this is a supposed end to science. Perhaps, from the perspective of media culture, the problem would be that science is basically being nurtured by candidates with 20 mediocre publications rather than by candidates with brilliant ideas. All these data clearly show a decline in new scientific ideas, in the working and economic conditions of scientists and in the crisis of scientific vocations.

6

Western Intellectuality Against Science

In 2011, the President of the United States, Barack Obama, together with the country's leading business leaders, launched the STEM—Science, Technology, Engineering and Mathematics—initiative, with the aim of graduating 10,000 more scientists and engineers each year in the United States, as well as obtaining 100,000 new STEM teachers. The proposal was picked up by the media.[1] The goal was that the United States—and the West in general—should not lag behind the emerging Asian countries in science and technology. Why do young Americans—and Europeans—decide against studying science, technology, engineering and mathematics (STEM subjects) and enrol instead in social science or media studies, and does the popular culture of the mainstream have anything to do with it? This was one of my lines of research when I applied to the Harvard Visiting Scholar programme.

One of the greatest paradoxes of the twenty-first century is reflected in the following fact: while science has advanced by leaps and bounds, public opinion is less and less interested in it. The historian Eric Hobsbawm describes it clearly in his book, *The Age of Extremes: the short twentieth century, 1914–1991*: 'No period in history has been more penetrated by and more dependent on the natural sciences than the twentieth century. Yet no period, since Galileo's recantation, has been less at ease with it. This is the paradox with which the historian of the century must grapple'.[2]

Another apparent contradiction is even more worrying: never before have there been so many journalists or news reporters professionally assigned to

---

[1] Among others, *New York Times*, 4 November 2011: 'Why science majors change their minds (it's just so darn hard)', http://www.nytimes.com/2011/11/06/education/edlife/why-science-majors-change-their-mind-its-just-so-darn-hard.html?pagewanted=all&_r=0.

[2] Eric Hobsbawm. (1995). *The Age of Extremes: the short twentieth century, 1914-1991* London: Abacus.

© Springer Nature Switzerland AG 2019
C. Elías, *Science on the Ropes*, https://doi.org/10.1007/978-3-030-12978-1_6

covering scientific issues; and yet scientific vocations are steadily decreasing in Westernized societies (which include countries such as Japan and South Korea). All these countries have two elements in common: dominance by the prevailing mainstream media culture and a declining interest in science.

The philosopher Mario Bunge explains perfectly in his book, *Pseudoscience and Ideology*,[3] how scientific progress and the advancement of knowledge are not linear: throughout the history of humanity there have been times when science has flourished and others when it has almost disappeared. He mentions that, for example, the fall of the Roman Empire did not mean the loss of the scientific work developed by the Athenians, because science was no longer relevant to Rome at the end of its decline: 'It is very likely,' says Bunge, 'that Alaric did not interrupt the demonstration of a single theorem, the recording of any natural process or the conception of a single original scientific or philosophical theory. He entered Rome at a time when few individuals were engaged in intellectual work, and those few were more consumers than producers' (Bunge 2013, 218).

Drawing a parallel with the current situation, Bunge believes that we may be witnessing a new decline in science, as happened in the opulent—but unscientific—Roman Empire: 'It is possible that we are at the beginning of a crisis in basic science that, if followed, would lead to a New Middle Ages' (Bunge 2013, 224). He adds:

> If we are to avoid this catastrophe, we must do something to change the 'public image' of science so that it can continue to attract some of the smartest young people and continue to deserve the support of enlightened administrations and politicians without promising them what it cannot give. Let us stop painting science as a provider of wealth, well-being or power; let us instead paint it as what it is, namely, the most successful effort to understand the world and to understand ourselves. (Bunge 2013, 224)

Bunge stresses that the loss of faith by young people in basic science is due to the fact that in some cases science is blamed for the nuclear and ecological crises, and in others it is seen as 'the ideology of capitalism' (Bunge 2013, 222).

I agree with Bunge. Only, in my view, the current crisis of science, 'the public image of science' and, in short, the evils attributed to it by a part of society are not due to religion or political power, as in the past, but to something more liquid, permeable and pernicious: the media culture of the mainstream (of Hollywood or Madison Avenue) and how it deals with science. Young people acknowledge this. This media culture does not come out of nowhere:

---

[3]M. Bunge. (2013). *Pseudociencia e ideología*. Pamplona: Laetoli.

it has been created in the universities, where those who will later take charge are taught not only about the mass media but of other social aspects, such as the curriculum or the narrative of what science has meant in the history of humanity.

My position on this point is clear. In terms of the lack of interest on the part of young people and, above all, the lack of vocations, as well as the increase in magical thinking as evidenced in all surveys and the framing of media content, the decline of science is not so much due to erroneous educational approaches or political decisions. Given that the phenomenon appears in all Westernized countries—which have very different educational policies and educational cultures—it is due to what unites them: the mainstream media culture and the cult of celebrity.

I repeat, we cannot lose sight of the second part of the equation: in Western universities, especially since World War II—won by the Allies, entirely thanks to science—there has been an enormous proliferation of professors and researchers in the humanities and social studies fields with the clear, although seldom admitted, objective of intellectually discrediting the natural sciences. For decades, these professors, departments and faculties have been graduating students who, without knowing about science, are highly critical of scientific disciplines. Especially dangerous in this respect are the students and teachers of philosophy, sociology, communication, political science, history and education studies. They take on the role that the Spanish Inquisition had played in the Counter-Reformation, crushing any research that advanced the knowledge of modern science begun by Galileo.

In the twenty-first century, the attacks on science are not so much from theology or papacy as from the secular philosophy, sociology and mass communication taught in universities (especially those suffering from the harmful influence of French academia): together, they have built cultural paradigms that promote values that are contrary to rational thought, scientific knowledge of nature and rigorous work. On the contrary, the cult of fame, rapid success, irrationality and magical thinking (which are highly literary) is increasingly extolled in the media.

The case of France is curious: its rise as a country, as a culture and as a language coincided with its firm support for the ideals of the Enlightenment; its decline in every sense—not only cultural but economic—has coincided with uncritical acceptance of irrational thought in its universities.

As the Venezuelan philosopher Gabriel Andrade (2013) so well maintains, it is ironic that, in just two centuries, France has gone from being the country that promoted the ideals of the Enlightenment to become the country that attacks them the most. France is the country of origin of the illustrious:

Voltaire, Diderot, D'Holbach, D'Alembert and Montesquieu. But it is also the country of origin of the great postmodernist gurus: Lyotard, Baudrillard, Foucault, Derrida, Deleuze (Andrade is critical of these anti-scientific approaches, because he believes that they have spread throughout Latin America and are one of the causes of these countries' scientific and technological (i.e. economic) backwardness).

However, the problem of the lack of vocations and interest in science is not only French but has spread throughout the West. In Latin American countries—including France—it is barely on the agenda. However, in English-speaking countries it represents one of the areas of concern, with a notable impact not only on media discourse but also on politics. In Great Britain, the Royal Society considers it to be the main educational problem.

The Influence of the French Philosophers of the French Theory

There is one thing that I will always thank France for: the *Encyclopaedia* and the Enlightenment. And there is one thing that I will never forgive it for: that two centuries later its universities let in the irrational thinking of French postmodern gurus such as Lyotard (1924–1998), Baudrillard (1929–2007), Foucault (1926–1984), Derrida (1930–2004) and Deleuze (1925–1995), among others, and accorded them the same status as physicists, chemists and biologists. The origin of their anti-scientific mentality is unclear: maybe it is because they were born in the 1920s, and were at a formative stage' during World War II. Yet others were unaffected. The problem lies not with them (mental disorders have always been present, even in academia) but with the university as an institution, considering them seriously. Above all, the problem is that biologists, chemists, physicists, mathematicians, geologists, engineers and serious philosophers would find it impossible to thrive in this academic climate.

That was the end of a part of the French university (it is now almost irrelevant, in a world context), yet it contaminated heavily the American university, the best in the world and, from there, arts faculties across the entire Western world. Due to this influence, 'anything goes'. Since 2016, when Donald Trump won the presidential elections and there was talk of anticipation, or alternative facts, there has been a shift: Trump's advisors have been trained in social studies or humanities departments, which take these authors seriously. Fortunately, Oxford and Cambridge are immune (not totally, as we shall see) to postmodern philosophy, but not the American universities of Harvard, Yale, Stanford, and so on.

The French Encyclopaedia—the *Encyclopaedia*—by Diderot and D'Alembert was a triumph of free thought, of the secular principle and of private enterprise. It foreshadowed the triumph of the French Revolution and of the enlightened values in favour of science, technology and, especially, of reason over irrationality.

Encyclopaedias are an old aspiration of human knowledge, from the cuneiform tablets in the archives of the kings of Mesopotamia (668 BC) to the *Natural History* of Pliny the Elder and many other lost Greeks and Romans. In the sixth century, Chinese emperors commissioned their colossal plethora of officials with an immense encyclopaedia, culminating in 1726 with the *Gujin tushu jicheng*'s 745 hefty volumes. Yet this had no influence, as the Chinese officials did not circulate it. Likewise, the French *Encyclopaedia* was produced by a private company, and its sale and business profit involved publishing the volumes, as we know, not all at once but alphabetically: one letter per volume, one at a time. There were ground-breaking decisions: sorting the entries alphabetically, which put 'king' (*roi*) below 'rat'; ranking the amount of text in each entry; and selecting what was defined and what was not, who was cited and who was not.

As the entry for 'soul', the classical and authoritative definitions by Plato, Aristotle and St. Augustine were inserted. Skilfully, as though to refute these, the encyclopaedia also added those of other, more 'disputed' intellectuals, such as Epicurus, Hobbes and Spinoza. These quotes ended the supremacy of religion. Obviously, there was much criticism—the *Encyclopaedia* was outlawed by the Inquisition. The monarchy and the clergy put pressure on the publishers to change their attitude, but the bourgeoisie, who bought the encyclopaedia, wanted more science and technology and less theology. This led to the fall of the old regime. D'Alembert brilliantly defended the choice that he had made:

> One will not find in this work... neither the life of the saints nor the genealogy of the noble houses, but the genealogy of the most valuable sciences for those who can think... not the conquerors who devastated the Earth, but the immortal geniuses who have illustrated it... because this Encyclopédie owes everything to talents, not to titles [aristocratic], everything to the history of the human spirit and nothing to the vanity of men. (Prologue to the *Encyclopaedia*, in Blom 2004, 186)[4]

Apart from these beautiful ideals, which changed Western mentality, the *Encyclopaedia* had another characteristic: its editors were proud of the intellectual qualifications of the authors of the entries. These ranged from Rousseau

---

[4]P. Blom. (2004). *Encyclopédie. The triumph of reason in an unreasonable age.* London/New York: Fourth State.

and Voltaire to others who were less well known yet eminent in their time, such as the chemist Gabriel Venel, the physicist Louis Guillaume Le Monnier, the mathematician Guillaume Le Blond and the architect Blondel, among others. They knew the huge difference between being a specialist and just a self-taught amateur. And this is not trivial in our times, when the dominant encyclopaedia is Wikipedia, with its attendant problems.

Many articles were read and edited at meetings at the home of Baron D'Holbach, patron of the *Encyclopaedia* who, under a pseudonym, is the author of a controversial book in defence of atheism, *Christianity Unveiled: Being an Examination of the Principles and Effects of the Christian Religion*. In this book, published in 1761, he accused Christianity of oppressing humanity. Besides the elite Parisian progressive intelligentsia, foreigners of the stature of David Hume or Benjamin Franklin visited his home. And besides the focus of the articles (progressive, secular, rational, concerned with mankind...), even decisions on the inclusion of which entries were inserted were discussed. It was no small matter to include an entry for 'atheism' and its then provocative definition: 'The opinion of those who deny the existence of God in the world.'. In other words, the entry states that it is not that the existence is unknown, but that, knowing the notion of God, they reject it. This is the origin of the Church–State separation. That entry possibly changed much of the history of the West.

Baron Holbach wrote *Le Système de la Nature* ('The System of Nature') under the pseudonym of Jean-Baptiste de Mirabaud, the secretary of the French Academy of Sciences who had died 10 years earlier. He described the universe as just matter in motion, guided by the inexorable natural laws of cause and effect. There is, he wrote, 'no necessity to have recourse to supernatural powers to account for the formation of things'[5] By studying matter, its composition and the laws of cause and effect, one would understand the universe and, with it, human beings. The publication of this book in 1797 led ultimately, intellectually, to the nineteenth century's consolidation of science as the way of understanding.

The encyclopaedists were not neutral, but took sides in favour of science, reason, progress and economic development as the source of prosperity, and against slavery and the Inquisition. Now, it seems obvious, but in the eighteenth century this was an act of heroism.

---

[5]Paul Henri Thiry, Baron d'Holbach. (1797). *System of Nature; or, the Laws of the Moral and Physical World*, vol. 1, p. 25.

Against the Enlightenment

In the middle of the twentieth century, an intellectual movement began in Europe and then spread throughout the United States. This movement despises rationality—the rationalist hegemony of the West, its followers said—and promulgated a return to mediaeval thoughts of myths, romanticism and alternative facts. By turning its back on rationality, it obviated the need for data and arguments: it wanted followers who were emotional and with little affection for data. Until then, irrationality had always belonged to the ultra-religious Right (nothing else has done more harm to religion and magic than scientific and rational thought). What is interesting now is that, while rationality had been responsible for progress, the new mystics were from the political Left and claimed that scientific thought was totalizing. They blamed rationality for the Nazi holocaust and science and technology for labour exploitation.

There the decline of the Left began as progressive yet rational thinking, using the same arguments as the far Right. But the decline of the West and the rise of the East also began. China accepted the values illustrated: free trade, rational thinking and training in physics, chemistry, biology and mathematics as the basis for solving any social, economic, political or philosophical problem. It is true that China is not a democratic country, yet democracy—which I, as a journalist, will always defend—is part of the European cultural tradition (from ancient Greece), and is not exclusive to either scientific tradition or the Enlightenment.

It is far from clear that scientific and technological progress and democracy go hand in hand. In fact, it was in Nazi Germany that two of the most important contributions to physics were made—quantum mechanics and the theory of relativity—and technological progress has also come from Soviet communism. But wanting to introduce political ideas into science created a scientific setback. A clear case in point is that of the Russian biologist Trofim Lysenko, whose merit was to create a bizarre genetic theory that discredited Mendel (an Austrian Catholic monk), which basically claimed that the genetic laws discovered by Western biologists were irrelevant. This politicized genetic science, pointing out that the inheritance of acquired traits allows the cause of improved society to be fixed in the effects of the revolution. Every Soviet citizen, regardless of his or her genetic past, once he or she has experienced the improvement of the revolution, may feel superior to citizens of decadent and bourgeois environments.

Lysenko's genetic theory was nonsense, but he was Stalin's friend. Applied to plant genetics, it ruined Russian agriculture. The Soviet power appointed him in 1940 as director of the Institute of Genetics at the USSR Academy of

Sciences. He was twice the winner of the Stalin Prize and once of the Order of Lenin, a hero of the USSR and Vice-President of the Supreme Soviet.

Dictatorships, such as that of Soviet communism, led to the gradual disappearance of the other, highly prestigious, Russian (and obviously Mendelian) geneticists. In 1933, Chetverykov, Ferry and Efroimson were sent to Siberia, and Levitsky to an Arctic labour camp. In 1936, the geneticist Agol was executed for having 'converted to Mendelism'; that is, for having made it clear that science is not political. The most shocking case was that of Nikolai Vavilov, the best Russian geneticist, member of the Royal Society and who formulated the laws of the homologous series of genetic variation. In 1940, claiming that Lysenko's theory had no scientific basis, he was sent to prison for defending Western genetics: 'a bourgeois pseudoscience', according to the Supreme Soviet. And there, in prison, he died of malnutrition. Russian biology never recovered. This is the danger that dictatorships pose to the advancement of science; it is not that democracy is necessary, but that dictatorships can govern madly and eliminate scientists simply because they do not agree with certain political ideas.

The counter-movement in the West began in the Enlightenment itself, and came not only from the Church or the Inquisition, as might be expected, but from philosophers such as Rousseau. Faced with Diderot's and D'Alembert's (both scientists) optimism for science and technology, Rousseau pointed out that science cures some diseases yet generates more ills than it solves. Another idea was also his: that the general will must be defended against dissidents (and that they must be crushed by society), which would make science—from Galileo to Darwin to Einstein—a social mistake. Science has always been intellectual dissidence, but Rousseau's idea left a strong legacy among literary intellectuals and triggered the so-called romanticism movement, promoting a return to myth and popular religiosity. Through literature, Rousseau idealized the life of the Middle Ages precisely because it was unscientific.

In his controversial essay, 'Three Critics of the Enlightenment: Vico, Hamann, Herder', Oxford professor Isaiah Berlin argues that the counter-movement comprised those who were, above all, anti-rationalists and relativists. That would have favoured irrational romanticism and nationalist totalitarianism. Irrationalism came not only from the ultra-religious (the clearest example would be De Maistre) but atheists such as Nietzsche, opposed to all kinds of rule including that of scientific method.

Scientific thought, together with secularism, egalitarianism, materialism and Renaissance humanism, is part of the most radical of the French Enlightenment (Diderot, Condorcet, Mirabeau, etc.). There were the more moderates (D'Alembert, Montesquieu, Rousseau and Voltaire) who, in reality, were reformers of the Old Regime, but they did not believe in the pro-scientific manifesto that Condorcet gave in his speech of admission to the French Academy in 1782. In this, he affirmed that the 'moral' (social) sciences should follow the same methods and acquire the same exact and precise language and the same degree of certainty as the physical (natural) sciences.

This led to a declaration of war by philosophers such as the German, Wilhelm Dilthey (1883), who, influenced by the idealists Kant and Hegel, held the thesis that everything social should be 'spiritual' rather than 'material'. In a way, it was a reproduction of Spanish mysticism as a way of understanding the world: St. Teresa of Jesus before Bacon, Newton or Dalton.

In fact, according to the science historian Phillip Ball, German physicists during the Weimar Republic (1918–1933) wanted to encourage a quasi-mystical perspective of quantum theory in the face of the growing rejection of the supposed evils of materialism: commercialism, greed and the invasion of technology (Ball 2013, 49).[6] Science was associated with matter (it is actually the study of matter), and many literary intellectuals linked it to these supposedly degenerate values. The literary intellectuals of the Weimar era considered that the aspirations of science were inferior and could not be compared with the 'noble aspirations' of art and 'high culture'.

There is debate on whether the emphasis on metaphysical aspects of quantum mechanics was cultivated to free physics from materialism. The blame was not laid on quantum theory but on that of the microscopic probability of matter developed by the Scottish physicist James Maxwell (1831–1879) and the Austrian, Ludwig Boltzmann (1844–1906). The statistical distribution of the molecules proposed by Maxwell and Boltzmann to explain the kinetic theory of gases laid the foundation for fields such as special relativity and quantum mechanics. Also, by renouncing a precise and deterministic description of atomic movements, there was an abundance of renunciation of causality and a rise in indeterminacy.

The German philosopher and historian Oswald Spengler, in his influential book *Der Untergang des Abendlandes* ('The Downfall of the Occident', 1918), pointed out that the doubts of the physicists of his time (referring to Maxwell and Boltzmann) about causality were a symptom of what he considered to be the moribund nature of science.

This was the background, even though it was somewhat forgotten during World War II. It gained momentum in the mid-twentieth century. There was a movement that linked the atomic bomb to science. To be unscientific was to be concerned with the humanities, and the postmodern philosophers (French and German) began a far more continuous attack on science than the Roman-

---

[6]Philip Ball. (2013). *Serving the Reich. The Struggle for the Soul of Physics under Hitler.* Bodley Head.

tics had carried out. Alternatively, and this is my hypothesis, it was no more continuous, yet was able to have great influence when its ideas were received among the growing community of communication students, who would later dedicate themselves to journalism or cinema.

The famous analytic philosopher of the University of California, Berkeley, John Searle (one of the most eminent philosophers of the mind), maintains that French postmodern philosophy—Foucault, Derrida, Lacan, and so on—deliberately practises 'obscurantisme terroriste (terrorism of obscurantism)',[7] with the intention of making the ideas of charlatans look like genius. Let us briefly analyse some of these philosophers' ideas about science.

One of the most acclaimed is Roland Barthes (1915–1980). Studied extensively in non-English-speaking Western schools of communication, and also in media studies in the latter, he went so far as to say that clarity in writing is a 'bourgeois ideology' (for him, the bourgeoisie is something negative). The dark language allows us to affirm both one thing and its opposite, because everything is left to the discretion of interpretation. And it is this has been exploited by the movement of 'alternative facts' and post-truth.

It holds extravagant (but pleasing to the ears of obscure literary scholars) ideas whereby literary critics have the same level of creativity as the writers themselves. Barthes, in his book *Critique et vérité*, states that anyone who writes a dark essay on Shakespeare or Cervantes must also be worthy of artistic glory.

However, in my view, it is perhaps the Frenchman Jacques Derrida who has contributed the most to the concept of post-truth so beloved of today's populists. He has elevated irrational approaches to academic and seriousness. Derrida bases his work on an attack against what he calls logocentrism; that is, knowledge based on logic and reason. In his opinion, in the West, intellectual violence has been exercised by giving priority to the rational over the irrational. Derrida maintains that the emphasis on rationality favours the domination of men over women, and points out that logocentrism leads to phallocentrism.

Like all French postmodernists, Derrida is an extremely dark author, because logical reasoning (something that Derrida opposes) invites us to think that, if rational thinking or logocentrism leads to phallocentrism and male dominance, the way to obtain gender equality is to return to irrational thinking or magic. But for the postmodern French Theory, magic and science are at the same level. It would not be a problem if we were considering a hairdressing salon or a mechanic's workshop. The serious thing is that it is sustained in the university;

---

[7] Steven R. Postrel & Edward Feser (2000) 'Reality Principles: An Interview with John R. Searle'. *Reason*. https://reason.com/archives/2000/02/01/reality-principles-an-intervie.

the worst thing of all is that those who promote these anti-rational ideas are rewarded.

In fact, in 1992 there was a huge scandal when the University of Cambridge wanted to award Derrida an honorary doctorate and a group of 20 philosophers opposed it in a letter to the *Times*. One of the paragraphs of the letter states:

> Derrida describes himself as a philosopher, and his writings do indeed bear some of the marks of writings in that discipline. Their influence, however, has been to a striking degree almost entirely in fields outside philosophy – in departments of film studies, for example, or of French and English literature. In the eyes of philosophers, and certainly among those working in leading departments of philosophy throughout the world, M. Derrida's work does not meet accepted standards of clarity and rigour.

The signatories saw a pernicious influence on the university in those departments for which magic is as valid—because it is a literary object—as science: the departments of film studies, French and English literature. That letter did not prevent Derrida from being awarded an honorary degree by the University of Cambridge (1992), following a vote of 336 to 204 in favour. Perhaps that was the turning point in the decline of the Western university: that the university where Newton taught or where the structure of DNA was discovered had awarded Derrida a doctorate. It is proof that in the university of the late twentieth century, the arts (literary studies) had power over science and rigorous philosophy. This is not because they were right, but because they had more students, and therefore more teachers who voted. If a majority of teachers voted that Derrida's thought was valid, then it was valid.

Under that premise, Galileo could never have succeeded, not to mention Einstein. When Hitler wanted to discredit Einstein, he also displayed the populist spirit that most people are right. He recruited intellectuals to discredit the German physicist, compiling the opinions of 100 scientists who contradicted those of Einstein in *One Hundred Against Einstein* (in German, *Hundert Autoren Gegen Einstein*), published in Leipzig, Germany, in 1931. When Einstein was asked about the book, his answer stated that he knew what science was: 'If I were wrong, only one would have been enough.'

The post-truth and alternative facts that have brought Trump to power are nourished by another variant of French anti-scientific and irrational philosophy: relativism. For its followers, truth does not exist, science does not lead to truth and truth is always relative to the one who enunciates it. If a person believes that the Earth is flat, it is his truth and it is as true as that of a professor of physics who claims that it is spherical.

One of the main French representatives of this current is Jean-François Lyotard. In his opinion, postmodernism is defined as the loss of credibility of the great narrative. Science, for Lyotard, is just one more narrative, although it is a great one: it is a great narrative. But since, in his opinion, the great narratives lost their legitimacy in the mid-twentieth century, science has likewise done so[8]. In any case, for Lyotard, it being only a narrative, science is really an ideology with which one can agree or disagree. This cultural relativism is not new. Already the Greek sophists, with Protagoras at the head and his famous phrase 'man is the measure of all things', have established that every human being has his truth. The sophists, who recognized that they used rhetoric with the intention of persuading, not finding the truth, were pre-scientific.

However, the French postmodern philosophers were able to say both that science was just another ideology and to say it on television, or to fly in an aeroplane, without taking into account that, if science were not true, there would be no television and the aeroplane would not fly. Inconsistency becomes a value. They were not afraid to use that aeroplane to fly to the United States to spread the stain of their irrationality—French Theory—by polluting American universities, the best in the world, and, from there, the whole West.

It is a fascinating story described by the communication expert François Cusset in his book, *French Theory: Foucault, Derrida, Deleuze & Co. Transformed the Intellectual Life in the United States* (2003).[9] It explains how these French theorists first entered the French language departments in order to reach the influential English literature and literary critics at universities such as Irvine or Cornell. It was the Modern Language Association, the most important forum for teachers and researchers in literature (founded in 1883), which from the 1970s and 1980s onwards went from analysing Shakespeare's theatre or the Baroque poetry of Calderón de la Barca to lectures with titles (influenced by French Theory) such as 'Clitoral Imaging and Masturbation by Emily Dickinson', 'Victorian Underwear and Representation of the Female Body' and 'The Sodomite Tourist'. And this was an academy, and they paid them the same salaries as in physics. It is the responsibility of scientists not to strike—as at Cornell, for example—but to find another university where they do not have to work alongside departments that accept French Theory. It is not enough to vote as, at the moment, rational scientists and philosophers are in a minority in the West because they have no students. And if they lack students, they have no power: not just economic power, but influence in the world.

---

[8]J. F. Lyotard. (1979). La Condition postmoderne: Rapport sur le savoir. Paris: Éditions de Minuit.
[9]François Cusset (2003). *French Theory: Foucault, Derrida, Deleuze & Co. Transformed the Intellectual Life in the United States*. Minneapolis: University of Minnesota Press.

These departments of literature have incorporated French philosophy into English literary theory and, from there, it has gone into cultural studies and film studies. Literature teachers (Paul de Man, Gayatri Spivak, etc.) promoted reading groups of French authors. It would not have spread beyond a group of eccentric university professors and appointments among their brotherhood of pedants if it were not for a widespread error in many universities of the West: the creation of faculties of communication. Cinema, journalism, advertising and television programmes are not staffed by professors of thermodynamics, genetics or chemistry, but by professors in film studies, cultural studies, media studies, semiotics or literary theory. When their students came to control mass media, both irrationality and post-truth spread like wildfire.

One of the most interesting episodes of this struggle in the United States was when they found anti-Semitic writings by Professor Paul de Man (1919–1983), considered the promoter of the Yale Deconstructive School. De Man invited Jacques Derrida from France to Yale every year and generously invited him to translate his work into English. In 1986, the *New York Times* unveiled Paul de Man's Nazi collaborationist past. While he was still living in Belgium (where he was born), he published an article (4 March 1941) in the Belgian newspaper *Le Soir* entitled 'Jews in Contemporary Literature', in which he stated that the expulsion of Jews from Europe, and of some of their personalities of mediocre artistic value, would not lead to any regrettable consequences for Western literary life.[10]

This controversy arose just as Martin Heidegger's concessions to the Nazi regime were being reviewed in Europe (a key reference for the deconstructivist philosophers of French Theory). The philosophers of French Theory had dazzled the Left, yet by some media they were associated with a Nazi ideological past. Many critics considered that the relativism they promoted, whereby there is no distinction between the true and the false, was to protect academics such as Paul de Man. In fact, Derrida himself, grateful for the trips that De Man (with his Yale funds) had arranged for him and other philosophers of French Theory, proposed 'deconstructing' the pro-Nazi articles of his friend De Man. One thing can be said as well as another; the same thing can be said and used to defend specific interests.

In any case, what is interesting in the twenty-first century is that this relativism and deconstructivism, which went from Heidegger to the Left of the French Theory that incorporated him into cultural studies, is now one of the main tools of the far Right: it is used by politicians from Marine Le Pen to Donald Trump and the defenders of Brexit in order to build realities that run

---

[10]Werner Hamacher (dir.). (1989). *Paul de Man. Wartime Journalism 1939–1943.* Lincoln: University of Nebraska Press. (in Cusset, 2003, op. cit).

parallel to the facts. It is used to elevate fiction to the same status as reality and facts, as long as it serves to construct persuasive stories.

During the NBC show *Meet the Press*, when Kellyanne Conway, Donald Trump's government advisor, told Chuck Todd that, despite the photographs and facts showing more people in Obama's inauguration than Trump's, she did not hesitate to use the deconstructivist concept of Derrida and the French Theory to talk about 'alternative facts' (Conway studied politics and took her doctor's degree in law at American universities): 'He (referring to Sean Spicer, who was responsible for Trump's communications at the time) presented alternative facts. There is no way to count people in a crowd accurately', Conway said. Yes, there is, actually, but physics and mathematics should be involved, not the creation of narratives, storytelling or literary theory that Conway and Spicer used. The same goes for Le Pen's economic, anti-immigration and anti-European statements.

I insist that literary theory or cultural studies would have no relevance (they would not appear on the walls of a darkened university department) unless they are taught to students who then have responsibilities for mass communication. And not only do these students go on work in the media, but a significant majority do so as campaign advisors or as public discourse creators. The influence of those French Theory-loving professors who later devoted themselves to the mass media is the real triumph of the irrational and relativistic intellectuals who saw science as the cultural and intellectual paradigm that defined the West in the mid-1950s. From French Theory (and, above all, its American version) there is also contamination of 'science studies', which, influenced by another Frenchman, Bruno Latour (and also by the British Steve Woolgar) have tried to deconstruct the scientific disciplines (from his influential book—first published in 1979—*Laboratory Life: The Construction of Scientific Facts*[11]).

Latour now (in 2018) worries that he has gone too far. Citing an op-ed in the *New York Times*, in which a Republican senator argued that the way to gain public support for climate change denial is to artificially maintain a controversy by continuing to 'make the lack of scientific certainty a primary issue', Latour notes 'I myself have spent some time in the past trying to show 'the lack of scientific certainty' inherent in the construction of facts'.[12]

---

[11] Latour, B and Woolgar, S (1979). *Laboratory Life: The Construction of Scientific Facts*. Beverly Hills: SAGE.

[12] Salmon, P. (2018). 'Have postmodernist thinkers deconstructed truth? Bruno Latour once sparked fury by questioning the way scientific facts are produced. Now he worries we've gone too far.' *New Humanist* (March). Available https://newhumanist.org.uk/articles/5296/have-postmodernist-thinkers-deconstructed-truth.

In the 1990s, many studies under the umbrella of 'social studies of science' have accused science of being imperialist and of defending reason and rationality, of a macho attitude. This has been taught at universities that hold high-ranking positions because of their findings in chemistry, physics and biology, not because of their studies in literature, film, sociology or media studies.

Relativists defend themselves by saying that the hegemony of some ideas—the scientific ones, for example—over others, those of magic, is an example of intolerance and, ultimately, of totalitarianism. But relativism is not the same as tolerance. Tolerance implies that although there is in existence a true idea, I tolerate your having one that is false. When Voltaire, in his *Treatise on Tolerance*, said, 'I do not agree with what you say, but I will defend your right to say it to the death', he was not defending relativism but the right freedom of expression for the wrong speaker. He did not mean that both were right.

Not only by philosophy, but also by anthropology—which sometimes defines itself as 'social science'—it has been argued that science does not have a monopoly on rationality or truth. For the influential French anthropologist Claude Levi-Straus (1908–2009), the ancestral beliefs of indigenous peoples, although unscientific, are perfectly rational because, in his opinion, rationality should be judged on the basis of consensual beliefs, not scientific data or experiments. But, without explaining very well why, he considers that since science is one of the great narratives and they have lost legitimacy, science has also lost it.

Science as the Source of All Evil

The intellectual critique of science in the twentieth century began with Germans or Austrians who were persecuted by the Nazi regime, linking science with totalitarianism (Nazism and Communism). They did so without realizing that it was precisely science that had allowed the Allies to defeat fascism and even communism (which destroyed biology, as we have said), from the mathematician Alan Turing and his code decryption machine (the origin of today's computing) to the atomic bomb itself and the space race.

One of the first intellectual criticisms of science after World War II came from a liberal economist, Friedrich von Hayek (Vienna, 1899–Fribourg, 1992). Hayek (Nobel Prize in Economics in 1974) was an Austrian who had to emigrate to England during World War II. He taught at the LSE and was famous for his disputes with the British social democratic economist, John Maynard Keynes (1883–1946). Keynes did not suffer under fascist repression, had studied mathematics before economics and was a great admirer of scientists such as Newton (he bought original editions by the English physicist and mathematician). Hayek, on the other hand, came from the Austrian school, where economists did not come from mathematics or physics but from law, political philosophy or history. This lack of scientific knowledge made them mediocre intellectuals and may explain his book, *The Counterrevolution of Science* (1955), published just after World War II. Hayek attacked the blind scientific optimism that, in his opinion, had been experienced since the Enlightenment. In his opinion, this scientific optimism had resulted in socialist planning and Soviet totalitarianism.[13]

Another German émigré, the political scientist Eric Voegelin (Cologne, 1901–Stanford, 1985), considered that the principle of objective science, in the sense of the Enlightenment or of Marx or Darwin, can be interpreted as a symptom of a fatal 'deification', of an alienation and dehumanization that, in his opinion, would allow Germany to draw a line from the monism of Haeckel or Ostwald to Hitler's Nazism.[14] Voegelin was trained in political philosophy, not physics or biochemistry.

However, neither Hayek nor Voegelin are authors who study communication. Journalists and filmmakers from film studies or cultural studies begin their anti-scientific sentiments by reading the German authors who did not study science, although they spoke about it: the members of the so-called Frankfurt School, which some term neo-Marxist. Possibly one of the most influential works in anti-scientific thinking is *Dialektik der Aufklärung* ('Dialectic of Enlightenment'), published in 1944 by two Jewish German philosophers who suffered Nazi persecution: Max Horkheimer (1895–1973) and Theodor Adorno (1903–1969). Together with other philosophers (mostly German and retaliatory Jews) such as Walter Benjamin (1892–1940), they founded the so-called Critical Theory (much discussed among teachers and students of communication).

Critical Theory is a current of thought that is closely involved with the commitment to social emancipation from the structures of Western capitalist society. Criticism from mass culture to fascism: in principle, that's an interesting link. Fascism, like all populisms, is based on mass culture. However, what is most bizarre is that it links the concept of reason and rationality to the modern social system (unjust to these thinkers; it may be true, yet it is also true that the social system was infinitely worse prior to the Enlightenment).

The *Dialectic of Enlightenment* (in a 1944 context) circulated underground during the 1950s and 1960s. By the end of the 1960s, it had become a leading text in media and cultural studies. Its central idea (remember that it was

---

[13] Friedrich Hayek. (1955). *The Counter Revolution of Science*. London: Collier-Macmillan.
[14] Eric Voegelin. (1948). 'The origins of scientism.' *Social Research*, 15, 462–494.

written by German Jews who survived Nazism) is that the Holocaust was not a coincidental event but an ideological consequence of the way that Western imperialism is constituted. A similar idea is found in Horkheimer's 1947 book, *Eclipse of Reason*, whose title is sufficiently eloquent: reason and science (as a discipline that wants to dominate nature) have led to barbarism in the West. But of course, they do not explain why there was no descent into barbarism in England or the United States (countries with an important scientific tradition and cult of reason equivalent to Germany's).

Other important German philosophers of Critical Theory, such as the German Jew Marcuse (Berlin, 1898–1979) or his disciple Habermas (Düsseldorf, 1929), who also suffered under Nazism, unceasingly held the belief that the progress of science and technology leads humanity to alienation and even fascism. In 2001, when Habermas received the Peace Prize in Germany, he spoke of the Holocaust and in the same speech pointed out that science, scientists and, in short, modernity have not led to increased well-being. The German faculties of chemistry or physics did not reply. His discourse against science was worthy of the Spanish counter-reform inquisitors.

All this would make sense in Germany, but it is incomprehensible that these philosophers were respected in England and the United States, countries that deny that science leads to fascism. Science served to defeat fascism, but this narrative would diminish the importance of arts departments. In the United States, another idea seduced us: the criticism of science served to allow departments of humanities and social sciences to stand up to the departments of natural sciences and engineering. They nourished discourse on how what was being done in social sciences (especially in media studies and cultural studies) was also important, and deserved both funding and students.

France is not to blame for this situation of a decline in rational thinking, vocations and scientific interest. It is difficult to know which came first: the culture of celebrity, the superficial and the media-spreading telebubble in the West; or its inspiration. Remember, those who produce media culture were university students—inspiration came from postmodern philosophy, whose greatest exponents such as Feyerabend, Derrida or Lyotard extended the absence of deep thought to the classroom and replaced it with bombastic phrases and frivolous analysis. The philosopher Gabriel Andrade points out the steps in his extraordinary book *Postmodernism: A Fool's Bargain!*[15]

Postmodernism has become one of the philosophical doctrines used as a spearhead by those who defend pseudosciences and irrational beliefs. Defenders of astrology, psychoanalysis or homeopathy often invoke the names of postmodern gurus

---

[15] Gabriel Andrade. (2013). *El posmodernismo ¡Vaya Timo!* Pamplona: Laetoli.

such as Feyerabend or Foucault to protest against scientific hegemony and thus proclaim the legitimacy of irrational disciplines and beliefs. Therefore, it is not enough to attack the specificities of each scam. It is also necessary to attack the pseudo-philosophical baggage in which these absurd disciplines and beliefs are protected. (Andrade 2013, 14)

One might think that these philosophers or sociologists who embrace post-modern philosophy are social outcasts living in hippie communes, isolated from the world. However, as Andrade well recalls, nothing could be further from the truth: they have colonized the Western university with their 'anticipation':

The defenders of postmodernism have university degrees. Most of them are professors in the best universities in the world (it must be acknowledged that, fortunately, two of the best universities in the world, Oxford and Cambridge in England, are very reluctant to accept defenders of postmodernism into their teaching staff). They write in the world's largest newspapers, are interviewed by the most famous personalities on TV, and governments frequently ask them for their opinions and advice on military, economic, political, cultural and other matters. And, of course, while there are fortunately almost no books in university bookstores that promote creationism or homeopathy, unfortunately there are plenty of books in those same bookstores that promote postmodernism, and even occupy the privileged shelves. Thus, postmodernism enjoys a prestige inside and outside the academy. The advocates of postmodernism have something that attracts, and it is not precisely the clarity and depth of their ideas. It is rather a kind of sex appeal that generates followers of all kinds. They are, so to speak, rock stars in the academic world. The young students would like to be like them. Many wear long hair, smoke pipes, wear exotic costumes; in short, they seem to have a concern for their image. In this, they are much more like artists than conventional university professors. (Andrade 2013, 5)

In fact, from my point of view, postmodern philosophers, educators and sociologists have not been interested in thinking, in a philosophical sense, but in taking on the vital aspects of a media culture: easy success; admission that image and form are more important than substance; discrediting the culture of effort; and valuing the extravagance of arguments as a lever for attracting media audiences (something that has grown exponentially in digital culture). And those premises have been introduced into the Western university.

Postmodern Philosophy from British and North American Universities

The French postmodernists had influence because there was a breeding ground in the US and UK universities among philosophers who had also lived through World War II. Some had fled to Great Britain—Karl Popper (Vienna, 1902–London, 1994) and Irme Lakatos (Lipschitz, Hungary, 1922–London, 1974). Others had not fled, but had influence in the US and UK academy—Paul Feyerabend (Vienna, 1924–Zurich, 1994), while yet others were exclusively English-speaking, such as the American Thomas Kuhn (Cincinnati, 1922–Cambridge, United States, 1996).

Popper, Lakatos and Kuhn had all studied physics. You can see this when you compare their work to Feyerabend's. The intellectual work of these philosophers (who are also postmodern) is far superior, intellectually speaking, to those of French Theory. They had much influence in the English-speaking world (including Feyerabend) and, in many cases, especially Popper and Kuhn have been misunderstood by literary intellectuals, who have twisted their ideas to make them conform to the discourse that science is not valid, when they never said any such thing.

The problem, however, lay in giving intellectual authority (in scientific matters) to Romantics such as Hegel or Nietzsche. While the philosophers of the French Theory did not consider themselves to be philosophers of science (science was marginal to their work) but philosophers and experts in literary theory, Kuhn, Popper, Lakatos and Feyerabend did belong to what has been called the 'new philosophy of science', which emerged in the 1960s with, among others, the American Thomas Kuhn (1922–1996).

For scientists, the respect of colleagues is almost as important as the air that they breathe. That does not mean that they are isolated or that do not care what the rest of society thinks of them, just that prestige in science is what guarantees that they dedicate all their efforts to this activity that, in practical terms, provides them with nothing else. However, this circumstance has been misinterpreted. The historian and philosopher of science, Thomas Kuhn, who was for decades the most cited author in absolute terms in the discipline of science, technology and society, denounced this interest in scientific honour and considered it to be the cause of his social isolation. In one of his most studied works, which has most influenced philosophers, sociologists and, by extension, the training of journalists and other graduates in the field of communication, Kuhn conflates prestige with isolation. In this work, *The Structure of Scientific Revolutions*,[16] first published in 1962, Kuhn wrote:

---

[16]T. S. Kuhn. (1962). *The Structure of Scientific Revolutions* (1971 edn). Chicago: University of Chicago.

Some of these are consequences of the unparalleled insulation of mature scientific communities from the demands of the laity and of everyday life. That insulation has never been complete – we are now discussing matters of degree. Nevertheless, there are no other professional communities in which individual creative work is so exclusively addressed to and evaluated by other members of the profession. The most esoteric of poets or the most abstract of theologians is far more concerned than the scientist with lay approbation of his creative work, though he may be even less concerned with approbation in general. That difference proves consequential. Just because he is working only for an audience of colleagues, an audience that shares his own values and beliefs, the scientist can take a single set of standards for granted. He need not worry about what some other group or school will think and can therefore dispose of one problem and get on to the next more quickly than those who work for a more heterodox group. Even more important, the insulation of the scientific community from society permits the individual scientist to concentrate his attention upon problems that he has good reason to believe he will be able to solve. (Kuhn 1962, 164)

Kuhn was revered by philosophers and sociologists of science, especially, in my opinion, by those who have never practised real science. His book has had more than 100,000 citations in its English edition and almost 4,000 in the Spanish, which are enormous figures for the humanities. However, from scientists' point of view, Kuhn is so far-fetched as to be ignored completely rather than refuted. If I quote Kuhn in this chapter, it is because of the extraordinary influence that he has had on teachers who will later train journalists, communicators and those in the humanities.

It can be said that, since Kuhn, science began to have a bad reputation. That's why it's important, at least, to know his work, because relations between scientists and journalists are based on the ideological substratum that this philosopher created. It is not enough to ignore it, as some scientists claim. It must be studied in all areas of science in order to train scientists who can challenge it, because, in my experience, most natural scientists have never even heard of this philosopher.

According to Kuhn, natural scientists do not respect social judgement, nor are they interested in showing how the research is important or practical. If they do, this interest may be circumstantial, to obtain funding, and there may even be some pretence about the dubious usefulness of the research. For Kuhn, scientists are interested only in the judgement of their colleagues, and the problems that they dare to face are those that they believe can be solved. This feature, Kuhn believes, is necessary to boost high scientific productivity.

Does Kuhn think that the mediaeval society of the magician Merlin is better than present-day society, and are scientists not addressing the major problems facing society, such as climate change or the search for effective new drugs to combat disease? What do Kuhn's followers want from science, that it should be devoted to studying the gender of angels? What's more, when society has a problem, such as obtaining an AIDS vaccine, it turns to scientists, not to poets or philosophers.

It is, however, revealing that in the faculties of science, medicine or engineering, at least in Spain, Kuhn is never studied, but as soon as you enter the humanities or social sciences you are bombarded with his work. In my case, in the five years that I studied chemistry he was never mentioned, but when I started studying journalism his work was explained to me, among other topics, in the topics of public opinion, in the theory of communication and even in journalistic ethics and deontology. If that's not worshipping him, I don't know what is. I believe that natural scientists are unaware of how much anti-science is taught in the faculties of humanities and social sciences because, in general, usually they do not meet it either as students—undergraduate and doctoral—or as teachers and researchers.

Let us continue with Kuhn because, in his opinion, the scientists' authority is afforded by their own colleagues, who allow them to isolate themselves from any problems that are not admitted by this community:

> The effects of insulation from the larger society are greatly intensified by another characteristic of the professional scientific community, the nature of its educational initiation. In music, the graphic arts, and literature, the practitioner gains his education by exposure to the works of other artists, principally earlier artists. Textbooks, except compendia of or handbooks to original creations, have only a secondary role. In history, philosophy, and the social sciences, textbook literature has a greater significance. But even in these fields the elementary college course employs parallel readings in original sources, some of them the 'classics' of the field, others the contemporary research reports that practitioners write for each other. As a result, the student in any one of these disciplines is constantly made aware of the immense variety of problems that the members of his future group have, in the course of time, attempted to solve. Even more important, he has constantly before him a number of competing and incommensurable solutions to these problems, solutions that he must ultimately evaluate for himself. (Kuhn 1962, 164–165)

But, according to Kuhn, this does not happen in science careers, in this sense denigrating these studies:

> Contrast this situation with that in at least the contemporary natural sciences. In these fields the student relies mainly on textbooks until, in his third or fourth year of graduate work, he begins his own research. Many science curricula do not ask even graduate students to read in works not written specially for students. The

few that do assign supplementary reading in research papers and monographs restrict such assignments to the most advanced courses and to materials that take up more or less where the available texts leave off. Until the very last stages in the education of a scientist, textbooks are systematically substituted for the creative scientific literature that made them possible. Given the confidence in their paradigms, which makes this educational technique possible, few scientists would wish to change it. Why, after all, should the student of physics, for example, read the works of Newton, Faraday, Einstein, or Schrödinger, when everything he needs to know about these works is recapitulated in a far briefer, more precise, and more systematic form in a number of up-to-date textbooks? (Kuhn 1962, 165)

Kuhn's idea that textbooks are systematically substituted for the 'creative scientific literature that made them possible' is very relevant, because I think that it is untrue. That is to say, that what student textbooks on chemistry, physics, biology or geology do, always in Kuhn's opinion, is to isolate some problems, point out others and recruit future scientists by separating them from society as a whole. This strategy, which may at first seem crazy from a didactic point of view, according to Kuhn is the secret of the effectiveness of science.

A scientist reading these lines might argue that what Kuhn mentions is so stupid and meaningless that it is not worth bothering with. But the scientist is wrong. Kuhn is revered by those who have never entered a research laboratory. As I have mentioned, for a considerable majority of today's philosophers (who basically work by teaching this and other disciplines in institutes and universities) or sociologists of science, Kuhn has undoubtedly been the most influential philosopher in the conception of science in the second half of the twentieth century. His ideas, according to these twentieth- and twenty-first-century philosophers, have transformed the philosophy, history and sociology of science. Moreover, these philosophers consider that Kuhn has established permanent links between these three disciplines that deal with science as a historical phenomenon. The reader should note that Kuhn's descriptions of scientists correspond almost exactly to the image that the media, especially films, propagate.

From my point of view, there is another big mistake by Kuhn that has had a great impact on the decline of science, on the poor image of science in the media and on the rise of irrationality. Kuhn is wrong when he points out that the advance of science is but the result of a sum of 'revolutions' in which scientists agree among themselves, establishing what Kuhn called a 'paradigm', that is to say a dominant consensus, to conclude that this is the scientific truth. In time, according to Kuhn, they will 'come together' to change paradigms.

This Kuhnean belief is now accepted by some historians, sociologists, philosophers and journalists who have academic standing yet lack training or research experience in the natural sciences to have experienced that Kuhn is wrong. These scholars agree with Kuhn, and often conclude that 'scientific truth' is no more true than 'artistic truth'. This hypothesis is widespread in information science degrees.

However, the idea of paradigm, that science advances by means of consensus and fashion, is false. As the science historian John Gribbin rightly argues, accepting Kuhn's approach implies that 'Einstein's theory of relativity might go out of style just as Victorian artists' painting went out of style over time'.[17] But Einstein, Newton, Darwin and Dalton will never go out of style. These tendencies towards 'intellectual fashions' arise in art, history and the social sciences, such as education studies, psychology, sociology and communication, but not in the natural sciences.

Fashion does not influence science because, for example, any description of the universe that goes beyond the scope of Einstein's theory must, on the one hand, go beyond the limitations of that theory and, on the other, include in itself all the successes of Einstein's own general theory; just as Einstein includes in his theory all the successes of Newton's theory of universal gravitation. Newton, who formulated his theory in the seventeenth century, is not outdated: in fact, his laws are studied by every schoolchild in the world. Galileo and Copernicus don't go out of fashion, either. That is the great potential of the natural sciences: their description of the world that they investigate is basically true and not subject to fashion as in the social sciences.

Although in the natural sciences there are impostors, as in all fields, unlike in the social sciences or the humanities intellectual they are easy to detect. Among other things, any work must withstand the test of experimentation. In fact, in my opinion, this tendency to feel that natural science is an absolute truth is the real reason why some scientists isolate themselves. It is not, as Kuhn argues, because scientists ignore social judgement, despise the rest of the world and just live for their colleagues. It is quite the opposite: they feel—as I did when I was researching in an inorganic chemistry laboratory—that they are on the path to something true and eternal. Why worry about politics or university management, if it takes time away from researching what really matters?

Scientists know of many examples in the history of science of researchers who could have made a greater contribution had it not been for 'wasting time' on other tasks. From Mendel, who abandoned his experiments with peas in which he established the laws of inheritance, upon being elected abbot of his monastery, to Hooke, who might have eclipsed Newton somewhat if he had

---

[17] John Gribbin. (2003). *Historia de la ciencia: 1543–2001 (Science: A History)*, p. 500. Barcelona: Crítica.

not devoted time to the management of the Royal Society. Newton himself spent much time on social work: he was the Director of the Mint, President of the Royal Society and a Member of the British Parliament. How much progress would science and humanity have made if, instead of 'wasting time' in parliament or the Royal Society, he had been more involved in research?

Well, probably not much more. And this is probably where many scientists make mistakes. Newton formulated most of his work as a young man, as did Einstein and many other great scientists. In fact, the famous Fields Medal (the world's highest award in mathematics) is awarded only to scientists under the age of 40. It is assumed that, after that age, few ideas can be radically innovative. So, from my point of view, Newton did well to become involved in society. For example, under his presidency, the Royal Society became the world's largest scientific institution. It still retains a prestige that is difficult to match. Without knowing Newton's exact role in parliament, the mere fact that in the seventeenth century there was a scientist in it says much about a country such as Britain. A scientist must know that society benefits greatly from his or her science and that it is not something that depends on a single person so, as well, he or she has a moral obligation to devote himself or herself to matters apart from science.

Scientists must resist the huge temptation to isolate themselves in search of scientific truth. Because, apart from other professional scientists, people will never understand that feeling, and it can be detrimental to both science itself and to its image. The scientist must aspire to all spheres of real power in order to mitigate the decline of science.

The World Ruled by the Arts

It is all perfectly clear to social scientists—educationalists, economists, researchers in law, journalism, sociology… unless they are at all conceited, they do not feel that their contributions are relevant. However, they do not fail to take any opportunities to gain power to put their 'social hypotheses' into practice. It is therefore rare for a professor of economics, law, journalism or sociology to regard being a deputy or a minister as a waste of time. On the contrary, they think that they are doing the right thing. And that is why the world of the twentieth or twenty-first century, in general, is governed by this type of professional rather than by scientists. Scientists have locked themselves away in their laboratories, evaluating the impact of their publications, and have not realized that the real impact is achieved through economic, political and media power.

From these spheres of power, natural scientists could extend the idea that the social sciences—and not the natural sciences—are the perfect example of Kuhn's notion of a paradigm. This includes how their methodologies are weak and their results are susceptible to manipulation, because rarely does anyone bother to do studies designed to replicate earlier experiments in order to verify the findings. These hypotheses and conclusions are subject to fashion.

In fact, in a social science in principle as sterile as economics, it is not surprising that a Nobel laureate in economics, such as Wassily Leontief, should say that 'technology always generates unemployment' and, at the same time, hear another Nobel laureate in economics, Gérard Debreu, state that 'there is no link between unemployment and technological innovation'.

Furio Colombo maintains in *Latest News on Journalism*, a book of headlines in the faculties of communication, that 'the problem of scientific news becomes especially serious when one enters the field of social sciences, being this field more exposed than ever to the influences of political thought, so that we witness dramatic changes in definitions, analyses and proposals' (Colombo 1997, 97)[18]. But this reflection rarely appears in the media, which affords social science studies the same credibility as natural science studies.

Colombo's book includes a curious example of a supposedly serious social science, sociology, at a supposedly important university, Princeton, and of an apparently competent researcher, Sarah McLanahan. Well, this researcher published a study in 1983 that stated that, 'for the happiness of children, the presence of one or two parents is indifferent'. She added: 'Better one good father than two bad or warring parents. The important thing is love.' It is important to note that at that time there was a general sense of recognition of the value, and even heroism, of single mothers. The study was published in all the media and quoted in the best social science journals.

Ten years later, a conservative wind was sweeping across the United States. The same sociologist, still at the same prestigious university, published new research: 'The more I study the cases of girls and boys who have only one father, the more I am convinced that there is nothing worse than an incomplete family,' she told the media. Data from the new study showed that children with only one parent 'are twice as likely to drop out of school early. If they're girls, they get pregnant sooner. And if they're kids, they give into violence.' The problem was in the methodology. In the first investigation, 100,000 children from well-to-do classes were used and, in the second case, the study system was reformatories. Furio Colombo uses this example to warn journalists that

---

[18]Furio Colombo. (1997). *Últimas noticias sobre periodismo* ('*Latest news about journalism*'). Barcelona: Anagrama.

scientific truths are not absolute. But of course, he omits to point out that what is not absolute is the truth in social sciences.

The example also serves to further explore this issue of the differences between the social and natural sciences. There's no way that McLanahan could have failed to miss her methodological flaw. The difference between her and a physicist is that she is aware that neither her first research nor her last are absolute truths, because in sociology there are no such truths as: 'The Earth is the one that revolves around the Sun and not the other way around.' Therefore, she will never really attach the same importance to her research as a physicist who investigates super-string theory.

McLanahan knows both that her events will never happen again and that in fifty years all her studies will be out of date. That is why in the social sciences researchers don't break their necks as they do in the natural sciences. That's why they are also susceptible to fashionable results. And that is why it is no trauma for them to stop researching and to engage instead in politics or writing for the media.

The difference between McLanahan and Galileo—in fact, the gulf—is the fact that showing how the Earth revolves around the Sun, not the other way around, almost cost Galileo his life. This is because what was 'fashionable' at the time was the view that the Sun revolves around the Earth, yet Galileo was sure of the outcome and the method. And he defended these not because he wanted to change his paradigm or his style but because he knew that his method—observation plus mathematics—led to the absolute truth, and that this was the truth. And so it goes on: spaceships only corroborate what he demonstrated with his humble spyglass.

This doesn't happen in social science. McLanahan used her position in the college to make a little extra money by acting as an expert to the media or advising politicians who agreed with her result. It is true that there are also natural scientists who adapt their methodologies to obtain results that satisfy politicians. In fact, I studied this phenomenon in my doctoral thesis.[19] But these 'industrialization does not play a role in climate change'-type results cannot be considered as science, even if some politicians pass them off as science. Serious scientific associations, such as the American Academy for the Advancement of Science or the Royal Society, constantly and publicly discredit scientists who are open to political interests.

Another major essential difference from the social sciences is that the natural sciences make it possible to conceive how a given system will evolve in the future. These predictions are confirmed by constructing rockets that reach their destination perfectly, formulating drugs that act as expected and producing

---

[19]Carlos Elías. (2000). *Flujos de información entre científicos y prensa.* Universidad de La Laguna.

materials that behave as we expect in theory. This doesn't happen in the social sciences. Not even in an economy that uses visible mathematical apparatus. From my perspective, I see comparisons and a mixing in of the word 'science', as if everything were the same. Above all, I think it is dangerous for the media culture to consider that all academic disciplines are of the same value. In fact, while the natural sciences make predictions, the social sciences only give prophecies. It's not bad, but prediction is not synonymous with prophecy.

This does not mean that the social sciences of academic life should be discarded. But we must take into account the causal inferences and the explanatory models that they propose, because they are easily proved wrong. Journalists and filmmakers should be aware that natural science results often express truths while social scientists have points of view. Neither the media nor the public authorities should consider as truths what are only points of view, and not the opposite, either: to consider as a point of view what is proven truth.

I am not saying that sociology or economics are not valid. What I am saying is that the social sciences and the humanities are narratives, and they use data to support their points of view, as a rhetorical technique. But that is not applicable to physics or chemistry, which have a totally different type of knowledge, and hugely superior. Therefore, it must be present in all higher studies, to form the mind. But, in the university of the twenty-first century, a professor of economics or sociology can be considered to have the same intellectual status as a professor of physics. And this is unacceptable. If it extends to 'junk' courses such as film studies or cultural studies (which are irrational narratives), we can say that the university is not reliable.

Scientists Against Kuhn and Postmodern Philosophers

In a review in *Science* of the book that brought Kuhn fame, *The Structure of Scientific Revolutions*,[20] the reviewer warned: 'Since Kuhn does not permit truth to be a criterion of scientific theories, he would presumably not claim his own theory to be true.'[21] If truth is not a criterion that supports a scientific theory, and a philosopher who was also a physicist says so, how can we demand that Trump say that climate change is true?

In the Western university, there are legion philosophers, sociologists, educationalists, journalists, filmmakers and writers who have specialized in science without knowing about it, with the aim of criticizing it. Moreover, philoso-

---

[20]T. S. Kuhn. (1970). *The Structure of Scientific Revolutions.* University of Chicago Press.
[21]N. Wade. (1977). *Science, 197,* 143–145.

phers or sociologists who defend rational thought and natural science are often ridiculed by their departments and have their academic careers castrated. By contrast, there are few cases of physicists, chemists or biologists who have criticized sociology, philosophy or education studies as lax academic disciplines that succumb easily to unreasonable arguments. Scientists look the other way, not taking into account how, while they have been shut away in their laboratories and papers, their colleagues in the social studies and humanities have been fighting for decades for humanity to return to perceiving sympathetically the magical and irrational thinking that the Enlightenment tried to eradicate.

I should like to relate one of the most relevant episodes of scientists' struggle against postmodern philosophers. It was in 1987, more than thirty years ago, that two British physicists, Theocharis and Psimopoulos, published a controversial essay in *Nature* entitled 'Where science has gone wrong',[22] subtitled 'The current predicament of British science is but one consequence of a deep and widespread malaise. In response, scientists must reassert the pre-eminence of the concepts of objectivity and truth.'

The article blamed four philosophers of science for the fact that today's society does not support scientific truth. This implicitly meant that philosophy, or at least part of it, prefers the world to go down the path of irrationality. The essay was illustrated by photographs of the four 'traitors to the truth': Thomas Kuhn (1922–1996); Karl Popper (1902–1994); Imre Lakatos (1922–1974); and Paul Feyerabend (1924–1994). It argued that these four, who created the so-called 'new philosophy of science', were intellectual impostors basically because they themselves refuted their sceptical ideas about scientific truth that they wanted to give out as true:

> For reasons known only to themselves, the advocates of the antitheses refuse to recognize that their ideas are flagrantly self-refuting – they negate and destroy themselves. And if a statement entails its own falsity, then it must be nonsense: the proposition 'there is no truth' contradicts itself, and hence some truths must exist after all. Similarly, the structure 'nothing is certain, everything is doubtful casts doubt, besides everything else, upon itself, and so some certain and undoubted things must exist. Likewise, if 'anything goes', then 'nothing goes' must go too. (Theocharis and Psimopoulos 1987, 596)

The two physicists, and scientists in general, first of all, blamed these four philosophers of science for never having practised science. As a consequence of not knowing how it works yet wanting to define it, the conclusions of the philosophers would be bound to be impregnated with resentment—very

---

[22]T. Theocharis and M. Psimopoulos. (1987). 'Where science has gone wrong.' *Nature* 329, 595–598.

common among arts intellectuals towards science—due to the assumption that they do not have the capacity to understand something that other humans can.

What can be said is that the theories of Kuhn, Popper, Lakatos and Feyerabend are understood and interpreted in a totally different way by those who have studied science at university and by those who have a literature-based background. As a consequence, Kuhn complained at the end of his days that he had been misunderstood. And from this also comes the danger of teaching these authors in the humanities and social sciences if, by then, a good basis in physics, chemistry or mathematics has not previously been established in students, something that is increasingly outlawed in these disciplines. If these theories are taught by arts teachers, they are used to denigrate science, and the unwary students do not have enough arguments to defend themselves.

Theocharis and Psimopoulos blame these philosophers for the science of decline that they say British science had already experienced by the late 1980s. In N*ature*'s article, they were outraged by the BBC documentary broadcast on 17 and 22 February 1986 in the *Horizon* series. Its title: 'Science… Fiction?' was revealing of what the screenwriter of the documentary thought of science. The documentary took seriously the idea of Kuhn, Popper, Lakatos and Feyerabend that science was neither objective nor true—a fact that outraged scientists. If it had been asserted in the media, it would have been seen as totally irresponsible.

I would like to stress the role played by the BBC, in theory the most serious and rigorous media corporation in the world. Possibly, by 1986 the BBC was already using journalists or filmmakers who had studied everything from French Theory to the new philosophy of science. This is the problem: the shortage of physicists, chemists, biologists and engineers among BBC editors, and, above all, that journalists, sociologists, philosophers and historians talk about science without having practised it.

The physicists, Theocharis and Psimopoulos, also lashed out in *Nature* against an article in the television magazine *The Listener* that described the *Horizon* programme. Its title, *The Fallacy of Scientific Objectivity*,[23] outraged them:

> As is evident from their titles, these were attacks against objectivity, truth and science. After rehashing the usual anti-science chestnuts, both the film and the article came to this conclusion: '*The gradual recognition of these arguments may affect the practice, the funding and the institutions of science*' (our italics). At least in Britain, the repercussions of these mistaken arguments are already happening. (Theocharis and Psimopoulos 1987, 595)

---

[23] H. Lawson. (1986). 'The fallacy of scientific objectivity.' *The Listener*, 115, 12–13, 20 February.

Theocharis and Psimopoulos also denounced how the British media are guilty of expanding the theories of the philosophers of science and, as a consequence, spreading society's distrust of it. Physicists criticize the fact that learned societies do not attack the media when they publish such content:

> Articles and programmes attacking the scientific theses and championing the antitheses are published and broadcast regularly by the British media. But oddly, the Royal Society (RS), Save British Science and the other scientific bodies remain silent and do not deploy their powerful corporate muscle to answer such attacks against science (and sometimes against the RS itself). As a result, it appears, the editors of the popular media have come to the conclusion that the RS has no satisfactory answer. (Theocharis and Psimopoulos 1987, 595)

In the article published in *Nature*, the authors bemoaned the fact that there are moderately intelligent people who take up Popper's theory of falsehood, and explain in the following way the basis of the theories of this philosopher of science:

> In 1919 Sir Karl Popper by his own account had taken a strong dislike to the theories of Marx, Freud and Adler, whose supporters maintained that they were scientific. The difficulty was that Popper could not find any obvious way to refute them conclusively. Having noticed that Einstein's theories made (what seemed to him) falsifiable predictions, Popper resolved all his difficulties simply by declaring: 'irrefutability is not a virtue of a theory (as people often think) but a vice (…) The criterion of the scientific status of a theory is its falsifiability' (Example: 'The Earth is (approximately) a sphere' is not a scientific statement because it is not falsifiable; whereas 'The Earth is a flat disk' is indeed scientific.) Popper also thought that observations are theory laden. He phrased it thus: 'Sense-data, untheoretical items of observation, simply do not exist. A scientific theory is an organ we develop outside our skin, while an organ is a theory we develop inside our skin.'[24] (Theocharis and Psimopoulos 1987, 595)

In this way, Popper was able to make statements such as that 'the vast majority of scientific theories turn out to be false, including Newton's works'. On the other hand, the story of Adam and Eve may be an absolute truth yet, if it is, it is not a scientific truth but another kind of truth.

However, Popper's criticism may be somewhat exaggerated. It is true that it made the great mistake of comparing theories in social sciences (which are still more or less documented opinions, such as psychoanalysis or Marxism) to those in the natural sciences. Newton is not Freud or Marx. They're not on

---

[24]K. Popper. (1972). *Conjectures and Refutations: The growth of scientific knowledge* (p. 33–37). London: Routledge.

the same level. However, from my point of view, Popper had examined science so hard because he wanted to improve it, not to eliminate it. Psychoanalysis or Marxism may be dispensed with but, without kinematics or the theory of gravity, we would not have anything, from satellites to elevators.

Popper's view that the natural sciences were far superior to the social sciences is interesting. Popper does not deny the possibility of establishing empirical laws for the study of social phenomena, laws that can be subjected to empirical tests. But, for him, they will always be laws of low generality. He gives examples of sociological laws—such as the fact that every social revolution brings about a reaction—or economic laws—there can be no full employment without inflation—but makes it clear that they cannot be generalized or universally applied, like physical or chemical laws. This idea of 'historicism', that is, that it can be predicted in the social behaviour of humanity, was denounced by Popper for lack of rigour. He criticized Hegel and his 'predictive' idea of the establishment of a perfect state or Marx and his 'prediction' of a classless society. He praised the Marxist analysis of capitalist society at the time, but considered that Marx became a quasi-religious prophet when he propagated the idea that a classless society would soon come into being.

We also owe to Popper the idea that, while the natural sciences make predictions (we know exactly when an asteroid will pass by or what happens if we mix hydrochloric acid with sodium hydroxide), the social sciences make prophecies. And he warned of the danger of making predictions that are only prophecies. This is precisely one of the foundations of populism. Popper greatly reproved the Frankfurt School (much followed by the teachers of communication) and its champions—especially Theodor Adorno and Jürgen Habermas—for criticizing capitalism based on a synthesis of Marxism and psychoanalysis, two disciplines that Popper clearly identified as pseudosciences. The Frankfurt School inspired many of the ideas of May '68.

Although the movement did not succeed economically, it did succeed intellectually—not in the whole university, but in certain departments of philosophy, sociology, politics, media studies, cultural studies or film studies. These are precisely the departments where those who will control the media, science policies and even the policies of a country are formed.

But Popper's criticisms of the natural sciences, in the sense that he also thought that the experimental observations themselves were loaded with theory and subjectivity, were answered by natural scientists. Theocharis and Psimopoulos replied: 'But if observations are theory-laden, this means that observations are simply theories, and then how can one theory falsify (never mind verify) another theory?'

Theocharis and Psimopoulos criticized Popper for not being able to resolve this obvious contradiction. The philosopher Irme Lakatos was the one who solved the problem, considering that a theory could be scientific whether or not it was verifiable or falsifiable; that is, according to Lakatos, 'if a theory is refuted, it is not necessarily false'.[25] As an example, both 'the Earth is round' and 'the Earth is flat' are neither verifiable nor falsifiable. Everything will depend, as Kuhn states, on the agreement that the scientists will establish. Because, let's not forget, for the philosophical followers of Popper the measure is irrelevant and full of subjectivity. So, according to Kuhn, if the scientific establishment decrees that fairies exist, they exist.

This idea that science is not the truth but that scientific theories are just a succession of passing fashions is widespread in the humanities and social studies. In fact, these circles use the term 'paradigm' with astonishing ease. 'So according to Kuhn, the business of science is not about truth and reality; rather, it is about transient *vagues*' (Theocharis and Psimopoulos 1987, 596).[26]

What scientists were not very aware of until 1987 is that these faculties serve as the spring that nourishes the producers of media culture. This is why the British physicists were so astonished by the article that was published in *Nature*: 'So now that the term "truth" has become a taboo, we know what must take its place: mass popularity and prevailing fashion'.

In this race to turn irrationality into science, led by many philosophers of science from the universities themselves, the British physicists Theocharis and Psimopoulos asked themselves: 'Kuhn's view, that a proposition is scientific if it is sanctioned by the scientific establishment, gives rise to the problematic question: what exactly makes an establishment "scientific"?'. They pointed out that this Gordian knot was untied by what is undoubtedly considered to be the greatest enemy—within the university—of science of any time, Paul Feyerabend. For this philosopher of science, who preferred to study drama and opera rather than science, 'any proposition is scientific': 'There is only one principle that can be defended under all circumstances and in all stages of human development. It is the principle: anything goes' (Feyerabend 1975, 28).

Paul Feyerabend had much influence in the 1970s and 1980s (and still today) in all social sciences. In fact, Feyerabend had good persuasion and oratory skills—he had studied drama—yet British physicists Theocharis and Psimopoulos complained that in universities, in the late twentieth century, according to Feyerabend it can be asserted without blushing that: '"There are

---

[25] I. Lakatos. (1968). *The Problem of Inductive Logic*, ed. I. Lakatos (p. 397). Amsterdam: North Holland.
[26] P. Feyerabend. (1975). *Against Methods. Outline of an Anarchistic Theory of Knowledge*. London: New Left Books.

one million fairies in Britain" is as scientific as "There are two hydrogen atoms in one water molecule".'

Feyerabend was an agnostic, but his fascination for literary intellectuals who hate science is such that even someone as far from agnosticism as Catholic theologian Joseph Ratzinger (now ex-Pope Benedict XVI), in arguing his thesis on 'The crisis of faith in science',[27] relied on Feyerabend's next statement:

> The Church of Galileo's time was more grounded in reason than Galileo himself, and also took into account the ethical and social consequences of Galilean doctrine. Its trial of Galileo was rational and fair, and its review can only be justified on grounds of political expediency.

In order for the reader to observe the extent of this visceral hatred of the sciences and, above all, the influence of this type of anti-scientific philosophy, I quote Ratzinger himself. At the time of writing the text cited—1989—he was Prefect for the (Catholic) Doctrine of the Faith, and here tells of his personal experience with journalists:

> To my great surprise, in a recent interview on the Galileo case, I was not asked a question of the kind: 'Why has the Church sought to hinder knowledge of the natural sciences', but the exact opposite: 'Why has it not taken a clearer stand against the disasters that were to result when Galileo opened Pandora's box?' (Ratzinger 1989)

In 2008, a group of physicists from the Italian university, La Sapienza, Rome, rejected the presence of then Pope Benedict XVI at the inauguration of the academic year. They asked for the visit to be cancelled of 'in the name of secularism and science' and, above all, their reflections on Galileo, with arguments from Feyerabend. They wrote a letter and sent it to Rector Renato Guarini, who was a professor of statistics.

Guarini (Italian, from the same country as Galileo) said that those who asked for the event to be cancelled 'are a minority, since La Sapienza has 4,500 academic staff, so those who have expressed their own dissent through this letter are a small number'. That explanation, that most people are right, is the clearest proof that the university is not up to the challenges of the twenty-first century and that the thinking of Kuhn, Feyerabend and Lyotard, among others, has contaminated the academy. In my opinion, these physicists should have separated themselves from La Sapienza. If this university—and the rest of them—has any prestige and credibility, it is mainly thanks to physics and the

---

[27] J. Ratzinger. (1989). 'Perspectivas y tareas del catolicismo en la actualidad y de cara al futuro', reproducido en Altar Mayor: Revista de la Hermandad del Valle de los Caídos, 104 (January 2006).

method that this discipline has exercised to know the truth, and has extended to other areas such as chemistry, biology, geology and all technology.

In 1979 *Science* (from the AAAS) published a four-page feature article on Feyerabend in which it defined him as 'the Salvador Dalí of philosophy'.[28] In this article, Feyerabend states: 'normal science is a fairy tale.' He adds that 'equal time should be given to competing avenues of knowledge such as astrology, acupuncture, and witchcraft'. Strangely enough, in that interview he omitted religion. The British physicists complained that this type of philosopher does not seek truth but applause through factual means and powers, as Feyerabend had previously put science on an equal footing with 'religion, prostitution and so on'.

The British scientists said in *Nature* that all these ideas have had an effect on Britain by cutting science budgets: why finance such an expensive 'fairy tale'? Politicians with no university education in science, but with studies in the humanities or social sciences, will wonder. However, they warned what form the ideas of these philosophers are assuming in America: 'in the United States the most conspicuous impact of the antitheses has been the alarming growth of religious fundamentalism.' No sociologist predicted this thirty years ago; however, these physicists did.

At the time, journalists working for scientific journals such as *Science* described Feyerabend and, in the words of Theocharis and Psimopoulos, his 'monstrous ideas', as follows: 'Compared with the stiff and sober work that is often done in the philosophy of science, Feyerabend's views are a breath of fresh air.'[29]

Theocharis and Psimopoulos were warning us, back in distant 1987:

It is the duty of those who want to save science, both intellectually and financially, to refute the antitheses, and to reassure their paymasters and the public that the genuine theories of science are of permanent value, and that scientific research does have a concrete and positive and useful aim – to discover the truth and establish reality.

Thirty years later, the United States has a president, Donald Trump, who won because he and many like him believe that Feyerabend is as valid as Einstein and, above all, that someone can be undertaking some kind of intellectual work more by studying Feyerabend than Newton or Einstein.

In the academic field, the extension of the idea that the natural sciences do not reflect the truth but are just one more type of knowledge can lead to

[28]W. J. Broad. (1979). 'Paul Feyerabend: Science and the anarchist.' *Science*, 206 (4418), 534–537.
[29]W. J. Broad. (1979). 'Paul Feyerabend: Science and the anarchist.' *Science*, 206 (4418), 534–537.

the downfall of students who are interested in them. It should be noted that adolescence is an age when ideas are not mature, and students are easily carried away by teachers. It should not be forgotten that all students, throughout their academic lives in secondary school, have plenty of contact with teachers of the arts and philosophy who consider these unscientific ideas seriously. More than once I have had to calm down my physics and chemistry students after classes given by a teacher of philosophy or literature.

When a professor of philosophy—especially in Latin countries, where it is not necessary to study science or mathematics at university in order to graduate as a philosopher—tells his students that there is no way to get to the truth or that 'there are a million fairies in Britain is as scientific as saying that the water molecule has two hydrogen and one oxygen atoms', the intellectual, critical and analytical momentum of young people is affected.

Theocharis and Psimopoulos complained bitterly that these authors are taught in universities, and describe the effect that they have on students:

> The hapless student is inevitably left to his or her own devices to pick up casually and randomly, from here and there, unorganized bits of the scientific method, as well as bits of unscientific methods. And when the student becomes the professional researcher, lacking proper tuition and training, he or she will grope haphazardly in the dark, follow expensive blind alleys and have recourse to such (unreliable) things as random guess, arbitrary conjecture, subjective hunch, casual intuition, raw instinct, crude imagination, pure chance, lucky accident and a planned trial – and invariable error. Can this be an adequate methodology by means of which to make new discoveries and beneficial applications? Of course not, yet this is the entire methodology that the exponents of the antitheses [Kuhn, Lakatos, Popper and Feyerabend] actually recommend to professional researchers. (Theocharis and Psimopoulos 1987, 597)

The problem is not so much for the natural sciences themselves—apart from, as we have seen and will continue to see, their media coverage—because, I repeat, scientists and science studies simply ignore these philosophers. The main tragedy is played out by the social scientists themselves, who believe that the natural sciences work not as they really do but as these philosophers have told them. Therefore, they design a whole 'scientific' social science apparatus that is based on what they believe are the sciences that have really advanced—but they do know how they really work. The consequence is pseudoscientific results in areas such as education studies, sociology, communication theory, economics and psychology, among many others.

Worse still, some of these results are applied to society in the form of laws (e.g. education), which politicians say are endorsed by 'scientific studies'. This would explain why there are more and more educationalists and research projects in education studies and yet, every day, schoolchildren know less and less. In other words, the results of these studies push back the field in which they are applied, contrary to what happens in the natural sciences.

In most cases, the 'researchers' who practise these social science disciplines have never studied a serious science such as physics, chemistry or biology. They have never been able to approach it from the primary sources and direct experiences, so what guarantees can their conclusions have? In the twenty-first century, many regulations are not based on common sense or tradition—which for me would be interesting elements to take into account. Instead, it is considered better to legislate using the kind of results from many social sciences. That this happens and no one protests is also a symptom of the decline of serious science.

For some social scientists, according to Feyerabend, esotericism or trickery is as relevant as science.

In March 2007, Dr. David Voas, a senior research fellow at the faculty of social sciences at the University of Manchester, appeared in the world press with his study showing that relationships between couples do not involve the conditioning factors of the different signs of the horoscope.[30] This is obvious, because horoscopes are a falsehood, although they are widespread in the media, so widespread that Dr. Voas felt that perhaps there was some truth in them. Obviously, Voas has never understood the law of universal gravitation of his fellow countryman Newton, or perhaps he considered it to be, as Kuhn maintains, a simple agreement between scientists now due to be revoked for the benefit of magicians and astrologers.

To conduct his study, Voas—a senior research fellow—identified 20 million husbands and wives in the British census and he determined that there is no relationship between the durability of their couple relationships and the premonitions of magicians.[31] Thus, in the face of astrological speculation that a Virgo woman cannot be paired with an Aries man, Voas showed that there are as many happy unions as there are of Pisces and Sagittarius (also banned by astrology) as of Cancer and Taurus (which are blessed by astrologers).

Someone may argue that it is precisely this kind of research that dismantles false beliefs such as astrology. But my question is this: is it worth spending the money to prove it? Did Voas ever think that the horoscope was something serious that deserved the slightest credibility? How did someone finance the research? How did a university give him time to study such a hoax? Could it

---

[30]The results were published in the British *Daily Mail* (27 March 2007) or *The Guardian* (27 March 2007), among others.

[31]D. Voas (2007). 'Ten million marriages: A test of astrological "love signs"'. https://astrologiaexperimental. files.wordpress.com/2014/05/voasastrology.pdf (retrieved July 2018).

be that in his degree Voas, who graduated in social sciences, had studied Kuhn or Feyerabend more than Newton or De Broglie?

If Voas continues along this path, the next step will be to investigate whether the Earth is flat or whether matter is made up of fire, air, water and earth. To what extent have the true scientists (physicists, chemists, biologists) of the University of Manchester lost prestige and credibility because of its decision to allow the social sciences the same academic status as the natural sciences? The *Daily Mail* had entitled its article 'Love isn't in the stars, say scientists',[32] and *The Guardian* 'Zodiac off the track for true love'.[33] The journalists had accorded Voas the status of a scientist or a researcher. To what extent is the existence of Professor Voas, as a member of the faculty of the University of Manchester, not proof of the decline of science?

Creationism or Intelligent Design

There are more clues to confirm the decline. All this philosophy that claims that science is just one more of the many possible versions is used by another dangerous collective: the one that claims that Darwin's theory of evolution has as much validity as the theory of intelligent design, which holds that a superior being—which its theorists do not define, but that many associate with God—has created the various animals and plants, and, therefore, these are not products of evolution. It is revealing that these groups are basing their arguments not on the results of experiments but on the weakness of the prevailing scientific method as defended by Popper, Feyerabend and Kuhn.

When serious scientists throw up their hands and ask about the evidence for intelligent design, its proponents argue that, according to Popper their theory is as scientific as Darwin's and, therefore, both should be taught in schools. Creationists have cited the philosophers of science in the US courts as proof that intelligent design is as valid as Darwin's theory of evolution: 'The official university itself maintains that science is just another type of explanation of reality and that this explanation is only the consensus of the scientific hierarchy. Consensus that is just a passing fad,' US lawyers argue in court. That is to say, based on the philosophers of science—the one most often cited in the courts is Popper—and on what they consider to be 'scientific relativism', a door is being opened to teach a lie—such as creationism—as if it were science.

---

[32]'Love isn't in the stars, say scientists,' http://www.dailymail.co.uk/news/article-444821/Love-isnt-stars-say-scientists.html.

[33]'Zodiac off the track for true love'. (*The Guardian*, 27 March 2007), https://www.theguardian.com/education/2007/mar/27/highereducation.uk2.

In case the reader is not familiar with the theory of intelligent design, also known as creationism—I can advance some postulates to prove that it is a fairy tale. The theory holds that Noah moved the dinosaurs into his ark, that planet Earth is only 6,010 years old—an age obtained by calculating backwards the ages of the protagonists of the Old Testament—and that all the animals that lived in the Garden of Eden were herbivores. It maintains that crocodiles' sharp teeth are 'so that they can break a watermelon'. However, like any theory, there are also uncertainties. For example, theorists are unclear as to whether Noah, to gain space in his ark, chose adult dinosaurs, young or just fertilized eggs…

But how to design an experiment to prove that a supreme being designed plants and animals? There is no way to prove that the nonsense of intelligent design theory is either true or that it is false so, even from a Popperian point of view, the theory of intelligent design is still a hoax. However, this hoax, which even the Vatican itself has described as such, is gaining followers in the United States, a country where half the population declare themselves to be 'strict creationists' and where, according to a survey by the demographic studies company, Zogby, '78% of Americans want biology teachers to teach Darwin's theory of evolution in class, but also the scientific arguments against it'.[34]

But what are the scientific arguments against it, if even molecular biology itself supports Darwin? You can't take up a neutral position between science and anti-science, because that's what the Jesuits wanted with Galileo: to expose his heliocentric model as just another model. If, in the United States, they are considering teaching both creationism and evolutionism, it is not surprising that someone such as Donald Trump has won the presidency of the country there.

The worst part is that the idea is coming to be accepted by science students themselves. A survey published by the British daily, *The Guardian*[35] in 2006 indicated that one in six biology students in London colleges would be in favour of this theory. Aware and alarmed by this problem, the Royal Society organized a competition with the slogan: 'Why is Creationism wrong?' The winner was geneticist Steve Jones, professor at University College of London. His statements to *The Guardian* are conclusive about the decline of science and what that would mean: 'There is an insidious and growing problem,' said Professor Jones. 'It's a step back from rationality. They (the creationists) don't have a problem with science, they have a problem with argument. And irrationality is a very infectious disease, as we see from the United States'.[36]

---

[34] Zogby polls on teaching evolution, Summary from 2001 to 2009, https://freescience.today/2009/12/19/zogby-polls/.

[35] *The Guardian*. 'Academics fight rise of creationism at universities,' 21 February 2006, 11.

[36] https://www.theguardian.com/world/2006/feb/21/religion.highereducation.

During my stay in London I saw how the Royal Society and other scientific institutions are developing many initiatives to promote rational and scientific thinking, from lectures to 'scientific cafés' where researchers chat with young people. Scientists denounce that philosophy and the media are looking the other way because they are not interested in a cultural supremacy of science and rational thought.

A significant percentage of scientists are unsettled and surprised in this twenty-first century that, suddenly, everything is now considered science, including intelligent design. There is no explanation for why a country like the United States—which has 55 Nobel Prize winners in chemistry, 41 in physics and 74 in medicine—should put the theory of evolution—for which experiments can be developed to refute it, although it usually comes out more than successful—on the same level as that of creationism.

My hypothesis is that the fault lies partly with these philosophers, but above all with their extraordinary influence in the universities where those responsible for the prevailing media culture are trained. If it weren't for the fact that most journalists, filmmakers, political scientists and policy-makers come from studios where these philosophers are taken seriously, there would be no problem. They'd just be a brotherhood of self-appointed college pedants. Social networks have magnified this problem, because trenches are dug and we receive information only from one side, promoting ideological hooliganism.

The Sokal Scandal: War from Science

The next attack by scientists on the intellectual dross of literary intellectuals who have been seduced by French Theory or the postmodern Kuhn, Popper or Feyerabend came from another physicist: Alan Sokal.

Alan Sokal! It is a cursed name, unmentionable in any university department of social sciences or humanities. However, he is the hero of experimental scientists, because he 'brought to light the intellectual bullshit that is plotted in those other departments that are not of pure science'.

Let us remember why this thorny matter reopens so many wounds. Sokal jokingly wrote a smug article that was nothing more than an accumulation of grotesque nonsense and obvious falsehoods. He wrote it in the style in which some philosophers and sociologists of science write postmodern critiques of physics, chemistry or biology. He gave it a pompous title, as befits something that he wanted to pretend it is not: 'Transgressing the boundaries: Towards a transformative hermeneutics of quantum gravity[37]'. Sokal sent this to a

---

[37] https://physics.nyu.edu/sokal/transgress_v2_noafterword.pdf.

periodical known as an impact journal; that is to say, publishing there serves to increase the store of professors of humanities and social sciences. The one in question was *Social Text*. The article was approved by the editors and published in April 1996.

But, in the name of experimental scientists, Sokal had prepared his own particular revenge against social scientists and academics in the humanities. The day after the publication in *Social Text*, Sokal revealed on the cover of the *New York Times*, no less, that it had all been a joke. In his opinion, he had thus exposed the incompetence and lack of level and rigour of many publications in which social and 'humanities' researchers reveal their work and, if such a thing had happened in science, a catastrophe would have been unleashed. But in the departments of social affairs and humanities, people just decided not to talk about 'the Sokal affair'.

Two years later, Sokal and Jean Bricmont rubbed more salt into the wound and published *Fashionable Nonsense: Postmodern Intellectuals: Abuse of Science*.[38] This book brings together all kinds of quotations from pretentious intellectuals who, from their literature-based chairs, have analysed everything from Lacan's identification of the penis with the square root of one, to the criticism of the equation of relativity (that energy is equal to mass for the square of the speed of light) to 'privileging the speed of light over other speeds with the same rights'.

The book unravels the lies of many social science intellectuals. One of the most interesting chapters, especially from the point of view of those of us who teach journalism, is how he dismantles all the work on the 'expert' in linguistics and semiotics, Julia Kristeva. This linguist, of Bulgarian origin, was awarded the Chair of Linguistics at the prestigious University of Paris VII after publishing two books following a totally absurd doctoral thesis entitled 'La revolution du langage poétique' in 1974, in which, among other nonsense, she attempted to demonstrate that the poetic language 'is a formal system whose theorization can be based on [mathematical] set theory' and, on the other hand, said in a footnote that this is 'only metaphorical'.

Sokal and Bricmont fiercely attacked this conception of passing off as something serious what are merely laborious cogitations, or outbursts like those of St. Teresa when, unbalanced by the lack of food (she had undergone hard fasting as penance), she elaborated her mystical work. In Kristeva's case, they point out:

> Kristeva's early writings relied strongly (and abusively) on mathematics, but she abandoned this approach more than twenty years ago; we criticize them here

---

[38] Alan Sokal and Jean Bricmont. (1998). *Fashionable Nonsense: Postmodern intellectuals' abuse of science*. New York: Picador.

because we consider them symptomatic of a certain intellectual style. The other authors, by contrast, have all invoked science extensively in their work. Latour's writings provide considerable grist for the mill of contemporary relativism and are based on an allegedly rigorous analysis of scientific practice. The works of Baudrillard, Deleuze, Guattari and Virilio are filled with seemingly erudite references to relativity, quantum mechanics, chaos theory, etc. So we are by no means splitting hairs in establishing that their scientific erudition is exceedingly superficial. Moreover, for several authors, we shall supply references to additional texts where the reader can find numerous further abuses. You don't understand the context. Defenders of Lacan, Deleuze et al. might argue that their invocations of scientific concepts are valid and even profound, and that our criticisms miss the point because we fail to understand the context. After all, we readily admit that we do not always understand the rest of these authors' work. Mightn't we be arrogant and narrowminded scientists, missing something subtle and deep? (Sokal and Bricmont 1988, 8–9)

Both mathematicians and physicists later proved that Kristeva had no idea about mathematics, much less ensemble theory, as her assumptions would not have allowed her to pass even a high school diploma test. There are massive errors, like everything concerning Gödel's theorem. Her level of mathematics was negligible, yet she considered that it was not necessary to know mathematics for her thesis, although the subject was an important part of it.

But none of those supervising her doctoral thesis understood a word that Kristeva was saying and, rather than be regarded as ignorant, they decided to approve it. The same thing happened with those who awarded her a chair. She next abandoned her approach, because she realized that everything was absurd. But she didn't give up her professorship, and neither did her semiotic followers, most of whom make their living—teaching—in faculties of communication, including Spanish ones.

The semiotic Roland Barthes, another representative of the French Theory, neither knew of mathematics nor found it interesting enough to study. He went on to write in the 'prestigious' journal *La Quinzaine Littéraire* (May 1970, 94, 1–15; 19–20) the following about Kristeva's intellectual work:

Julia Kristeva changes the order of things; she always destroys the latest preconception, the one we thought we could be comforted by, the one of which we could be proud: what she displaces is the already-said, that is to say, the insistence of the signified, that is to say, silliness; what she subverts is the authority of monologic science and of filiation. Her work is entirely new and precise (...). (Barthes 1970, 19)[39]

---

[39] R. Barthes. (1970). 'L'étrangére'. La Quinzaine Littéraire. (1–15 May) pp. 19–20.

Aside from the fact that what he says isn't true, it's important to note how Barthes uses difficult and meaningless language so that it looks like he is writing about something important when it is pure rubbish. But Barthes is followed and quoted with devotion by professors of journalism and audiovisual communication. Ignacio Ramonet, editor of the influential newspaper *Le Monde Diplomatique* and professor of media studies at several universities, describes as a personal milestone 'having been a disciple of Roland Barthes'.

It should be noted that, in *Fashionable Nonsense*, neither Sokal nor Bricmont are so much denouncing journals for letting their hoax/joke through their filters as disgusted at the intellectual weakness of the theories that sustain many social sciences:

> But what was all the fuss about? Media hype notwithstanding, the mere fact the parody was published proves little in itself; at most it reveals something about the intellectual standards of one trendy journal. More interesting conclusions can be derived, however, by examining the content of the parody. On close inspection, one sees that the parody was constructed around quotations from eminent French and American intellectuals about the alleged philosophical and social implications of mathematics and the natural sciences. The passages may be absurd or meaningless, but they are nonetheless authentic. In fact, Sokal's only contribution was to provide a 'glue' (the 'logic' of which is admittedly whimsical) to join these quotations together and praise them. The authors in question form a veritable pantheon of contemporary 'French theory': Gilles Deleuze, Jacques Derrida, Felix Guattari, Luce Irigaray, Jacques Lacan, Bruno Latour, Jean-François Lyotard, Michel Serres, and Paul Virilio. The citations also include many prominent American academics in cultural studies and related fields; but these authors are often, at least in part, disciples of or commentators on the French masters. (Sokal and Bricmont 1998, 3)

Another suggestive part of the book is when fun is made of Lacan's attempt to mathematize psychoanalysis, first because psychoanalysis is not science (psychology became science when it left philosophy and went into neurochemistry and neurobiology); second, because, as the physicists Sokal and Bricmont demonstrate, Lacan has no idea of mathematics:

> What should we make of Lacan's mathematics? Commentators disagree about Lacan's intentions: to what extent was he aiming to 'mathematize' psychoanalysis? We are unable to give any definitive answer to this question-which, in any case, does not matter much, since Lacan's 'mathematics' are so bizarre that they cannot play a fruitful role in any serious psychological analysis. (…) The most striking aspect of Lacan and his disciples is probably their attitude towards science, and the extreme privilege they accord to 'theory' (in actual fact, to formalism and word-

play) at the expense of observations and experiments. After all, psychoanalysis, assuming that it has a scientific basis, is a rather young science. Before launching into vast theoretical generalizations, it might be prudent to check the empirical adequacy of at least some of its propositions. But, in Lacan's writings, one finds mainly quotations and analyses of texts and concepts. Lacan's defenders (as well as those of the other authors discussed here) tend to respond to these criticisms by resorting to a strategy that we shall call 'neither/nor': these writings should be evaluated neither as science, nor as philosophy, nor as poetry, nor .... One is then faced with what could be called a 'secular mysticism': mysticism because the discourse aims at producing mental effects that are not purely aesthetic, but without addressing itself to reason; secular because the cultural references (Kant, Hegel, Marx, Freud, mathematics, contemporary literature...) have nothing to do with traditional religions and are attractive to the modem reader. (Sokal and Bricmont 1998, 36–37)

What Sokal criticizes, although very common, is the general tone. There are great works in social sciences and humanities, just as there may be some false scientific work in supposedly pure areas such as physics or mathematics. The difference, from my point of view, is that in the natural sciences it is the scientists themselves who detect the imposture and, when they do, tear to pieces those who dared to use the false methodology. They do this because they know that the credibility of the natural sciences is such that unmasking an intellectual impostor will result in greater credibility for science.

But the same is not true of the social sciences. They do not have the credibility of physics or chemistry, yet they do have much political credibility, and such as economics can involve much more than academic debate. That's why all social science professors keep quiet about flagrantly false studies, and media criticism of their methods is rare. I repeat that there may be serious social studies and scientists, but I have not seen the same criticism of physics directed at economics, sociology or education studies.

In this sense, the documentary *Inside Job*[40] analyses the causes of the financial crisis in the United States. The filmmakers denounce the role played by the salaries of bankers, rating agencies and politicians but, most interestingly, for the first time the professors of economics and business at universities and business schools. The documentary shows how the financial sector had corrupted the study of the economy. The first example is when we talk about the 'Nobel Prize in Economics' instead of its real name, 'the Bank of Sweden Prize in Economic Sciences in Memory of Alfred Nobel'. As we have already pointed out, the Bank of Sweden's interest is to equate economics (political philosophy, in which anything can be said) with physics or chemistry.

---

[40] Ferguson 2010 (its a documentary. It does not need more reference.

Inside Job, which won the Oscar for the best documentary, explains how many professors and researchers in economics supported the deregulation of markets, not as scientific knowledge but as a collaboration rewarded by many zeros on their consulting fee. Apart from the aforementioned documentary, there has been no catharsis in the economic sciences. What's worse, the number of young people who want to dedicate themselves to this discipline, so far removed from the rigour of physics yet which has not been criticized with the same harshness, has not fallen. Ferguson, the director of the documentary *Inside Job*, was not from film studies, cultural studies or media studies, but from something more solid: he graduated in mathematics at Berkeley.

Glenn Hubbard was dean of one of the world's most prestigious economics schools and researchers, Columbia Business School, and appears in the documentary. Hubbard comes over on film as haughty, defiant and visibly nervous when asked about his potential conflict of interest. The dean of Columbia was head of the economic council during the George W. Bush administration. He received $100,000 for testifying on behalf of the Bear Stearns fund managers accused of fraud, and wrote a report for Goldman Sachs in 2004 praising derivatives and the mortgage securitization chain as improving financial stability.

The press, specifically *El País* (31 May 2011),[41] echoed the finding of the corruption of economic science and recalled that the links between the financial industry and the academic world are not restricted to Hubbard:

> Martin Feldstein, Harvard professor of economics and advisor to Ronald Reagan, was a leading architect of financial deregulation and served on the AIG board; Laura Tyson, a professor at the University of California, Berkeley, became a director of Morgan Stanley after leaving the presidency of the National Economic Council during the Clinton administration; Ruth Simmons, president of Brown University, is a director of Goldman Sachs; Larry Summers, who pushed deregulation in the derivatives market during his time in office, is president of Harvard, a position that has not prevented him from earning millions of dollars by advising several hedge funds; Frederic Mishkin, who returned to teach at Columbia after working at the US Federal Reserve, wrote a report in 2006 funded by the Icelandic Chamber of Commerce (he received $124.000) praising the strength of this country's economy…

The newspaper asked to what extent the financial industry used its convincing chequebook to earn academic endorsement, providing intellectual support for its ultraliberal policies. Ángel Cabrera was in 2011 the only Spaniard to

---

[41] 'Académicos tras la especulación' ('The scholars behind the speculation'), (*El País*, 31 May 2011), https://elpais.com/diario/2011/05/31/sociedad/1306792801_850215.html.

head a business school in the United States, serving as President of the Thunderbird School of Global Management in Arizona. In his opinion, the role of the academic world in the lead-up to the crisis was not so much specific conflicts of interest but the fact that for decades the centres had been transmitting a series of values about the functioning of markets, risk management or human resources that have been shown to be wrong and harmful:

> The whole theory of market efficiency, for example, became a religion, and the universal conclusion was reached that any intervention was bad. Likewise, in retribution policies, if you treat people as opportunistic and selfish by nature, creating huge short-term incentives, you open the door for them to behave like this. (Cabrera, in *El País*)[42]

Cabrera added that the centres had been shaping the Wall Street value system, creating a 'platform of legitimacy' for certain behaviours. However, enrolment in economics has not fallen, nor have social scientists and academics in the humanities dedicated themselves to criticizing the view that economics is not science but only an opinion that is mathematized to give a halo of credibility to its prophecies. Physics has been attacked more than the economy. And in the West there are more young people studying a false science such as economics than a true science such as physics, chemistry or biology.

A social scientist will generally be aggrieved if another researcher attempts to repeat his or her experiment and its conditions (in the unlikely event that they are reproducible). It is also common for other researchers to have no access to the experiment's databases,[43] so it is not even possible to know whether, at best, the methodology was incorrect or, at worst, the results were invented. There is no way to prove it, or at least it is not as easy as it is in pure science. By contrast, a natural scientist will consider it an honour if his or her experiment is repeated as many times as necessary by as many researchers as wish to do so. Herein lies the strength of the natural sciences and the weakness of the social sciences.

This is an obvious advantage of the natural sciences for many, but not for information or communication science graduates. This is because their publications give the same credibility to the social sciences as to the natural sciences, or almost less to the latter, which are not usually taught in journalism or film studies schools. It is detrimental to science that academic structures

---

[42] https://elpais.com/diario/2011/05/31/sociedad/1306792801_850215.html.

[43] Some prestigious social science journals demand complete data, in order to approve articles. However, these data are still not reproducible, because the conditions (the state of society, established through surveys, interviews, statistics) are not replicable over time, as in the natural sciences.

give the same status to a professor of chemistry as to a professor of film studies, because they put both disciplines on the same intellectual level.

During 2018, three scholars—James Lindsay, Helen Pluckrose, and Peter Boghossian—wrote 20 fake papers using fashionable jargon to argue for ridiculous conclusions, and tried to get them placed in high-profile journals in fields including cultural studies. *The Wall Street Journal* began to investigate,[44] and the three authors went public with their project before it was finished. Their success rate was remarkable: by the time they took their experiment public (on October 2018), seven of their articles had been accepted for publication by ostensibly serious peer-reviewed journals. They describe their articles as 'shoddy, absurd, unethical and politically-biased'. Seven more were still going through various stages of the review process. Only six had been rejected. *The Atlantic* published a fascinating report entitled 'What an Audacious Hoax Reveals About Academia[45]'.

According to the authors of the fake articles 'each paper began with something absurd or deeply unethical (or both) that they wanted to forward or conclude'.

> We then made the existing peer-reviewed literature do our bidding in the attempt to get published in the academic canon… This is the primary point of the project: What we just described is not knowledge production; it's sophistry… The biggest difference between us and the scholarship we are studying by emulation is that we know we made things up.[46]

One paper about rape culture in dog parks (in which the writer claimed to have inspected the genitals of just under 10,000 dogs while asking their owners about their sexuality) was honoured for excellence as one of 12 exemplary pieces in feminist geography by highly ranked journal *Gender, Place and Culture*, which published the paper.

Their detailed account of the process for *Areo Magazine*, in an article called 'Academic Grievance Studies and the Corruption of Scholarship' is fascinating and important reading. This is the introduction:

---

[44] Jillian Kay Melchior (Oct. 5, 2018) 'Fake News Comes to Academia. How three scholars gulled academic journals to publish hoax papers on 'grievance studies'.' *The Wall Street Journal.* https://www.wsj.com/articles/fake-news-comes-to-academia-1538520950.

[45] Yascha Mounk (Oct. 5, 2018) 'What an Audacious Hoax Reveals About Academia. Three scholars wrote 20 fake papers using fashionable jargon to argue for ridiculous conclusions.' *The Atlantic.* https://www.theatlantic.com/ideas/archive/2018/10/new-sokal-hoax/572212/.

[46] https://www.irishtimes.com/life-and-style/people/hoax-papers-the-shoddy-absurd-and-unethical-side-of-academia-1.3655500.

Something has gone wrong in the university—especially in certain fields within the humanities. Scholarship based less upon finding truth and more upon attending to social grievances has become firmly established, if not fully dominant, within these fields, and their scholars increasingly bully students, administrators, and other departments into adhering to their worldview. This worldview is not scientific, and it is not rigorous. For many, this problem has been growing increasingly obvious, but strong evidence has been lacking. For this reason, the three of us just spent a year working inside the scholarship we see as an intrinsic part of this problem. We spent that time writing academic papers and publishing them in respected peer-reviewed journals associated with fields of scholarship loosely known as "cultural studies" or "identity studies" (for example, gender studies) or "critical theory" because it is rooted in that postmodern brand of "theory" which arose in the late sixties. As a result of this work, we have come to call these fields "grievance studies" in shorthand because of their common goal of problematizing aspects of culture in minute detail in order to attempt diagnoses of power imbalances and oppression rooted in identity.[47]

In my opinion, the lack of seriousness of many social scientists may be demonstrated, for example when they publish their results or surveys in the media rather than in prestigious journals. Likewise, they may choose to publish their results in a book—which they edit themselves, using their research budget—rather than subject their data to the scrutiny of other colleagues in rigorous, blind evaluations. What is the scientific rigour of books financed by public or private institutions? None, for me.

I'm not talking about books of essays. The social sciences were born of a type of volume about analysis and reflection that a researcher constructs around an issue. Books such as Adam Smith's *The Wealth of Nations* (1776) were the origin of the social sciences, long after Machiavelli wrote *The Prince* (1513) and long before Karl Marx published *Das Kapital,* also known as *Capital: Critique of political economy* (1867). Nor do I criticize scholarly books by academics in the humanities. Francisco Rico's critiques of *Don Quixote* represent a magnificent contribution to Culture with a capital letter, and their rigour and seriousness are praiseworthy. In that kind of book there is nothing to object to.

But there is a widespread trend among some social scientists, who cannot aspire to this type of essay but supplement their publication with hundreds of surveys or content analyses with an economy of scientific rigour. This kind of research can also damage science's image among journalists and filmmakers, especially if students are involved. As a student of journalism, I remember suffering the tyranny of the surveys that many teachers gave out to us as

---

[47] Available at: https://areomagazine.com/2018/10/02/academic-grievance-studies-and-the-corruption-of-scholarship/.

compulsory classwork, to pass on to people. At other times, they hired pollsters who, in turn, had the students distribute the surveys. The same is true of some 'brainy' content analyses that rely on data collection by unsuspecting students. Now, as a professor, I still see how many of my colleagues continue with these pseudo methodologies of research. The use of undergraduates or interviewers not directly linked to the research to conduct studies in which they will not actively participate (their names will not appear as co-authors of the research) is one of the greatest sources of intellectual corruption of our time, as the counterfeiting cannot be detected and the data cannot be reproduced.

I remember in my student days how we used to falsify data to obtain our teacher's approval or earn money from the pollsters. I cannot forget how, afterwards, many teachers published those results as if they were their own. Our form of revenge—widespread in the social sciences—was to falsify the data. I laugh when people wonder why the polls never predict the election correctly. An analysis of how and by whom the data are collected is sufficient explanation. If physics were as wrong as polls, civilization would still be in the dark ages. The fact that the polls are wrong is irrelevant: if they were never undertaken, the world would not change.

But students of communication have the idea that science is an impostor, from the very moment of data collection and sampling. This is because they do not have any other form of comparison since they do not study, for example, the natural sciences. I remember many cafeteria conversations in my days as a journalism student, trying to explain that which we all tampered with, as a way to annoy some professors—or survey companies—who wanted to take advantage of students, was not science. However, this experience, that science is something false, will never be erased from the minds of young students of communication when they see how their teachers later boast of those results. The works of many social science 'experts' contain little input besides these surveys or content analyses, which are likely to have been falsified, will hardly ever be developed by them, and which they attempt to explain as if they were the most excellent and objective of data.

My experience tells me that this practice, while not generalizable, is common in some social disciplines. Many social scientists are using the endorsement afforded by a knowledge of the natural sciences to something called the university, to propagate from that same platform a doctrine supposedly endorsed by science, scientific method and rigour. That this happens, and the natural scientists of that university do not criticize the practitioner of that unscientific procedure, is another symptom of the decline of science. This does not happen in the humanities. In the humanities, academics write essays for edited volumes but don't pretend that their idea is scientific truth; it is an opinion, a point of view, a synthesis of a problem. The talent of a humanities intellectual is to give other versions of reality that are attractive and to contextualize a complex world; they never pretend, as economists and educationalists sometimes do, that their version is the truth in the same way that Newton's laws and the existence of mitochondria are.

False Results in Natural Sciences

Someone may argue that false results may be published in the natural sciences also. It's true. And we'll talk about it next. But I will first try to illustrate to the reader that what happens in the natural sciences is unlike what happens in the social sciences. In the natural sciences, fraud is always detected and is proof of the strength of the method. In social science, precisely because of its method there is no attempt even to look for fraud. This starts right from the assumption of the investigator's honesty; there is nothing to say that the investigator is necessarily honest, especially if you know that any deception cannot be proved.

In the natural sciences, a multitude of cases of scientific fraud have been demonstrated by repeating the experiment. The good thing, moreover, is that these reprehensible acts are aired by the media, which acts as a barrier to publishing these fraudulent results because researchers know that they can lose their credibility. There might be really false news, like the Raelian sect's alleged cloning of a baby.[48] Even though the announcement was made by a doctor of biochemistry, everyone knew that it was a fraud. First of all, it was obvious because he had gone to the media to give a press conference rather than undergo blind peer review in a prestigious scientific journal. And, secondly, he did not give free access to his experiments, nor did he offer sufficient information to reproduce his data. Even so, the media published the news, and this says a great deal about the media culture.

But at least the media alerted us that the result could be false, a warning that never arrives when a result is from the social sciences, even if it is disclosed to the media in the same circumstances as in the Raelian claim—without previously appearing in major journals and without giving free access to the data for others to repeat the experiment before reporting the finding at a press conference.

In any event, the fact that a claim appears in a scientific journal of excellence does not guarantee much either, because the news may be false, as demonstrated by the case of the physicist Jan Hendrik Schön, author of 80 articles in scientific journals yet who falsified the data. He published three articles in *Nature* on

---

[48] In January 2003 this false story was published in most of the world's media, including prestigious media.

superconductivity, all false and all spread by the media. *Science* published one in its electronic edition, but it did not go to print because of something that, I repeat, the natural sciences have that other fields of knowledge do not: in the summer of 2002, many researchers began to detect that the results were irreproducible and, therefore, false. Schön had even published in *Physical Review Letters* and, already by the age of 31, he was a Nobel Prize candidate (2002).

But how does one prove a similar fraud in the social sciences? It is impossible. That is why it is so dangerous to make political decisions that are contrary to common sense yet are supported by social research, as in the field of education studies. In education, for example, the experience of teachers who have taught for many years may be more valuable than the research produced in faculties. But few leaders listen to the voice of teachers. They prefer the opinion of education researchers who have never taught children. The same happens in communication, journalism, law, business, cinema and politics…

Continuing with the natural sciences, in 1997 the German biologist Friehelm Herrmann published numerous false articles on leukaemia in the best journals for medicine and biology. This shows that there is fraud in such scientific journals, even though it represents a problem for Sokal's stance. However, in their defence, these frauds are almost always detected.

One of the cases that has had the greatest impact was the publication in 1998 by the prestigious medical journal, *The Lancet*, of a study that linked the trivirix vaccine—measles, mumps and rubella—to cases of autism. Six years later, in 2004, the author of that study, Dr. Andrew Wakefield, was found to have collected €82,000 from an institution—the Legal Aid Board—that wanted to sue the laboratories that produced the vaccines. The news was revealed by a newspaper—*Sunday Times*—after a process of journalistic investigation. In other words, in this case, journalism was more rigorous than science. *The Lancet* referred to the study as 'fatally flawed'.

The perversion in specialized medical journals has reached such an extreme that it is not uncommon to find headlines such as: 'A large medical journal apologizes for evaluating drugs with experts paid by their manufacturers' (*El País*, 25 February 2000, 44). In this news item, those responsible for the *New England Journal of Medicine*, one of the most frequently cited specialized journals, for example in the health supplements of *El País*, *El Mundo* and *ABC*, apologized because 19 scientific articles published in the last three years had been written by researchers belonging to certain pharmaceutical laboratories yet who had concealed it.

The scientists who wrote the articles praised the properties of certain drugs over others, depending on the funding that they received from the laboratories.

The companies were accused of using the 'purchase' method of a medical researcher to comment, with the appearance of scientific objectivity, on the benefits of a drug in this specialized journal which, in turn, disseminates the content across the world from its press office. Among the multinationals were: Roche, Glaxo Wellcome, Pfizer, Procter and Gamble, Bristol-Myers Squibb, Merk & Co. and Wyeth-Ayerst. The 'purchased' scientific studies reported favourably on the significant results of certain drugs for diseases such as viral hepatitis, meningitis and osteoporosis. However, in the social sciences it is common, and everyone assumes it, to receive funding from political parties or business or ideological lobbies to carry out 'biased studies', which appear scientific and neutral—although, 'coincidentally', they happen to endorse the funders' postulates. What is alarming is that the media often do not report this and give the study the same—or even greater—credibility as in the natural sciences.

Big Science Scandals: Do They Contribute to Decline or Reinforce Science?

An interesting case of science–media relations is that of the great scientific frauds, frauds that, I repeat, in the case of the natural sciences are always relatively easy to detect. There have been two spectacular and paradigmatic cases of scientific fraud. One is because of its duration—Piltdown Man—and the other because of its media, scientific and political repercussions: human cloning. Both have in common that they were a result of the pressure to which the scientists were subjected. And here's another difference from the social sciences, where the pressures are usually political or ideological; in the natural sciences, the pressures are often from the way that the scientific community is organized much do with the media, another proof of how science can be subject to and conditioned by media culture.

Piltdown Man

Piltdown Man has undoubtedly been one of the greatest scientific frauds in history, not only in its magnitude but in its duration of the fraud until it was exposed. The case of Piltdown Man teaches us how damaging media pressure, coupled with patriotism, can be to science. The fraud, however, was not perpetrated in a hard science such as physics or chemistry, but in one of the so-called 'soft sciences', palaeontology, a science in which it is difficult to reproduce the experiment because the raw material—the skull, in this case—usually belongs to the research centre that discovers it.

It all started, as I mentioned, with pressure from the media. At the beginning of the twentieth century, palaeontology was in a golden age. The discoveries followed one another and it was common for the names of Neanderthal and Cro-Magnon to appear in the newspapers, while any new bone was presented as a real social and media event in the best scientific societies.

The media-scientific-patriotic war was served, because the French had their Cro-Magnon and the Germans had their Neanderthal. But… the British? With science as well developed as in Britain then, how could there be no 'English prehistoric man', the newspapers wondered.

And the man from Piltdown wanted to be that 'English *Homo*'. Its discoverer was a certain Charles Dawson, an amateur palaeontologist who, he told the press, received the remains in 1908 from workers mining gravel at the Piltdown Common quarry in Sussex, England.

However, the news began to spread wider in 1912 when the remains reached Arthur Smith Woodward, a respected English geologist who worked at the Natural History Museum in London. The remains of the box represented the perfect missing link in evolution: among the bones were a typically human skull and a jaw with the classic features of apes. All you had to do was bring them together to have a new *Homo*. And that's what Dawson's amateur did: he rebuilt the skull to fit the jaw and stuck the two parts together. Then he named it *Eoanthropus dawsoni*, something like 'Dawson's initial man'.

Everything was therefore ready to be presented to the press as the 'first Englishman'. The event took place on 18 December 1912 at the prestigious Royal Society of Geology in London. Newspapers such as the *Illustrated London News* covered the story with great lines such as: 'The first predecessor of man is British' or 'The capital of European prehistory is no longer in Les Eyzies' (France). The photo of the skull went around the world.

The journalists didn't notice the fraud but, what's worse, neither did the palaeontologists. They accepted the specimen without criticism and came to the conclusion that 'Piltdown Man was the common ancestor of modern humans'. Some dissidents, such as Sir Arthur Keith, shyly said that they thought they saw the features of a simple Neanderthal, but British newspapers and scientists almost devoured him: reality could not be allowed to spoil good news, both journalistic and scientific.

But the natural sciences, unlike the social sciences, have weapons to detect fraud and the illusion of *Homo English* lasted only until 1949. Chemistry discovered the fraud through a very simple test that palaeontologists had already incorporated into their discipline.

Fluorine is a chemical element that is present in the environment. When bones are formed, they do not contain much fluoride, but they absorb it from the environment over time. In other words, the older a bone is, the more fluoride it has, so by measuring that amount we can know its age. It was found that the Piltdown remains had minimal amounts of fluoride, so it was clear that they had been buried recently. The news was sensational, especially in countries such as France and Germany, although, it has to be said, also in Great Britain and the United States, the latter a country where there were scientists who had doubted the English 'presapiens'. Yet many supporters of Piltdown Man—journalists and palaeontologists—refused to acknowledge the evidence provided by the fluoride. It was in 1953 that other analyses revealed that the remains had been dyed to give them an antique appearance and, worst of all, the teeth of the jaw, which turned out to be from an orangutan, had been filed down to give them a human look.

Hwang's Human Clone

Perhaps, as readers go through this book, the fraud that they remember is the false human cloning by the South Korean scientist, Hwang Woo-suk. This case attracted much media coverage (including covers and editorials), but scientifically it was less relevant because it took just a year to be exposed, which says a great deal about the good health of the natural sciences in this century.

The Hwang case has analogies with the Piltdown case, such as the high media coverage—both of both findings and of the fraud—and the scientific pressure to which the researchers were subjected. In both cases the impact was due to the fact that they dealt with controversial issues related to the human species, which is always journalistic news. With regard to pressure, there is a patriotic element that needs to be removed from science if we are not to make serious mistakes.

In the case of Piltdown Man we commented on the importance given to the fact that Great Britain did not have its own particular *Homo*. In the case of Hwang, the pressure came from the government of South Korea, a techno-logical country which, in his view, found it unbelievable that by 2004 it did not yet have a scientific Nobel Prize.

However, comparing the two frauds allows us to see how science, and the public communication of science, evolved in the twentieth century. The veteri-narian, who, according to press reports, slept only four hours and came from a humble family, astonished the world in June 2005 when he published in the prestigious journal *Science* data that supposedly confirmed the cloning of the first human embryos and what is even more important: the subsequent derivation of cell lines from these clones.

Shortly after the publication, the South Korean's media fame was unstoppable. But the media also began to report numerous irregularities: from payments and the pressure on his team of researchers to donate human eggs—something that is forbidden—to reasonable doubt about how the human cloning technique could be achieved so quickly by a person who, only a few years earlier, was not at the forefront of this type of research.

Until then, Hwang had been adored by the media. He was delighted, because he handled the coverage at will to achieve his goals. In this way, the results published in *Science* were presented at a major press conference held at the historic Royal Institution in London. I draw the reader's attention to how Hwang preferred meeting a group of journalists to a group of experts in the field. If he had given experts all the data from the experiment and allowed them to ask questions, they might have detected his fraud immediately. With the questions that journalists would ask, he knew that his manipulation was safe.

Hwang, like many not very good scientists, preferred to move among journalists, not colleagues. Hwang's fascination with the press is attested to by the *Financial Times'* account of the fraud:

> When two *Financial Times* journalists visited his laboratory on a public holiday last May, his reputation was approaching its zenith and his workload overwhelming. But he set aside the entire morning to entertain them and seemed disappointed when his invitation to lunch was refused. (*Financial Times*, 19 January 2006, 15)[49]

Let's remember that in November 2001 the first human embryo had been frontpage news in all the media when Robert Lanza, vice-president of the US biotech company ACT, announced that he had succeeded. However, this discovery did not exceed six cells, so no one considered it to be a true embryo, generating a boo from the scientific community. However, these six cells were indeed from a clone; but in 2005 no one remembered the news of 2001.

In December 2005, a few months after Hwang's press conference at the Royal Institution, an investigation at Seoul National University itself (where the scientist was working) acknowledged that it had all been a fraud. The university's release note clearly stated that 'Hwang did not produce a single one of the 11 cell lines from cloned human embryos that his team attributed to itself in *Science* journal'. It is important to clarify that it was the university

---

[49]'Korea's rapid development has hard scientific limits', https://www.ft.com/content/15cf77d0-885a-11da-a25e-0000779e2340.

itself—and some of Hwang's American collaborators (in particular Gerald Schatten of the University of Pittsburgh)—who announced the fraud. Natural science is always above the prestige of a particular researcher.

What Hwang basically did was to take a cell line from MizMedi Hospital in Seoul and a few adult tissues that had already been generated from it. It was a cell line that the hospital had generated long ago from an embryo of one of the many leftovers from in vitro fertilizations. But it wasn't a clone. Hwang sent the pictures to the journal, and the rest is history. How could they have known that the photos were faked and did not correspond to the experiment? Many scientists criticized *Science* and warned that journals are not foolproof. That's right: I myself do. We have already mentioned this and will develop it further. However, the method of attempting to replicate the same experiment in other laboratories is quite safe and, in this instance, it was the only thing that allowed the fraud to be exposed.

The attitude of the University of Seoul, resolving to conduct an investigation against one of its most prominent scientists, may be considered a healthy move that never would have happened in the social sciences. If such a case had arisen in education studies or sociology, the usual thing would have been to try to cover it up 'so as not to damage the prestige of the institution or the area'. Also, this would be done because, I repeat, in social sciences, studies to verify earlier experiments are seldom undertaken. If it is a collaborator who uncovers a fraud—for example, that the surveys were carried out by inexperienced students or that there is a mistake in a key methodological parameter—in the social sciences, in my experience the collaborator is labelled as disloyal. Instead of being seen as a good thing for science, he or she is treated as an enemy.

In the natural sciences, this excuse of calling the person who criticizes the wrong results an enemy is only available to a fraudulent scientist. Thus, in the *Financial Times*, Hwang said goodbye to the journalists with the words, 'I have so many enemies in the world'. The difference is that while in the natural sciences everyone laughs at phrases like these, in the social sciences it is not so. A corporatism covers up the miseries, because people know that, in most cases, the entire intellectual edifice of these social sciences is founded on sand, and will crumble at any serious criticism.

However, here the social sciences have a great ally: it is very difficult for a study with a sample of people—such as is used in sociology, education studies, mass communication science, and so on—to be reproduced in the same way by another research group to validate the data. Therefore, frauds, which can range from using incorrect methodologies to completely invented results, are almost undetectable. And this the journalist must know.

I am not saying that social science researchers invent data and doctoral theses. I say that, if they did, there are not many mechanisms to detect it. Nor can the social sciences make predictions, like the natural sciences, to check whether the theories are correct. Even economics, the most prestigious of them all, is defined as 'the science that explains the next day what happened the day before'. One of my chemistry teachers said that the natural sciences make predictions and social sciences prophecies. Maybe he wasn't too far off the mark.

This means that, from the point of view of journalism, which seeks to seek the truth, social sciences and their conclusions must always be considered at a level that is absolutely inferior to that of the natural sciences. This is an obvious statement for many academics, both in the natural and social sciences, but I believe that some journalists do not take it into account but put them on a par. Giving social science studies the same media validity as natural science studies is undermining the prestige of natural science in the eyes of the public, which may think that both have a similar endorsement to explain reality.

What Does the Sokal Case Prove?

We have seen how fraud is organized in the natural sciences and how it is discovered, and at the same time how poorly organized are the social sciences (especially those using surveys), in general, in preventing the detection of fraud. However, I believe that the Sokal case does not prove as much as some scientists have claimed. It does not do so precisely because scientific methodology was not used to prove whether what Sokal alluded to made sense, not from a mathematical or physical point of view but from a social, linguistic or cultural point of view, as the authors intended.

The Sokal case, although it was given much importance in the media—and, in fact, it was the first approach to discovering intellectual fraud in the humanities—is not scientific or philosophical proof of anything. At most, it is an 'exemplary action' which, normally, only serves to reaffirm its postulates in those who carry it out, but in no case does it give reason.

The way to show that the social sciences—especially the unmathematized ones—would be an 'intellectual imposture' would have been to dedicate themselves to them. Reproduce the experiments. Carefully analyse the methodologies and obtain new results. If the conclusions were different, then one could speak of an intellectual 'sham'. This is an interesting line of research because, in my opinion, the social sciences and the humanities can make substantial progress if natural scientists or researchers with strong training in mathematics and the natural sciences are involved.

After the repercussions of the Sokal scandal, some of the arts-based intellectuals met up who had founded their careers on quoting and praising those criticized by Sokal and Bricmont as intellectual impostors. It was in the summer of 1997, in Paris—of course—in a country that, curiously enough, has many followers in discrediting science. Perhaps this is because, despite France's remarkable contribution to modern science in its early days—and to rational thinking—French intellectuals, most of whom are now civil servants, consider that knowledge of the natural sciences has gone out of their hands and that, at the moment, it is more in the US/UK domain. Thus, for the French, one way to discredit the English-speaking cultural hegemony—and, at the same time, to deplore the loss of French cultural influence—is to discredit science. The intellectuals critical of Sokal met in what they called an informal seminar entitled 'Around the Sokal case'.

At that meeting, the organizers acknowledged that there was much anger at the physicists' criticism of the social scientists and philosophers: 'This meeting revealed many tensions within the group, as well as latent violence that resulted in aggressive attitudes and behaviour', Baudouin Jurdant, the editor, wrote in the foreword to the book the minutes of the meeting. This was entitled—it could not be otherwise—'Scientific Impostors: The misunderstandings of the Sokal case'.[50] Although he called for 'reconciliation between philosophers and physicists to end the war of science', he did nothing but activate it. Above all, this was because, at heart, the aforementioned people recognized that 'one does not have to know about science to speak, to inform journalistically or to philosophize about science' and this is not acceptable for a scientist, because then one would return to 'anything goes' and lead a thought that is not based on rationality and logic.

The discussion of the book edited by Jurdant does not clarify the veracity of what social scientists have said who have used physics, mathematics, chemistry or biology, without knowing about them, to demonstrate cultural or social trends. And by failing to do so it can be concluded that using scientific ideas out of context is not bad practice but the opposite: a symptom of intellectual flexibility. And this idea, which may be interesting, although I do not share its sentiments, needs further conceptual and methodological study in order not to appear irrational.

Intellectual flexibility can be very dangerous, and it is what led alchemy to fail to obtain results in proportion to the work and effort dedicated to its study. Are Lacan, Kristeva, Latour, Deleuze and Virilio right? Is it possible to advance in the social sciences thanks to research using natural science concepts yet

---

[50] Originally published as *Impostures scientifiques. Les malentendus de l'affaire Sokal*, ed. Baudouin Jurdant (1998). There is a Spanish edition by Cátedra (2003).

without its methodological rigour? What have the social sciences contributed to cultural debate? Have the world and society advanced by virtue of the funds invested in social sciences and the teaching salaries paid to Latour, Kristeva, Lacan, Deleuze and Virilio? I believe that intelligent answers to these questions, including some criticisms, would have improved the view of natural sciences among social scientists.

There is unanimity shown in the book *Scientific Impostures*, calling Sokal's action a 'trap', while there is a lack of self-criticism of some practices in social sciences and the humanities, an obvious fact, that clearly leaves room for improvement. This keeps the book *Intellectual Impostures*, written by Sokal and Bricmont, at the same level of confrontation as that book, edited by philosophers such as Jurdant and to which Jean-Marc Lévy-Leblond, among others, contributed.

These are two trenches that no one dares to climb out of. Perhaps it will take an entire generation to recover Aristotle, to enlighten us at this cultural crossroads of the West. I might propose that we do what he did: to study and investigate with the same passion and interest matters related to zoology or physics as to ethics, metaphysics, rhetoric, logic and poetry. That is to say, that the intellectuals of literature-based studies should read and research natural sciences, and vice versa.

Let us look at this issue from the perspective of media culture. Very few journalists have enough of a cultural and scientific background to enlighten public opinion on these controversies. For most, what comes from scientific sources—universities or journals—must necessarily be serious—and this can be a big mistake. What interests us, from the point of view of this book on the decline of science, is that journals, whether *Social Text*, *Nature* or the journal of the Royal Academy of Sciences, can be dangerous sources for journalists. They generate what Furio Colombo[51] has called the phenomenon of 'compliance news'. This is annoying to a journalist with a background in literature-based studies, because he or she feels humiliated and therefore tends to have an anti-scientific attitude.

This 'compliance news,' says Colombo, 'comes when news reaches the newsroom with such a high degree of security that journalists drops their guard on verification and is easily the target of hidden interests.' Colombo points out that this phenomenon, 'which poses a very significant risk to the future of journalism', is present in all areas of journalism, yet he believes that the clearest example is scientific news:

---

[51] Furio Colombo. (1997). *Últimas noticias sobre periodismo* ('Latest News About Journalism'). Barcelona: Anagrama.

Journalism controls literature and, in general, everything that belongs to the humanist world, philosophical statements, political positions, theories about history and the world. There are severe critics, competent reviewers (and also incompetent no less severe), who have in common the use of critical reflection as an instrument and as a working environment. All this is not true of science. It is not produced by the general feeling of submission of the humanist to the scientist, characteristic more and more evident in the contemporary world (...) And it is not produced (the critical reflection) by the clumsiness and impossibility of the reporter, hindered to intervene also by the voracity of the 'already prepared' of the information instruments'. (Colombo 1998, 105–106)

The Possible Solution of the Third Culture

In this discussion on the two cultures and their impact on the decline of science in the West, I think it is necessary to insist on what the solution is: to unite the two cultures again, establishing a third. As Colombo rightly argues, the phenomenon of 'news compliance' is generated in science journalism 'by the general feeling of submission of the humanist to the scientist, a characteristic that is increasingly evident in the contemporary world'. And such submission can only be avoided if the reporter has a perfect knowledge of scientific language. In this sense, the new intellectual leaders of social sciences and the humanities must be blamed for their contempt for the study of scientific language—mathematics, chemistry and physics. This is evident in the Western curricula for careers such as law, philosophy, humanities or philology, among others. In them, there is little study of physics, botany, mathematics or genetics: 'Obviously it will not be by renouncing the main source of information we have (science) that we will get to know each other. Science has to be milked, not feared', says the science philosopher Jesús Mosterín.

The anthropocentrism that has taken root in today's academics from a humanities background, especially in Latin countries, according to Mosterín also constitutes 'a lack of moral sensitivity towards non-human creatures'. In fact, this moral perversion stems from the fact that in the Jewish, Christian and Islamic traditions, it is only people, that is, humans as a species, who are the object of moral consideration. However, this is not the case in the Eastern cultural tradition, from Chinese Taoism to Buddhists and Jains, with their moral concern not to cause harm to creatures.

In this excessively anthropocentric view of Western culture, nature is not only ignored but, worse, is conceived as a mere object of human exploitation. Nature exists only to serve man. And there is no moral crime in natural destruction. In this view, the natural sciences should not be understood but simply used. That is why engineers are more socially recognized, especially in Latin countries, than scientists. And so, think politicians and literature-based academics, the knowledge of behaviour and language that man has designed in order to understand nature need not apply to the humanities.

However, it should be noted that, despite this cultural and religious influence on European civilization, which would appear to be opposed to science, the reality is stubborn. Science, in the broad sense as we know it today, was born in Europe in the seventeenth century. It was a Briton, William Gilbert and, above all, an Italian, Galileo Galilei, who made possible the birth of what we now call science. This is based on something simple but powerful: the need to carry out precise and repeated experiments to test hypotheses, not to rely on the old 'philosophical' approach of trying to understand how the world works by using pure logic.

Aristotle (and Greek philosophy in general) laid the first stone, yet it was Galileo who founded modern science. This is the science that shows you that, although logic and the senses invite you to believe that it is the Sun that moves around the Earth—indeed, literary (and journalistic) language still speaks of 'sunset' or 'sunrise'—it is the Earth that moves around the Sun. Literary language is usually quick to modify our notions, but not this one. Science is anti-intuitive, and that is its greatness; but for intellectuals for whom passion and emotion are more relevant than arguments and reason, the fact that science is anti-intuitive is precisely the cause of their disdain.

Other people may have had the technology and even made scientific explanations of natural events, but this cannot be considered science. In Europe—in Greek culture—a giant step was taken: the rational explanation of what happens to man. There are even those who can pinpoint the exact moment when humanity took this great step: in the transition from Herodotus (480–420 BC) to Thucydides (455–400 BC). Both Greek historians coexisted, although Thucydides was younger—and, in a way, the successor—to Herodotus.

It is not known if they ever met. What can be appreciated is that Thucydides highly valued precision and accuracy. Above all, he eliminated the gods from explanations of events. Both characters described similar historical events but, where Herodotus saw the manifestation of divine justice in action, Thucydides does not even speak of it, and gives only a political explanation.[52] Where Herodotus speaks of dreams and prophecies, or of the belief that everyone who goes too far ends up suffering a just revenge or divine punishment, Thucydides describes the abyss that separates expectations from results, intentions from reality.

---

[52]Thucydides 2.27.1, while Herodotus 6.91.1 speaks of a religious motive, cited in Fox, Robert Lane (2006), *The Classical World: An Epic History from Homer to Hadrian*. New York: Basic Books.

Instead of delving into myths and legends, as Herodotus did, the rational Thucydides explained the same events yet went deeper into the poor relations usually existing between justice and personal interests, between of the reality of power and the values of honesty. This opened up the possibility of predicting the events if characters with similar characteristics were crossed.

Thucydides took a giant step forward by understanding that what happens to man is only a product of man. No other ancient people had taken this step. 'Thucydides was aware of the difference between truth and rhetorical argument. He knew that what men said in public was not what they did in practice' (Fox 2006, 219–220). This supposes a great level of abstraction about what one appreciates with one's senses. But it is only a first step; we must expand it, from social relations among humans to knowledge and interaction with nature.

This formula of including observation in rational thinking and stimulating the capacity for abstraction was consolidated some four centuries ago by the birth of modern science and, remember, it was also consolidated in Europe. What was once divine punishment—a lightning bolt or a disease, like the plague—is now known not only for its true causes; science can also prevent man from suffering the consequences of nature, obviously more powerful than those of man.

But the science that was born in Europe is more than just an interpretation of nature: it is also an almost worldwide organization that publishes its discoveries and organizes conferences, and whose teachings are studied all over the world: heliocentric theory is not taught in one way in China and another in Spain or the United States. Science and rational thinking are the most universal and perennial part of human culture. And yet, despite being born in the West, this is where they are most criticized and suddenly seem unnecessary.

The great biologist and disseminator Richard Dawkins argues in his book, *Unweaving the Rainbow: Science, delusion and the appetite for wonder*,[53] that 'lawyers would make better lawyers, judges better judges, parliamentarians better parliamentarians and citizens better citizens, if they knew more science and what is more relevant, if they reasoned as scientists' (Dawkins 2000, 129).

The intellectuals of literature-based studies have renounced an understanding of reality in all its aspects, eliminating science as it provides the most answers. But what can these graduates contribute if they don't even know how to enunciate Newton's laws? How can they explain what climate change is? How can they explain the chemical basis of the carbon 14 analysis used to date

---

[53] Richard Dawkins. (1998). *Unweaving the Rainbow: Science, Delusion and the Appetite for Wonder.* London: Penguin Books.

works of art? And how can they explain what happens to man if he is a product of nature and continually interacts with it?

The Spanish philosopher José Ortega y Gasset,[54] a profound thinker on the role of the university, said that:

> The university consists, first and foremost, of the higher education that an average man should receive. We must make the average man, above all a cultured man – place him at the height of the times. Therefore, the primary and central function of the university is the teaching of the major cultural disciplines. (…) You have to make the average man a good professional. (Ortega and Gasset 1966, 335)

Ortega maintained that those 'great cultural disciplines' that must be taught in the university for a graduate to be considered educated are physics ('the physical image of the world'), biology ('fundamental themes of organic life'), history ('the historical process of the human species'), sociology ('the structure and functioning of the social structure') and philosophy ('the plane of the universe'). With this highly advanced study programme, Ortega's idea was to combine science and literature in the development of the average university student.

This is the way to access the third culture: teaching pure sciences in the faculties of literature and pure humanities in the faculties of science. The gulf between these two cultures—and, incidentally, the decline of the sciences—can be detected in the fact that if, for example, a minister of education were to propose today—in 2018—a common introductory course of these disciplines, the Spanish university would be relieved. The idea is inconceivable. I also believe that this idea's main enemy would be our current false academics from a humanities background, represented by literary intellectuals who, at present, would be quite unable to follow a normal university course in physics, chemistry or general biology.

I would recommend that these fake Western academics in the humanities visit Asia. Curiously enough, here a better synthesis of the two cultures is being made. From the East people obtain their literature, painting or history, and from the West science (they are already putting technology into practice). The Asian world view is likely to dominate the twenty-first century, and one of its values is to persevere in uniting the two cultures at all levels of education.

---

[54]José Ortega y Gasset. (1966). *Obras completas de José Ortega y Gasset*, vol. IV. ('Complete works by José Ortega y Gasset', vol. IV). Madrid: Ediciones Castilla (6th edn). Article published in *La revista de Occidente* in 1929.

From Irrationality to Journalism

Does a journalist's earlier scientific background influence how science news is approached? Are literary journalists more critical of science, and do they encourage anti-science?

In the nineteenth century, astrology was forgotten. It was the press that introduced horoscopes into the newspapers in the twentieth century. Nowadays, people read more about astrology than about astronomy. In this sense, the press has contributed to pseudoscience. Apart from the angles on this approach that have already been mentioned and clarified in the chapter on the decline of science, Martin W. Bauer provided some perspectives, not in terms of analysis of content but a huge amount of scientific news. According to Bauer, the last 150 years have seen various periods in which the popularization of science has been very much on the agenda: between 1870 and 1885; during the 1920s; from 1955 to 1965; and, recently, from the mid-1980s. This presence is seen not only in the number of news items but in the number of cover stories containing scientific news.[55]

This analysis is interesting and intensely laborious. However, from my point of view, the fact that science appears in the media more does not mean anything, in itself. This is because, if the news that is transmitted has a negative frame or focus on science, it is almost worse that many items are published, as the effect of anti-science is reinforced. In other words, the important thing is to determine whether the approach is negative and, above all, why there might be a negative approach.

This negative approach can perhaps be alleviated by knowing the causes. For example, it would be necessary to analyse journalists' attitudes towards science when reporting on climate change. But, in principle, the fact that climate change or cloning appears in the media a great deal is neither good nor bad, if you don't know the approach.

Bauer's maxims about the presence of science on the agenda do not explain the drastic drop in interest and, above all, in vocations. If we are living at a time when there is a maximum of scientific news and every worthwhile scientific organization or scientific journal has its own press office (a circumstance that has never happened in the history of science), why this decline in interest? Why this bad image?

From the empirical data that we have, we can state one thing: from the 1980s onwards, no one knows for sure whether more science news will appear as a result of the movement for the public understanding of science. There are now

---

[55]Martín W. Bauer. (1998). 'La longue durée of popular science, 1880–present', in D. Deveze-Berthet (ed.), *La promotion de la culture scientifique et technique: ses acteurs et leurs logiques* (pp. 75–92).

more journalists working with scientists in press offices, and it was not since the 1970s, especially the 1980s and 1990s, that the decline in vocations and interest in science began. In other words, the more journalists that there are working for science—in press offices—or reporting on science, the more people are scared off such vocations. It would be necessary to analyse whether the relationship between scientific information in the media and the recruitment of vocations is inversely proportional.

In 1988, John C. Burnham, professor of history at Ohio State University, published a magnificent book, *How Superstition Won and Science Lost: Popularizing Science and Health in the United States.*[56] Over time, this has proved to be the case. The book is dedicated to 'to the science teachers of the United States (for their role in fostering civilization)'. Burnham believes that the proportions of 'amount of scientific information in the media' and attracting interest are clearly inverse. In his opinion, the problem in the United States lies in a lack of scientific culture among American journalists. Especially among editors of small newspapers, this has led to the predominance of sensationalism, social vision and the 'agenda setting' phenomenon in their output of scientific information. In his opinion, all of this has contributed to superstition winning a victory over scientific information, and that science is increasingly disqualified. In other words, the question is not whether or not science news is published but the focus and the intellectual background of the writer.

Burnham believes that the lack of genuine professionals who are capable of making science attractive causes journalists to choose the path of superstition and makes newspaper leaders—and their lack of scientific culture—responsible for the fact that magic and hackneyed superstitions appear in the media with total impunity. Burnham also criticizes scientists who, he says, have left the dissemination of scientific theories in the hands of journalists and media offices, dedicating themselves exclusively to their laboratories and forgetting that they, too, have a moral obligation to spread the word.

In Burnham's opinion, this was not the case in the nineteenth century, when science was more prestigious than in the twentieth century. This he attributes to the fact that in the nineteenth century the newspapers employed prestigious contributors with extensive training in science to write their scientific information, and these were replaced in the twentieth century by professional journalists. It is precisely because of this fact that what Burnham describes below is happening. The blame, I repeat, lies with the scientists here, not the journalists:

---

[56]John C. Burnham. (1988). *How Superstition Won and Science Lost: Popularizing Science and Health in the United States.* Rutgers University Press.

People often assume that just as in laboratory and field of research, so in the field of popular enlightenment, the fight of science against superstition was won by forces of rationality and naturalism – at least won in the highly technical society of the post-industrial United States. But the record shows that changes in the way in which science and health come to be popularized in fact ultimately reduced and frustrated the cultural impact of both science and scientists. (Burnham 1987, 3–4)

This statement is from 1988, when social networks did not yet exist. Today, the picture is much worse. Burnham wonders where his country will be led by the deterioration in the popularization of science, produced by the media, which in its attempt to attract the public manages to fragment and distort scientific information. He reflects on the consequences for the United States and the rest of the Western world of the increase in the forces of irrationality in today's society. He blames not only journalists for this situation, but the deterioration that he believes the US education system has suffered since the nineteenth century, especially since the 1960s and especially the 1980s:[57]

The influence of nature study in the schools spilled over into other aspects of American life. In extracurricular endeavors similar to those of the Agassiz Association (which continued for a few years after 1900) and especially in the Boy Scouts, Girl Scouts, Campfire Girls, and similar youth organizations that flourished for several decades, the programs and leaders carried the message and the enthusiasm of the nature movement. (…) In 1900 a surprising number of students [at American High Schools] were taking physiology and physics. (…) At the same time, physics, chemistry, zoology and botany, not to mention mathematics, were increasingly coming into the domain of the specialist teachers and educators. (…) Although in fact knowledge continued to dominate in the classrooms, educators were talking, as earlier in the century, about adaptation to the environment, adaptation that they interpreted as applying to everyday life. The science of high schools' wrote an educator, 'needs to be taught as preparation for life, as a consumer science rather than a producer science'. (…) To begin with, the prestige of the scientists after 1945 was very great. they were moreover, by 1950s able to mobilize first foundation and then federal founding on a massive scale ($117 million was spent by the National Science Foundation alone). Finally, the educator themselves had long talked about the importance of teaching the scientific method, and one survey, for example, showed that this concern was showing up in chemistry and biology curricular discussions with greater frequency after 1945. (…) By the 1980s, scientists and science teachers had good reason to lament their unpopularity and to try once again to tie funding for

---

[57]John C. Burnham. (1988). *How Superstition Won and Science Lost: Popularizing Science and Health in the United States.* Rutgers University Press, pp. 181–188.

science education into Cold War fears. The basic problem in science education therefore remained the same throughout the century and over the years underlay the shifts in content and rationale. The problem was that grounding in science required tedious memorizing and thinking that no pedagogical innovation (…) ever succeeded in easing. (Burnham 1988, 181–188)

In most of the Western world, the hours spent on physics and chemistry or biology and geology have only decreased: in times of cuts, laboratories are expensive, they can be dangerous for our cosseted students, and there is a shortage of teachers in STEM and too many other areas. If we add to this the fact that statistical data are now a weapon used in political campaigns, it cannot go unnoticed that the level of difficulty of STEM subjects may contribute to increasing failure among school students. And no government wants that.

This whole situation, of underlining in secondary education the negative and dangerous aspects of science—in addition to the approach by philosophy, ethics, cinema or literature classes that 'it is just another interpretation', together with the negative image that cinema and television offer of scientific research—explains the great decline in vocations. It is also in science journalism itself, because society itself does not want to read or hear news that is favourable to science.

Jon Franklin, the first director of the first department of science journalism in the United States and Pulitzer Prize winner, explains in a famous article entitled 'the end of science Writing'[58] how in the United States as well as in Great Britain, from the 1960s onwards the word 'science' began to have negative connotations and that, although science is occupying an increasing percentage of column inches in newspapers, the number of specialized science journalists is decreasing as the human tenor of the news takes precedence over the science.

In Franklin's opinion, this caused the vast majority of science news in the United States published since the late 1990s to be inaccurate, in tone, approach or context if not factually, something that did not happen so often in the 1950s or 1960s. He points out, however, that this inaccuracy is barely detected by American newspaper editors, as they have little scientific knowledge; a survey cited by Franklin in his article states that two out of three believed that men and dinosaurs lived at the same time.

Another survey, also cited in Franklin (2007) and whose field of study was journalism at Columbia University (the most prestigious in the world in the area of media communication and the one that awards the Pulitzer

---

[58] Jon Franklin. (1997) 'The End of Science Writing', The Alfred and Julia Hill Lecture. University of Tennessee. Published in Spanish, ed. Bauer and Bucchi, in *Quark, Ciencia, Medicina, Cultura y Comunicación*, 11, May–June 1998 (pp. 53–63) (ed) (2007), op. cit. (pp. 142–156).

Prize), concluded that 57% of journalism students believed in extrasensory communications, 57% in spiritual cleansing, 47% in aura reading and 25% in the lost continent of Atlantis. If this happens at Columbia, the best university of journalism in the United States, with the highest number of Nobel laureates and scientific research in the world, what can we expect from the rest of the Western world? Franklin commented on the situation in the late 1990s (although he updated it in 2007). In his words, it can be seen that in the West rationality and science were losing the battle.

This article by Jon Franklin was the result of a lecture given at the University of Tennessee in 1997.[59] And the truth is that, above all, because Franklin was speaking from his own forty-year experience as a science journalist in the United States—from the 1950s to the 1990s—he has had a great impact on science journalism scholars around the world. From the experience of working on the frontline, Franklin speaks about how the journalistic profession has evolved in the United States and, by extension, in the world. It is therefore worth reproducing some of the paragraphs from his talk about the end of science journalism for the reader to evaluate how journalism is effectively witnessing a decline in science. It's a long quote from a superb article that Franklin wrote that explains the situation at first hand:

> By 1960 it was palpable even at an academic reception. The rift was definitely there, and it was definitely increasing, and while we may argue about the social seismology involved there is one thing that any science writer can tell you for certain: the laboratory was on one side of the fault line, and the newsroom the other. (…) Scientists saw the world as theory and fact, observation and prediction, statistical significance and probability. The rest of the world titrated experience in terms of motive and morality, right and justice, miracles and fate.
>
> Liberals had backed science from the very beginning of the Enlightenment, and conservatives had come aboard because of the Cold War. Scientists, innocents that they were, confused being in political favor with being apolitical. It is useful to think of science as the faith of the Enlightenment. Scientists hate this. (…)
>
> By the 1970s, when I went to work in Baltimore, Snow's cultural gap had become a chasm. Earlier science writers had found ignorance to be a problem; now there was hostility. You had to be an ostrich not to notice it. Many journalists turned against science and were articulate about it. Animal rights activists called you at 3 am and told you what dress your daughter had worn to class that day. I am aware that most Americans still tell pollsters that they believe in science, but talk to those people and the so-called 'science' that they believe turns out to include astrology, yoga and extrasensory perception. They don't know what science is. In one study of American science literacy, half did not understand that

---

[59]Text extracted from Jon Franklin. (2007). 'The end of science journalism'. In *Journalism, Science and Society*, ed. Bauer and Bucchi, pp. 143–156. New York: Routledge.

the Earth travels around the Sun. Only 6 per cent of adults could be considered scientifically literate. More than half the students at Hollins College believed in ghosts and mental telepathy.

In the late 1970s, I was forced to rethink my journalistic strategy. I had been reporting and explaining discoveries, yet my stories were not being widely read. I generally used the word 'science' early in the story, thinking that this would attract readers. The word generally ended up in the headline. But I now realized that the effect was to tell general readers what to avoid. In theory they might trust science, but in practice it had bad personal associations – it confused them, made them feel negative about themselves. Science pages ghettoized science news, and gave people a whole section that they could throw away unread.

There was also something more sinister afoot. As attitudes changed, editors started wanting a certain negative spin on science stories. If you didn't comply, you were not detailed to cover outside stories or your existence otherwise made uncomfortable. Some science writers, especially those who identified with the ecology movement, saw hostility to science as a path to success. Many reporters, outspokenly neutral on other topics, found it easy to align themselves with the anti-science faction. This was often couched in terms that favoured plurality and an openness towards 'other ways of knowing'.

It was said, then, that there were more scientists working than had existed in all the years since Newton. Many were training up graduate students and postdocs, so that there were 10 scientists to take the place of each one who retired. This had been going on for decades, and there was an expectation of continued expansion.

They were wonderful years for me, too. Once I started down the road of leaving the word 'science' out of my stories, I wrote about science as though it were a normal human activity. That sold surprisingly well. Pretty soon, I was concentrating on essays and narrated stories, and getting a good slice of the readership. I won some prizes, which makes newsroom life easier, and I started thinking about books. I really loved the life. It gave me access to all these great minds on the cutting edge of knowledge. Once, I asked a Nobel Prize winner for some advice. He was having a meeting at the time with several senior scientists. He shooed them out and spent the next three hours explaining restriction enzymes to me. Isn't that an amazing story? Yet it's true. It happened all the time.

Journalism, meanwhile, was changing. It became difficult, and then impossible, to secure the time and space that good science writing required. I had enough clout to continue my own narrative work, at least for the moment, but the pressure was for 'harder' coverage of investigative stories about science. Science writers who had a pugnacious attitude towards science had an edge in terms of assignments and promotion.

I had enough clout to continue my own narrative work, at least for the moment, but the pressure was for 'harder' coverage investigative stories about science. Science writers who were pugnacious towards science had an edge in assignments and promotions. The 'gotcha' story, so conspicuously absent from science coverage, now arrived: reports surfaced about scientific malfeasance, misappropriation and dishonesty. The clinch was the story about the misuse of scientific overheads at Stanford. Later, the *Chicago Tribune* did a major story on the contradictory claims over the discovery of the AIDS virus: very, very dirty laundry. Science was a sitting duck. Scientists were accustomed to solicitous, if perhaps inaccurate, treatment by the press. They had dealt with science writers. Now, there were science reporters on their trail, and that was another thing entirely: It had never occurred to many scientists that their grant requests, their reports, their memos—this stuff is all public record. Science is a muckrakers' paradise, like shooting fish in a barrel, and I predict that you are going to see a whole lot more of it in the future. (Franklin 2007, 145–151)

7

The Two Cultures and Their Influence on Thought

Another most interesting aspect of the decline of science in the West is not so much the criticism from certain disciplines, but something far worse: ignorance. In Spain, for example, people can consider themselves educated without knowing anything of chemistry, physics or genetics, but they would not if they were unaware of authors of fiction or certain historical facts, most of which would be taken out of context.

The science baccalaureate in Spain has always included literature-based subjects with a level of knowledge similar to that of social studies (and humanities). This is why it is common for its students into pursue careers in science or literature—or both—without prejudice. However, the arts baccalaureate has outlawed physics or biology, as if these areas were not a product of human intelligence and, therefore, the subject of study in the humanities. In 2009, the subject 'Science for the Contemporary World' was introduced into Spanish pre-university studies, a subject also studied in literature courses yet not examined by the university entrance exams. In addition, the subject proscribes scientific language—chemical formulae or mathematical equations—and is content to spread science in a literary way, 'so that those of letters can approve it,' said an educational official.

In the aforementioned book *The End of Science: Facing the limits of knowledge in the twilight of the scientific age* (1996), by science journalist John Horgan, it is pointed out that we may be facing the end of science because, precisely because it is true, there are fewer and fewer gaps to investigate. In his opinion, the scientists' complaint about the supposed power of the four philosophers' (Kuhn's, Lakatos', Popper's and Feyerabend's) anti-scientific attitudes is not

© Springer Nature Switzerland AG 2019
C. Elías, *Science on the Ropes*, https://doi.org/10.1007/978-3-030-12978-1_7

true.[1] The scepticism of a few philosophy teachers has never really posed a serious threat to the solid and well-subsidized bureaucracy of science.

This may not be true for science, but it is for its dissemination, because science journalists (in fact, Horgan is a journalist with a degree from Columbia), especially from Latin countries, come from the humanities, an area in which only these authors are taught. And, what is worse, by teaching them in faculties of communication and not philosophy, they are studied in an elementary way, which generates the idea that, according to the philosophy of science, any theory—including that the Queen of England is an alien—is scientific or, at least, deserves the same intellectual or media respect.

The problem, therefore, is not in the hands of the philosophers of science, but in those responsible for educational policy, who believe that a lawyer can competently practise—let alone have political responsibilities—without university knowledge of genetics or chemistry. That is to say, the problem in the West is a division of the two cultures: into literature-based subjects and sciences. This is not the case in Asia: in the Chinese university entrance exam——the famous *gaokao*—mathematics is common to all students.[2] In recent years, Gaokao has been reformed to eliminate the division between science and humanities: 'Now, Gaokao includes the three main subjects (Chinese, foreign language, and math) and three elective subjects that can be chosen by the student (politics, history, physics, chemistry, and biology)[3]'.

The idea of the two cultures first appeared in 1956 as the title of an article that Sir Charles Percy Snow published in the *New Statesman*. This was the germ of an idea for a conference at the University of Cambridge in 1959 entitled 'The Two Cultures and the Scientific Revolution', and in the same year came the controversial book, *The Two Cultures*, published by Cambridge University Press. By 1961 it was already in its seventh edition.[4] The argument of the book is well known: that there are two cultures, one of literature-based subjects and the other of science, and both are not only separate, but worse: they are not interested in meeting.

---

[1] J. Horgan. (1996). *The End of Science: Facing the limits of knowledge in the twilight of the scientific age.* New York: Helix Books.

[2] Mini Gu. (2016). 'The Gaokao: History, Reform, and Rising International Significance of China's National College Entrance Examination'. *World Education News+Reviews.* https://wenr.wes.org/2016/05/the-gaokao-history-reform-and-international-significance-of-chinas-national-college-entrance-examination.

[3] Katerina Roskina. (2018). 'Educational System of China'. *NAFSA: Association of International Educators IEM spotlight newsletter, VOL. 16, Fall ISSUE.* https://www.nafsa.org/Professional_Resources/Browse_by_Interest/International_Students_and_Scholars/Network_Resources/International_Enrollment_Management/Educational_System_of_China/.

[4] C. P. Snow. (1959). *The Two Cultures and the Scientific Revolution.* Cambridge University Press.

Snow approached the problem from his own personal experience, as he was both a physicist and a novelist and therefore frequented both worlds. As a physicist, he not only worked at a university but had been recruited by the British government during World War II to participate in military programmes. Moreover, as he acknowledged, he had lived in the period of maximum growth of physics as a discipline that actually explains the material world. Moreover, he loved literature and, in fact, abandoned his scientific career to devote himself entirely to writing. According to experts, though, his novels are unlikely to go down in the history of literature, while some have received good reviews.

Snow describes in his book anecdotes that illustrate the science/arts mismatch. In this sense, he describes how at university or embassy parties people used to get together with people from both the arts and from the sciences. He also describes his experience at Cambridge, 'a university where scientists and non-scientists meet every night at dinner'. This was the case in the 1950s. However, it is now rare for university professors from different disciplines to dine together every night.

From reading the book, it seems that he went from one group to another. He criticized both. With the literary intellectuals he knew, he denounced their absolute ignorance of science. With the scientists, he criticized their general literary ignorance: 'most of the rest, when one tried to probe for what books they had read, would modestly confess, 'Well, I've tried a bit of Dickens', rather as though Dickens were an extraordinarily esoteric, tangled and dubiously rewarding writer, something like Rainer Maria Rilke (Snow 1959, 13)'. However, Snow clarified: 'But of course, in reading him [Dickens], in reading almost any writer whom we should value, they are just touching their caps to the traditional culture' (Snow 1959, 13).

However, Snow considers scientific culture to be far superior to that of literature:

They [the scientists] have their own culture, intensive, rigorous, and constantly in action. This culture contains a great deal of argument, usually much more rigorous, and almost always at a higher conceptual level, than literary persons' arguments even though the scientists do cheerfully use words in senses which literary persons don't recognise, the senses are exact ones, and when they talk about 'subjective', 'objective', 'philosophy' or 'progressive', they know what they mean, even though it isn't what one is accustomed to expect. Remember, these are very intelligent men. (Snow 1959, 13–14)

Snow defined literary intellectuals as 'natural Luddites' (p. 23), a term that refers to the Luddites, a movement born in Great Britain in 1811 characterized by the fact that its followers had the mission of destroying the machines of the incipient industrial revolution. They did not like the future that they saw with technology, and revered the agricultural and pastoral past. Loyalists attribute all the evils of civilization to machines, technology and science. The movement became so important that the British parliament passed a law that the death penalty would be imposed on anyone who destroyed industrial machinery. The narrow-mindedness of playfulness saw no guilt in harming the owner of the machine or in the owner of the machine, nor in the politician or monarch who protected the machine, but just the machine itself. Nevertheless, the movement spread in Great Britain, and writers of the stature of Lord Byron supported it.

Curiously, from the 1970s onwards, a movement called neo-Luddism began to spread through areas of knowledge such as science, technology and society, dominated by philosophers of literature-based subjects. The neo-Luddists blame evils such as global warming or nuclear proliferation on techno-science and, of course, blame scientists and engineers. A simple glance shows that those guilty of such decisions are graduates in art history, philosophy, law, economics or political science, who are the professionals in the parliaments and boards of directors of countries and companies that misuse technology precisely because they cannot understand how it works. Or does anyone think that the members of the US Congress that approved the use of the atomic bomb in World War II were primarily physicists, chemists and biologists? Saddam Hussein used chemical weapons to annihilate the Kurds, but whose fault was it: the chemists who investigated the substances or Saddam Hussein, who was a graduate in ancient history? The decisions to use science against nature come, in most cases, from politicians or businessmen with a degree in literature or social studies. It should not be forgotten that the decisions of boards of directors and parliaments around the world are taken by people from these disciplines. There are hardly any scientists or engineers among politicians or economists.

Mario Bunge defines it much better:

> The applied scientist or technologist—especially the latter—is responsible for what may result from their efforts because they can sell or refrain from selling their professional knowledge. It is clear that whoever buys this knowledge for evil purposes is the main culprit (not only responsible). Indeed, it is he who orders or allows his expert to go ahead with a project that can only serve reprehensible purposes. In short, the primary responsibility and blame for the social ills of our time lies with political and economic decision-makers. Let us hold them accountable and blame them, mainly for the arms race and the unemployment it causes, for acid rain and the destruction of forests, and for poor-quality commercial and cultural products. Applied scientists and technologists involved in these processes are but accessories to crime, even though they often display reprehensible enthusiasm. Let's face it: not because they are instruments, they are

not without responsibility. They have it, but less than their employers. (Bunge 2013, 227)[5]

In the face of this interplay, Snow commented that scientists 'had the future in their bones' (p. 11). That worried him, because it increases the tension between the two trenches:

I believe the pole of total incomprehension of science radiates its influence on all the rest. That total incomprehension gives, much more pervasively than we realise, living in it, an unscientific flavour to the whole 'traditional' culture, and that unscientific flavour is often, much more than we admit, on the point of turning anti-scientific. The feelings of one pole become the anti-feelings of the other. If the scientists have the future in their bones, then the traditional culture responds by wishing the future did not exist. (Snow 1959, 11–12)

This may be another explanation for why both filmmakers and writers assign such a negative role to science in their works. An example of such a desire would be George Orwell's apocalyptic novel, *Nineteen Hundred and Eighty-Four*, which has 'the most vehement desire possible that the future does not exist', Snow says. Finally, he warns of something that defines the situation of science in the West: 'It is the traditional culture, to an extent remarkably little diminished by the emergence of the scientific one, which manages the western world' (Snow 1959, 12).

As might be expected, the criticism of Snow's book was much fiercer on the side of literature than on the side of science. A literary critic, F. R. Leavis, published his attack in the *Spectator*:

Snow is, of course, a– no, I can't say that; he isn't: Snow thinks of himself as a novelist. His incapacity as a novelist is... total: as a novelist he doesn't exist; he doesn't begin to exist. He can't be said to know what a novel is. Snow is utterly without a glimmer of what creative literature is, or why it matters.[6]

'Not only is he not a genius', Leavis concluded, 'he is intellectually as undistinguished as it is possible to be'. Snow relates in his book that when he is at an event for those with a literature-based background, no one (apart from him) can explain what the second principle of thermodynamics establishes. Then he complains that ignoring that scientific law is tantamount to admitting never to

---

[5]M. Bunge. (2013). Pseudociencia e ideología [Pseudoscience and Ideology]. Pamplona: Laetoli.
[6]From Roger de Kimball. (1994). '*The Two Cultures* today: On the C. P. Snow–F. R. Leavis controversy.' *New Criterion* (12 February 1994), https://www.newcriterion.com/issues/1994/2/aoethe-two-culturesa-today.

have read a work by Shakespeare. But he is even harder on the literature-based academics:

I now believe that if I had asked an even simpler question – such as, What do you mean by mass, or acceleration, which is the scientific equivalent of saying, Can you read? not more than one in ten of the highly educated would have felt that I was speaking the same language. So the great edifice of modem physics goes up, and the majority of the cleverest people in the western world have about as much insight into it as their Neolithic ancestors would have had. (Snow 1959, 16)

Leavis replied furiously that:

There is no scientific equivalent of that question; equations between orders so disparate are meaningless. The second law of thermodynamics is a piece of specialized knowledge, useful or irrelevant depending on the job to be done; the works of Shakespeare provide a window into the soul of humanity: to read them is tantamount to acquiring self-knowledge. Snow seems oblivious to this distinction.

This answer from Leavis is typical of a literature-based writer. However, no one is arguing that Shakespeare is indispensable or that his works are not a view of the human soul. The idea is that, in addition to Shakespeare, it is also necessary to know the second law of thermodynamics, because it describes the world we live in.

Peter Atkins, Professor of Chemistry at Oxford and a great disseminator, suggests in his book *Galileo's Finger: The ten great ideas of science* that, precisely, the second law of thermodynamics is one of 10 revolutionary ideas that man has had and that have changed the world. In technical language (albeit without mathematics), this second law can be stated as that, in any isolated system (without exchange of matter or energy), entropy (i.e. molecular disorder) either increases or is maintained but never decreases. In an interview that appears on the magnificent portal, The Science Network (defined with the objective of building an online science and 'public square dedicated to the discussion of issues at the intersection of science and social policy'), Atkins explains why the second law is so important to human culture:

Well, I think when I begin my lectures to my undergraduates at Oxford on the on the second law, I begin by saying that in my view, no other law has contributed more to the liberation of the human spirit. And of course, they giggle because they think that's hyperbole and so on. But I actually think it is true, and I think the second law of thermodynamics, which is this example of difficulty and complexity and erudition in order to understand it, is none of those things. What it is a way of looking at the world that enables you to understand why anything happens at all, and I think at the very minimum people ought to know in the world why anything happens because they're surrounded by things that are happening. And it's extraordinarily simple, too. All it says, basically, in a simple way, is that things get worse. And everyone knows that anyway. But to be slightly more precise, it implies that matter and energy spreads in disorder, and so it gets worse in the sense of the quality of energy is declining because it is less useful when it is dispersed than when it is collected in a small region. But the extraordinary thing is, and really the beauty of the second law is that you can tap into that dispersal of energy with gear wheels, pistons, or biochemical processes, metabolic processes and so on and use the dispersal of energy, this sort of corruption of energy, to drive other things into existence. So we eat and the metabolic processes inside us really disperse the products of digesting the food and the dispersal of the energy. And they link that to other processes that go on inside us, like the joining together of amino acid molecules into a protein. So as we eat, so we grow. And so you can understand, you know, our growth. But we can also understand why an internal combustion engine works as well; why a steam engine works. Why anything works. And we can also really begin to understand what drives the mental processes in our brains as well because that could be seen to be a way in which the dispersal of energy through metabolism and so on, linked by metabolic processes, organizes the more or less random electrical and chemical currents in our brain and turns these random currents into coherent currents. And that is manifest as works of art, acts of valour, acts of stupidity, acts of understanding – whatever. And so you really do find the springs of creativity by looking at this extraordinarily simple law. (Atkins, on *thesciencenetwork*[7])

That is to say, while Shakespeare speaks to us of human feelings such as love, revenge, loyalty or family relationships (feelings, on the other hand, that we do not know are exclusive to humans or if animals can also have them), the second law of thermodynamics has something truly human: the ability to ask and answer questions about why things happen and, with those answers, to transform the world. That is the true capacity that has made *Homo sapiens* the triumphant species of the planet. Obtaining the second law of thermodynamics totally transformed our world, since it is the origin of the steam engine and, therefore, of the industrial revolution.

Leavis' critique, which he himself does not understand, concluding that it is specialized knowledge to be played down, is overly simplistic. The current problem in Western culture is that a professor of thermodynamics can usually understand Shakespeare, but a professor of literature may not be able to

---

[7] http://thesciencenetwork.org/media/videos/3/Transcript.pdf.

understand the scope of the second law of thermodynamics. But both teachers are considered to be at the same level by the academy. And that is a symptom of the decline of science: that a person can be considered to be educated or to assume responsibilities of interpretation of the world without knowing science—without studying it at university level.

Science Versus Literature: The History of a Struggle Throughout Western Culture

This separation between science and literature that Snow denounces is not new, and deserves a small epigraph because, from my point of view, it explains Western culture and the current decline of scientific vocations, as well as the struggle between reason and truth and magic, narrative, alternative facts and post-truth.

As I mentioned in Chap. 1, in Greek times there was a division between what the sophists (who would be the present relativists, defenders of post-truth and alternative facts) considered education to be and what philosophers, much more advanced and serious than the giants, Socrates, Plato or Aristotle, thought. Bear in mind that, for example, one of the greatest exponents of the sophists, Protagoras (481–411 BC), is credited with the phrase, 'man is the measure of all things'. This is as if man were not part of nature, or as if nature were also subject to man. Plato criticized the attitude of Protagoras (and all sophists), in that this preferred the weapons of rhetoric or persuasion to the pursuit of truth.

Above all, Plato criticized Protagoras for seeking worldly success in the face of true knowledge: Protagoras travelled throughout Greece, charging high fees for teaching the perversion of words as a weapon of mass persuasion. In other words, Protagoras was what we would today call an expert in mass communication or corporate communication, or a speechwriter. Like all sophists, he was a relativist: that is, a follower of what we would today call 'alternative facts', post-truth or post-factual or, why not just say it, postmodern philosophy. The truth that counts is the truth that convinces the most people. It's that simple, but it's also that dangerous. The difference from the present time is that Protagoras and, in general, the sophists were faced with philosophers who were the greats milestones of Greek thought. First there was Socrates (470–399 BC) and his disciple Plato (427–347 BC), who criticized this type of professional who, in Plato's words, preferred applause for exciting and often dangerous adventures to the power of the search for true knowledge.

Plato had printed on the front wall of his academy the famous slogan, 'Let no one enter here without knowing geometry', which indicates the high value that Plato placed on mathematics as a discipline for training logic and for attaining knowledge. His most brilliant student, Aristotle (384–322 BC), devoted himself to physics and biology as well as to logic, rhetoric, poetry, ethics and politics. His idea was that, in order to access metaphysics (or rhetoric or poetry), you first have to know physics: how matter is constituted, how bodies move, what life is, why animals reproduce and how they evolve... Aristotle wrote more about physics and biology than about rhetoric and politics. Being a great thinker, he made a great mistake: he believed that one could speak of physics as metaphysics; that is, with simple logical induction, and he did not apply the scientific method of measuring data, conducting experiments and establishing general laws.

Aristotle thought of physics as he would metaphysics and rhetoric. His mistake was that in rhetoric, politics, art, poetry or ethics, anyone can give their opinion and all opinions can be valid, but in science only those that are supported by mathematics (the most sublime and exact form of thought) and experimentation are acceptable. This was the great revolution of Galileo and his *The Assayer* (1623), possibly one of the books that has changed humanity the most.

By the time of *The Assayer*, Galileo already had a telescope, thus instrumental measurements, and was able to determine that comets do not move under the sphere of the Moon, as Aristotle had claimed, but all over the sky. The question was not trivial, for the Church had accepted Aristotle and his idea that everything above the Moon, that is, the sky, was something unchanging where God dwelt. However, Galileo's real revolution was that in the seventeenth century he rescued the tradition of the Pythagoreans, Euclid and the great Archimedes, placing it above that of Aristotle. He claimed, like the Greek mathematicians, that to know the world you have to do it in mathematical language. This was something that infuriated the College of Cardinals in Rome. Like many politicians and intellectuals today, its members believed that they could participate in the discussion about what the world is like without knowing about physics or mathematics. *The Assayer* is summed up in the famous quote from Galileo:

I seem to detect in Sarsi a firm belief that, in philosophizing, it is necessary to depend on the opinions of some famous author, as if our minds should remain completely sterile and barren, when not wedded to the reasonings of someone else. Perhaps he thinks that philosophy is a book of fiction written by some man, like the Iliad, or Orlando Furioso – books in which the least important thing is whether what is written there is true. Sarsi, this is not how the matter stands. Philosophy is written in this vast book, which continuously lies upon before our eyes (I mean the universe). But it cannot be understood unless you have first learned to understand the language and recognize the characters in which it is written. It is written in the language of mathematics, and the characters are triangles, circles, and other geometrical figures. Without such means, it is impossible for us humans to understand a word of it, and to be without them is to wander around in vain through a dark labyrinth. (Galileo, *The Assayer*, 1623)

This paragraph changed our world, but it filled the intellectuals (who were in the majority) of the time with anger. They conspired against Galileo by promoting the famous trial whose sentence took away his freedom: he was imprisoned in perpetuity and forbidden to publish. In fact, it was the struggle that took place in mediaeval education between the *Trivium* and the *Quadrivium* that, in my opinion, is but the antecedent of the current war between Snow's two cultures.

Trivium, a Latin word, means 'three ways or paths' of knowledge. So the students studied disciplines relating to eloquence, according to the Latin maxim *Gram. loquitur, Dia. vera docet, Rhet. verba colorat* (that is, they studied grammar, which helps one to speak; dialectics, which helps one to find truth; and rhetoric, which colours words). Indeed, this could be the basis for current communications degrees.

But there was an elite, and this added mathematics (arithmetic and geometry), cosmology and music to the *trivium* (grammar, dialectics and rhetoric): it now constituted the *quadrivium,* or 'four paths'. According to the Latin maxim *Ar. numerat, Geo. ponderat, As. colit astra, Mus. canit,* the students studied arithmetic 'that numbers', geometry 'that weighs', astronomy 'that cultivates the stars' and music that 'sings'. Tensions over whether a person could be cultured with the *trivium* alone were unleashed on Galileo because, in a way, he was a forerunner of Snow, revealing to the cardinals that they could say little about the reality of the world if they did not know mathematics or physics. This is especially evident in his book *Dialogue Concerning the Two Chief World Systems* (1632).

At the time of the *trivium* and *quadrivium*, it was thought that the former could be studied and then the latter, or just the first one. But not just the second one. That is to say, it was possible to study literature-based subjects and sciences or only the former; but not only sciences. And that made the science people superior (as now), because they knew the two pillars of the culture of the time, while the *trivium* people knew only one. Of the seven liberal arts that made up the *trivium* and *quadrivium*—grammar, dialectics, rhetoric, arithmetic, geometry, astronomy and music—three were of literature, three of science and one of art. From geometry and astronomy, modern science developed.

But Galileo didn't just appear out of nowhere. While he synthesized an era, the conflict had been brewing since the very beginning of the Renaissance. The dilemma was what you had to know to be like the classicists. There was intense debate about whether the classicists considered it necessary to study the seven liberal arts in order to be educated and to understand the world, or whether knowing just the three literature-based subjects was valid.

There is no clear consensus on when and why the Renaissance and humanism began. One of the aspects on which there is consensus is that the turning point was the Fall of Constantinople in 1453. The last stronghold of the Eastern Roman Empire fell to the Arabs of the Ottoman Empire, and the city was renamed Istanbul. What in principle would seem to be a tragedy for Western culture was, however, the beginning of its rise. It led to the flight of intellectuals and, above all, the arrival of books from that city, on the border between Asia and Europe with Western Europe. This influence affected especially the easternmost parts of Europe, such as Venice and Florence in Italy.

It was not only a question of books, but of languages: among the emigrants after the capture of Constantinople were scholars who understood classical Greek, and this meant that Europe could return to the original Greek sources. During the Middle Ages what circulated were translations from Greek into Arabic, from Arabic into the Romance languages and, from these, into Latin. This produced enormous distortions in the texts. For example, at the Toledo School of Translators (twelfth and thirteenth centuries) it was forbidden to translate directly from Arabic into Latin (a language that could only be used by the clergy), so that the only way to understand each other was a four-hand translation where the intermediate language was Spanish, the language common to the three cultures that inhabited the city: Arabs, Jews and Christians.

Petrarch Despised Science

The situation in Italy before the Fall of Constantinople was truly catastrophic. The great poet Petrarch (1304–1374), considered to be the father of the Renaissance humanist movement, always contrasted *studia humanitatis* with naturalist research in his writings. That is, of the seven liberal arts of antiquity, he considered that only the studies of the *trivium* are valid, not those of the *quadrivium*. The former, according to Petrarch, serve to form us as human beings, the latter only to know, 'although without any security', how the things of nature are.

In the letter dedicated to his *On the Life of Solitude*, Petrarch said that man is an evil, disloyal, hypocritical, savage and bloodthirsty animal unless he has learned to dress in humanity and shed his fierceness, and that this is achieved through the cultivation of human literature-based subjects, 'because these are the only ones that serve to humanize man'. For this reason, as Paradinas (2007, 146) pointed out, Petrarch was not interested in the knowledge provided by the natural sciences, the naturalist or the scientist. And they did not interest him because, according to Petrarch, they were not true in most cases and, furthermore, they do not help to know what really matters: what is man, what is his origin and where is his destiny?[8] From today's point of view, if there are disciplines that really help to know the origin and destiny of man, it is science. But Petrarch regarded them with a great deal of contempt:

> Our man knows a great deal about wild beasts, birds and fish: how many manes the lion has on his head, how many feathers the sparrowhawk has on his tail and how many rings the octopus forms to hold the shipwrecked one; he also knows that the elephants mate on their backs... that the phoenix is consumed in an aromatic fire and is reborn from its ashes; that the moles are blind and the bees are deaf... This news is, of course, for the most part false; and, furthermore, even if all this were true, it would serve no purpose for a happy life. For, tell me, what good is it to know the nature of wild beasts, birds, fish and snakes and to ignore or despise, instead, the nature of man, without wondering why we were born or where we came from or where we are going? (Petrarch, *The Life of Solitude*, 1360)[9]

Petrarch's problem, however, is that in his day Constantinople had not yet fallen, so there were hardly any Greek texts in Europe. Those that there were either poorly translated or were by Aristotle who, in fact, was quite mistaken in matters of physics and biology. After Petrarch, other Italian Renaissance humanists, such as Salutati (1331–1406) and Bruni (1370–1444), insisted on the idea, from my point of view erroneous, that it is the *studia humanitatis*, also called *humanae litterae* or 'the humanities', that humanize man because they make him a true citizen, a free and responsible being capable of acting in the civil community. For them, the natural sciences, the knowledge that does not deal with man but with nature, serve to instruct him in certain practical knowledge to earn a living as a professional, but they do not help him to live in society. This is because they do not deal with truth and falsehood, good and evil, justice and injustice, convenience and inconvenience, etc. That is to say, they do not serve to found and maintain political life (Paradinas 2007, 147).

---

[8] Jesús Paradinas. (2007). 'Pasado, presente y futuro del modelo humanístico de educación (Past, present and future of the humanistic model of education).' *Eikasia. Revista de Filosofía*, II (11), 133–170.

[9] Petrarch, *De la ignorancia de sí mismo y de la de muchos otros* ('From ignorance of himself and many others'). In F. Petrarch, *Work. I. Prosa*, p. 167. Madrid: Ediciones Alfaguara, 1978.

This caused the upper classes to study literature, not science and technology. They did it for two reasons: one was that they didn't need their studies to gain employment and, secondly, because literature-based subjects took less effort—and we know that the well-off classes tend to flee from effort. Obviously, there are exceptions, but this is still the case: those who study literature or social sciences in the West often belong to the economic elite. The study of literature-based humanities, these Renaissance humanists considered, develops that which is the most characteristic of man, what distinguishes us from the animal: the *logos*, the word. In other words, for them, mathematics, physics or technology are not something that needs *logos* or is specifically humane. Underlying this idea is the division of cultures.

And here appears an absolutely wrong notion that has contaminated the West. Many believe, even today, that one can be educated only by knowing literature, art or history, and without knowing mathematics, physics, chemistry or biology. Consequently, what most humanizes humans are the language-based subjects and grammar, which, according to these early Italian Renaissance humanists, are the door to all knowledge; plus rhetoric, which teaches us to persuade and convince others. None of the three are any good at finding out the truth. It's as if scientists are less human than others. The belief is that rhetoric and poetry facilitate power, given the important role that passions and emotions play in political life. And this is right, only our problem is that we need more rationality and less passion.

The Turn of the Fall of Constantinople

Fortunately, as I mentioned, for the West all this took a turn with the Fall of Constantinople in 1453. Throughout the fifteenth century, the number of Greek scientific texts available to scholars in Italy increased, as did the number of people able to read these works in their original language. Knowing classical Greek became essential because it allowed one to think like the Greeks, when reading them without erroneous or selfishly subjective translations. This new situation would awaken the interest of many Renaissance humanists in the study of nature. And that would change the destiny of Europe and the world.

One of the young people who was dazzled by the new books of Greek mathematics that appeared in Italy was Filippo Brunelleschi (1377–1446). His passion for mathematics was the origin of Renaissance art. He invented the principles of linear perspective with the use of mirrors around 1415. And this revolutionized Western painting because he developed an optical principle: the apparent size of an object decreases with increasing distance from the eye.

His experiments on physics (optics) and geometry are not preserved. With them Western art changed: in gothic painting there was no perspective; in the Renaissance objects are painted in the most approximate way to reality thanks to the laws of perspective devised by Brunelleschi. His major works include the dome of the Cathedral of Santa Maria del Fiore (1418–1436), known as the *Duomo*, in Florence. It was the largest dome built since Roman times. Brunelleschi was not trained as an architect, but as a mathematician: that is why he revolutionized art. But this idea is difficult to find in a faculty of arts.

The Ionian scientific tradition (Greek province, in present-day Turkey) was recovered, which allowed for a deeper knowledge of Greek mathematics and the atomistic physics of Democritus, which was so different from the Aristotelian tradition. This is no small thing, because the Arabic translations of the mathematics books were obscure and complex. In Greek works, mathematics was much better understood and this encouraged Italian humanists to become interested in this discipline not as a tool for accounting or architecture but as a philosophy in itself. In the Middle Ages, knowledge of Greek mathematics was limited to the works of Euclid (325–265 BC) and Archimedes (287–212 BC), whose translations, as I have mentioned, had errors and the mathematics was confusing. The Italian humanists had the Greek texts of both authors at their disposal.

This is especially true of the work of Archimedes. Murdered by the Romans, his work became known when it was compiled in the seventh century by Isidore of Miletus, the famous mathematician and architect and author of the famous cathedral of Hagia Sophia (in Greek divine wisdom) in Istanbul, thus he was better known in the East. Begun in 532, this cathedral, with its 30-m diameter dome, remains one of the most impressive monuments built by mankind. Isidore recovered the writings following a tradition of his city, Miletus—in Greek Ionia, belonging to present-day Turkey. It was here, a thousand years earlier, that an impressive route towards humanity had begun under Thales of Miletus (639–547 BC), considered to be the father of both the inductive method and the study of nature (it is known that he wrote a book on nature, but has not survived).

Thales of Miletus, whom some consider the first scientist, was, of course, one of the greatest Greek mathematicians: his studies covered the area of geometry, linear algebra, the geometry of space and some branches of physics such as static, dynamic and optical. He founded the famous School of Miletus from which several disciples would emerge to complete their master's work. One was Anaximander (610–546 BC), who also wrote a book on nature. He is now lost in the mists of time, yet through other philosophers, especially Aristotle, we know that he was interested in cosmology and the origin of fire, earth and stars.

Another, Anaximenes (590–525 BC), for the first time appreciated the possibility of transforming some elements into others. He wrote a book about nature, his *Peri Physeos* ('On Nature'), which is also lost, although comments and references have survived. He analysed breathing and the importance of air (hence, he was ahead of his teacher Thales and his partner Anaximander by valuing air, not earth or water, as the most important element). He studied, for example, the process of condensation, so important to future alchemy. He was the first to observe that the planet (which he mistakenly considered to be flat) revolved around a common point: the pole star.

However, Thales' most important disciple was undoubtedly Pythagoras (569–475 BC). I still remember that in primary school our teacher explained the famous Pythagorean theorem. He told us that we were going to study it exactly as Greek children had studied it 2,500 years before—just like Aristotle or Alexander the Great had studied it. And it ended with a phrase that I later, over the years, came to know was inspired by G. H. Hardy's mathematician's book, *A Mathematician's Apology* (1940).[10] A theorem, my primary school mathematics teacher pointed out, is the only human construction that can aspire to be eternal. We studied it as the Greeks did 2,500 years ago and as our descendants will study it within another 2,500 years, if the world still exists. Actually, what Hardy said was:

> The 'real' mathematics of Fermat and other great mathematicians, the mathematics which has permanent aesthetic value, as for example the best Greek mathematics has, the mathematics which is eternal because the best of it may, like the best literature, continue to cause intense emotional satisfaction to thousands of people after thousands of years. These men were all primarily pure mathematicians (though the distinction was naturally a good deal less sharp in their days than it is now); but I was not thinking only of pure mathematics. I count Maxwell and Einstein, Eddington and Dirac, among 'real' mathematicians. The great modern achievements of applied mathematics have been in relativity and quantum mechanics, and these subjects are, at present at any rate, almost as 'useless' as the theory of numbers. (Hardy 1940, 39)

Founder of the Pythagorean school, Pythagoras considered mathematics to be the centre of all knowledge. Apart from his contributions to music (the Pythagorean scale) or cosmology (he established the sphericity of the Earth),

---

[10] G. H. Hardy. (1940). *A Mathematician's Apology.* University of Alberta Mathematical Sciences Society. Available at http://www.math.ualberta.ca/mss/.

Pythagoras had an enormous impact on the Greek thought that Plato, and Aristotle himself, recognized.

Alexandria: African Science

If the Fall of Constantinople meant the return of Ionian knowledge to Western Europe, it also meant the entry of ideas from another school far from Europe, namely from the North African city of Alexandria (in present-day Egypt). The other great mathematician of antiquity, Euclid (324–265 BC), wrote what is undoubtedly another of the great books of Western history, the 13 volumes, of which only seven have reached us, of the *Elements of Euclid*.

This book is one of the most powerful instruments of deductive reasoning made by human beings. Euclid founded geometry and the theory of divisibility. Apart from his work's influence on thought, it has been extremely useful in many fields of knowledge such as physics, astronomy, chemistry and various types of engineering.

The Alexandria school (which would have derived much from the Miletus school) had other mathematicians who came to Europe only after the Fall of Constantinople, and possessed original versions of Greek works. One of the most important was Diophantus (250–334 BC) and his work *Arithmetica*, a treatise of 13 books of which only the first six are known. This work was the precursor of algebraic theory. It was translated into Arabic, but the first translation into Latin from Greek was not made until 1575 from a Greek original found in Venice by the German mathematician, Johann Müller Regiomontano, around 1464.

A curious fact about Diophantus' book is that it is like a series of problems to be solved. In particular, Problem VIII in Book II states that 'One cannot write a cube as a sum of two cubes, a fourth power as a sum of two fourth powers, and more generally a perfect power as a sum of two like powers'.[11] The history of mathematics tells how the great French mathematician Pierre de Fermat (1601–1665) found a solution: 'I have discovered a truly marvellous demonstration of this theorem, which this margin is too narrow to contain', wrote Fermat in a 1621 edition (in Greek and Latin) of the *Arithmetic*. This mathematical enigma, which has to do with the theory of numbers, so important in today's big data, had kept mathematicians on their toes for more than three centuries. It is said that, frustrated, the great Leonhard Euler (1707–1783)—the chief mathematician of the eighteenth century—implored

---

[11] *Cubum autem in duos cubos, aut quadrato-quadratum in duos quadrato-quadratos, et generaliter nullam in infinitum ultra quadratum, potestatem in duos ejusdem nominis fas est dividere. Cujus rei demonstrationem mirabilem sane detexi. Hanc marginis exiguitas non caperet.*

a friend to search Fermat's house from top to bottom in search of proof. He had no luck: humanity had to wait until the end of the twentieth century—in 1995—for the British mathematician and Oxford professor, Andrew Wiles, finally to be able to demonstrate Fermat's theorem, as set forth earlier by Diophantus. Wiles used mathematical tools that emerged long after Fermat's death, so Fermat must have found the solution in another way, if at all. In any case, he was right: it could be solved. New thought was already ahead of the Greeks.

Wiles' proof exemplifies, better than literature, art, politics or law, a fascinating element of scientific culture: the demonstration of this theorem represents the uninterrupted history of thought. A problem posed by the Greek Diophantus, possibly from Euclid or even from the Pythagorean school, was rescued in Venice by Regiomontano (the founder of trigonometry) after the Renaissance, to be solved in the seventeenth century in France by Fermat (although his solution was lost) and, finally, almost at the end of the twentieth century, it was solved in England, at the oldest university of that country, with theoretical contributions from two Japanese mathematicians (Gorō Shimura and Yutaka Taniyama). In 2016, Wiles was awarded the Abel Prize (considered the Nobel Prize in mathematics) 'for his impressive demonstration of Fermat's latest theorem by conjecture of modularity for semi-stable elliptic curves, ushering in a new era in number theory'.

It is also worth looking at other mathematicians from Alexandria, such as Apollonius of Perga (262–190 BC), known as 'the great geometrician' for his book *Conicorum* ('From the Conical Sections'). He was born in Perga (present-day Turkey), but trained and died in Alexandria. He is hardly known in the West (where poets such as Homer are given more importance), but in the *Conicorum* of Apollonius there are concepts studied by children from all over the world, such as the ellipse and parabola, among others. Homer is for deeply learned people. Apollonius is for everyone, as is his solution to the general equation of the second degree by means of conical geometry. He is also attributed with the hypothesis of eccentric orbits, or epicycles theory, to explain the apparent motion of the planets and the variable speed of the Moon. His extensive work on flat curves and squaring of areas is also relevant.

Another mathematician from Alexandria was Pappus (290–350 BC), translated and read from the Greek from the Renaissance on. His theorems are used to find the surface area and volume of any rotating body. Highly relevant to engineering, his works reformulated by the Swiss mathematician Paul Guldin, or Guldinus (1577–1643) from Greek copies.

The last of the great inspirations for the Renaissance who came from Alexandria was the great Hero (1st century AD), considered to be one of the greatest engineers of all time: he invented the first steam engine, known as the 'eolipile'. He is the author of numerous treatises on mechanics, such as *Pneumatica* (πνευματικά), in which he studies hydraulics, and *Automata* (Αυτοματοποιητική), considered the first robotics book in history. He was a mathematical genius: he wrote *Metrica* (Μετρικά), in which he studied the areas of surfaces and volumes of bodies. Hero also developed calculation techniques taken from the Babylonians and Egyptians, such as calculating square roots using iteration.

The city of Alexandria, founded by Alexander the Great in 331 BC, would be one of the beacons of ancient knowledge with its famous library. It is estimated that the following number of books were deposited in the library: 200,000 volumes at the time of Ptolemy I; 400,000 at the time of Ptolemy II; 700,000 in 48 BC, with Julius Caesar, and 900,000 when Mark Antony offered 200,000 volumes to Cleopatra, taken from the Pergamonese Library.

Now, in the twenty-first century, Alexandria is a city of about five million inhabitants, and goes unnoticed among the hundreds of similar size in the world. It has fewer tourists than the pyramids. But Western science was built in that city: Hipparchus of Nicca developed trigonometry, and defended the geocentric vision of the universe; Aristarchus, who defended the opposite—that is, the heliocentric system—centuries before Copernicus; Eratosthenes, who wrote a geography and composed a fairly accurate map of the known world; Chalcedonian Herophile, a physiologist who concluded that intelligence is not in the heart but in the brain; and the great physician Galen, who developed anatomy and medicine in general.

There were women scientists, such as Mary of Alexandria. Her year of birth is not known exactly, but was between the first and third centuries after Christ. Regarded as the founder of alchemy, in her work, which has reached us through *Dialogue of Mary* and *Rings*, she described procedures that would later become its basis, such as leukaemia (bleaching) and xanthosis (yellowing). One was made by crushing and the other by calcination. She described acetic acid, invented complicated apparatus for the distillation and sublimation of chemicals (such as the tribikos, a kind of still), and the famous 'bain Marie', a method still used today in industry (pharmaceuticals, cosmetics, food and canning), the chemical laboratory and even the kitchen to keep a liquid or solid substance at a uniform temperature.

The last-known scientist in Alexandria was a woman, Hypatia (355–426). She wrote about geometry and algebra and improved the astrolabe. Her death, at the hands of a mob of fanatical Christians, shows us very well how a society is doomed to failure if irrationality and religion are accorded a status above science and mathematics. Before the murder of Hypatia (who was dismembered and cremated), Alexandria's library had been looted then completely destroyed.

Egypt had already become an important Christian country by the fifth century and, now that they were in power, Christians were uncomfortable with paganism and the Greek and Latin culture of the city. Classical scholar Catherine Nixey describes in her book *The Darkening Age: The Christian destruction of the classical world* (Macmillan, 2017) how early Christian fanaticism swept through classical culture. As its title suggests, Nixey's book presents the progress of Christianity as a triumph only in the military sense of a victory parade: 'Culturally, it was genocide: a kind of anti-Enlightenment, a darkening, during which, while annihilating the old religions, the rampaging evangelists carried out "the largest destruction of art that human history had ever seen".'[12]

As she points out, there are some dark dates: in 312, Emperor Constantine converted after Christianity helped him to defeat his enemies; in 330, Christians began desecrating pagan temples; in 385, Christians sacked the temple of Athena at Palmyra, decapitating the goddess's statue; in 392, Bishop Theophilus destroyed the temple of Serapis in Alexandria; in 415, the Greek mathematician Hypatia was murdered by Christians; in 529, Emperor Justinian banned non-Christians from teaching; in 529, the Academy in Athens closed its doors, concluding a 900-year philosophical tradition.

The episode shows that, no matter how important the culture of an area may be—Miletus or Alexandria—decline may come. And that all this knowledge is either lost or transferred to other places: from Turkey and Egypt it was taken to Europe (Italy, France and finally England), and from there to the United States. Perhaps the twenty-first century is the time of China and India.

De Rerum Natura: Atomism Made Poetry

I do not wish to end this section without mentioning an episode that shows that collaboration between science and literature is essential to progress. It would seem that the furthest thing from physics or engineering would be a dark, scholarly poem in Latin. However, this wonderful poem contains the essence of physics and, with it, the essence of who we are, where we come from, where we are going and how we should live our lives. Written more than 2,000 years ago, with ideas from 2,500 years ago, all we have done since in science is to demonstrate, with facts, what the poem proclaims. It is all here.

I refer to the magnificent *De Rerum Natura* ('On the Nature of Things' or 'On the Nature of the Universe'), written in the first century BC by the Roman poet Titus Lucretius Caro. It is in hexameter verse in a Latin so sublime that

---

[12]Thomas W. Hodgkinson. (2017). 'Islamic State are not the first to attack classical Palmyra: The Darkening Age reviewed.' *Spectator* (September).

the poem has survived, since it serves as an exemplar of what that language can achieve. It is a literary masterpiece. But if the literary form is brilliant, the background is pure dynamite. In sublime poetry, it speaks of the atomic physics of Democritus and, from it, points out that the only thing that is truly immortal is an atom and that, if we really are atoms, a universe without gods could be proclaimed since it would be 'only' the evolution of atomic aggregate forms different from others.

The first time I heard about it was from my professor of inorganic chemistry at university. He introduced the atomic models and, as always, he spoke of Democritus. But one day he went a little further and quoted the poem… At the same time he told us about something fascinating: nucleosynthesis; that is, how the elementary particles that emerged after the big bang become hydrogen atoms, and how these become the heaviest elements of the periodic table in the stars. He told us of something even more seductive: our atoms did not form on Earth (this is obvious, but many first-year students confuse chemical elements with minerals), but are the product of stars, such as the Sun, that have exploded or consumed themselves, of galaxies that have collided, of supernovae in their colossal explosions, of the exchange of oxygen in our blood. The professor paraphrased the astronomer and great disseminator, Carl Sagan: 'We are,' he said, 'star dust, and we aspire to return to that phase. Stars and planets emerged before the solar system, the product of violent explosions that form atoms that have reconnected to form another solar system with a planet like Earth whose geological processes have combined the chemical elements into minerals and into a curious type of substance, called life, that can create (reproduce) itself.' No idea had ever seemed so fascinating to me as that.

It was not until 1957 that *Reviews of Modern Physics* published a now almost legendary article, 'Synthesis of the Elements in Stars', written by Democritus' intellectual descendants, scientists Margaret Burbidge, Geoffrey Burbidge, William A. Fowler and Fred Hoyle. The article is also known as 'B2FH' (the initials of the authors' surnames) and, in my opinion, it is one of the most important cultural treatises of humanity as it perfectly responds to where we come from and where we are going. But few academics with a humanities background or social scientists are aware of Democritus.

I repeat, the Greeks had sensed all that Democritus and a poet such as Lucretius expressed in Latin in a kind of didactic poem. Why are Greek mathematics and physics not taught on classical studies courses, and why is classical philosophy not taught in physics courses, in Western universities? This is a relevant question that explains much of what happens in the West. However, the scientists (B2FH) of the twentieth century did not dare to, any more. They published their paper and considered that it was just science, not philosophy.

The Greeks and the Roman Lucretius had been more daring: if we are only atoms that are formed in the stars, there is no need to fear death because it is simply the passage from another atomic combination to another. This poem, *De Rerum Natura*, which transformed Western mentality, is considered to be the greatest poetic work of Roman culture, thus the classical humanities. Its vocabulary is Latin, but it is obviously pure Greek thought—that of Democritus and Epicurus—and it is possibly the greatest attempt to understand the physical reality of the world, also that of human beings. What is certain is that it is the greatest source of atheism in the Greco-Latin world, and it does not come from a book on ethics or literature, but from physics.

In 2011, a wonderful book appeared by Stephen Greenblatt, John Cogan University Professor of Humanities at Harvard University, entitled *The Swerve: How the World Became Modern* (Johns Hopkins University Press, 2011) (Pulitzer Prize for non-fiction 2011). Greenblatt describes the discovery of this lost poem: the journey from the Middle Ages to the modern, scientific, secular world that we know today.

While today's physics and chemistry professors are tiptoeing around the philosophical significance of the fact that matter is composed of atoms, Lucretius did not flinch: 'Death is nothing to us', he wrote. He added that to spend one's life gripped by the anguish of death is madness. It's a sure way for life to slip through one's fingers without having lived it fully and without enjoying it. Lucretius addresses here the philosophy of Epicurus (341–270 BC), which can be summarized in a few lines: religious myths embitter the lives of men, that the aim of human life is to procure pleasure and avoid pain but always in a rational way, avoiding excess, since this causes later suffering.

Elsewhere in the poem Lucretius returns to the physics of Democritus (460–370 BC). The matter of which the universe is composed, he proposed, based on Democritus' atomic theory, is an infinite number of atoms that move randomly through space, like specks of dust in a ray of sunshine, colliding, hooking up with each other, forming complex structures and separating again in an incessant process of creation and destruction. And there is no escape from such a process. The French philosopher and science historian Michel Serres (1930) stated in his book, *La naissance de la physique dans le texte de Lucrèce* (Minuit, Paris, 1977), that Lucretius' poem is not a text of metaphysics or moral philosophy but, essentially, of physics. What's more, it's not just that the poem is a mathematical and experimental physics like ours—with models, experiences and applications—it's that it is exactly our physics, not so much the one that opens with Galileo and culminates with Newton, but rather the one that we're starting to do today, based on experiences like those of Einstein, Heisenberg or Prigogine.

Another part of the poem states that when we look at the night sky and, feeling an ineffable emotion, are amazed at the infinite number of stars, we are not seeing the work of the gods or a crystalline sphere separated from our ephemeral world. We are seeing the very material world of which we are part and of which we are made: there is no master plan, no divine architect, no intelligent plan.

Darwin was possibly inspired by the verses that indicate that all things, including the species to which we belong, have evolved over very long periods of time. Evolution is fortuitous, although in the case of living beings it involves a principle of natural selection. That is, species that are able to survive and reproduce properly persist, at least for a while; those that are not quickly become extinct. But nothing—from our species to the very planet on which we live or the sun that shines on our days—lasts forever. Only atoms, Lucretius reminds us, are immortal.

The Church considered *De Rerum Natura* to be the most dangerous book in history, because it was the start of the emancipation of the individual from the supposed infallibility of the dominant religion. And it did it in terms of physics. In fact, its enormous impact lies in the fact that, until its discovery, anyone who wanted to explain how the world was created without taking into account the presence of gods had no argument. They did not have the capacity to refute the vast literature that supported the idea of God, because it had either been lost or deliberately destroyed after the Christian takeover of the Roman Empire.

In 1982, the Italian science historian Pietro Redondi discovered in the archives of the Holy Office the document that changed the view for which Galileo was condemned for insisting on the heliocentric theory. The Church had held that it was the Earth that was at the centre of the universe and it was the Sun that was moving, 'otherwise Joshua could not tell the Sun to stop, as the Bible says'. Redondi's paper on the Galileo trial shows that his inquisitor was concerned at the heresies in *The Assayer*, because he saw evidence of atomism. Atomism, his inquisitor insisted, is incompatible with the Council of Trent, which described the dogma of the Eucharist and the phenomenon of transubstantiation (the theological doctrine defended by Catholics and Orthodoxy, according to which, after the priest's consecration, the bread and wine of the Eucharist become the body and blood of Jesus):

> If this theory is accepted as true [Galileo's atomism], the document of the Inquisition points out, when the terms and objects of the sense of touch, sight, taste, etc., accidents or features of bread and wine are found in the Blessed Sacrament, it must also be said, according to the same theory, that these features are produced in our senses by tiny particles. And so we must conclude that in the Sacrament

there must be substantial parts of bread and wine, a conclusion that constitutes a flagrant heresy.[13]

This was a major problem for the Catholic hierarchy, as it demonstrated the profound irrationality of theology. An obvious issue is that anyone who had tasted the host before and after it was consecrated had found that it tasted, smelled and looked exactly the same. Therefore, at the Council of Trent it was determined that it was an 'experimental' issue, of philosophical and theological value, and that after consecration the permanence of warmth, colour, taste, smell and other sensory attributes of the bread and wine, which had had all their substance transformed into the body and blood of Christ, was miraculous.

The explanation by Galileo or the Greek atomists, suggesting that the taste, smell, touch, and so on, were due to atoms and bonds between atoms, was an anathema to the Church. And it still is, despite the fact that there are teachers who see no incompatibility between teaching chemistry and being Catholic (Pope Francis himself studied chemistry,[14] yet has not eliminated the existence of Eucharist).

To avoid this problem (and to be able to continue to investigate atomism and the structure of matter), Protestants do not believe in transubstantiation but in consubstantiation (a theological doctrine defended by Lutherans, whereby the substance of the bread and wine of the Eucharist coexists with the body and blood of Jesus after the priest's consecration). Few historians in the West explain the conflict of religions in Europe from the atomic theory of matter, but it would be an interesting perspective of union of the two cultures.

In the fifteenth century, the poem could have some influence, because the discovery of the sole copy, which had survived for 1,500 years (although the books of Epicurus were lost, for example) coincided with the invention of the printing press. At first, most of the print editions included some kind of retraction. A 1563 edition reads, 'It is true that Lucretius is against the immortality of the soul, denies Divine Providence and affirms that pleasure is the greatest good,' but then underlines 'although the poem itself is alien to our religion because of its affirmations, it is no less a poem'.[15]

Copernicus, Galileo, Newton, Descartes, Gassendi, Montaigne and many other intellectuals were dazzled by this poem. Also in the literary field, the great French playwright Molière made a translation which, unfortunately, has been lost. One of the most recent intellectuals to fall in love with the poem

---

[13] Pietro Redondi. (1983). *Galileo Herético*. Madrid: Alianza Editorial.

[14] Alex Knapp. (2013). 'Pope Francis, Scientist'. *Forbes* Available at: https://www.forbes.com/sites/alexknapp/2013/03/12/pope-francis-scientist-2/#722d91f14486.

[15] Stephen Greenblatt. (2011). *The Swerve. How the World Became Modern*. New York: Johns Hopkins University Press.

and its ideas (and to try to apply them to the social sciences) was Karl Marx (1818–1883). Before writing his famous *The Capital* (1867), in which he describes the mechanisms of economic production or social domination in terms of the ownership of matter, the German philosopher wrote a doctoral thesis (read in 1847) on a subject that, at this point in the nineteenth century, seemed to have already been decided: 'The Difference between the Democritean and Epicurean Philosophy of Nature'. If we consider his chapter titles, we can see Marx's interest in physics (Chapter 1: The Declination of the Atom from the Straight Line; Chapter 2: The Qualities of the Atom; Chapter 3; *Atomoi archai* and *atoma stoicheia*; Chapter 4: Time; Chapter 5: Meteors).

According to Marx, Democritus considers empirical support to be fundamental to overcoming the illusion of the senses and thus reaching the knowledge of reality, while Epicurus considers that it can be grasped through the senses and philosophical thought. Democritus defends the view that reality is deterministic and that its understanding is based on principles or causes. Epicurus, on the other hand, considers that reality occurs by chance, without the facts being explained by specific causes. For Marx, the fundamental difference in arriving at such disparate conclusions from an atomistic physics would lie in the postulate introduced only by Epicurus, which states that the atoms in their falling movement deviate from the rectilinear movement.

Marx uses Hegel's dialectics to analyse this movement and to construct Epicurus' worldview, as a consequence. Many argue that, when Marx prefers Epicurus' idealism to Democritus' scientism, he is being more mystical than scientific. In any case, this shows that the study of physics is highly relevant to understanding economic, social and historical phenomena. The two cultures are back together again.

Perhaps the latest example of this symbiosis is the book by quantum physicist César Hidalgo (currently at MIT Media Lab), *Why Information Grows: The evolution of order, from atoms to economies* (Basic Books, 2015), where he describes economics, politics and current communication in terms of quantum physics. But of course, if there is one place in the world in the twenty-first century where it is clear that the two cultures must come together, it is the Massachusetts Institute of Technology. However, in other Western universities—especially in the faculties of literature—physics and chemistry are not seen as starting points for trying to understand the world.

8

The Effect of a Humanities Culture on the Mainstream Media Industry

There are two keys to understanding the impact on science by cinema or television—the main agents of mainstream culture: a denigration of the figure of the scientist and, on the other hand, an exaltation of myth and magic; that is, of the irrational and unscientific. But this is not new: it has its roots in the culture of literature-based humanities. Cinema or television fiction are descended from literature and theatre. And, since the beginnings of Western culture, those who are 'artistic' have not only despised science and knowledge but vilified it. The Greek playwright Aristophanes (444–385 BC) in his comedy *The Clouds* (first performed in 423 BC) ridiculed none other than Socrates: in the play, he is presented as hanging from a basket and looking up at the sky. He is accused of demagoguery and is laughed at for his thirst for knowledge and interest in astronomy as a science. Aristophanes, like today's filmmakers, preferred myth to science.

This play by Aristophanes destroyed Socrates' reputation. When he was tried and condemned to death, the image of Socrates that was held by the citizens was derived more from the play than from the interesting questions on which Socrates forced them to think in depth. Let's not forget that, in Greece, the court rulings were dictated in the *agora* and then were voted upon. That is why his disciple, Plato, hated fiction (and democracy). The Socratic method of questioning everything has propelled Western critical thinking.

Much later, in European Gothic literature, we have the character of Faust. We all know his story: the man who sells his soul to the devil in order to find the answers to how the world works. The search for knowledge is related to diabolical evil. The title of the work in English—possibly translated from German by the great English playwright Christopher Marlowe (1564–1593)—is revealing: *The Tragicall Historie of the Life and Death of Doctor Faustus.* The

© Springer Nature Switzerland AG 2019
C. Elías, *Science on the Ropes*, https://doi.org/10.1007/978-3-030-12978-1_8

word 'doctor' is linked to the devilish. And, of course, we have the wonderful version of *Faust* (1808) by the great German poet Goethe (1749–1842). Goethe's Faust, also a scientist, is inspired by one of the greatest scientists of all time: Alexander von Humboldt (1769–1859), considered to be the first to see the Earth as a biological unit and, without doubt, the first scientific ecologist. Goethe was 20 years older than Humboldt, but they were friends and shared evenings of intense cultural debate (along with the other great German poet, Schiller (1759–1805), a medical doctor or physician).

It should not be forgotten that curiosity was a sin in the Catholic Church until the eighteenth century: it was due to curiosity, according to Judaeo-Christian doctrine, that Eve committed that original sin that we are still apparently paying for. The tree from which Eve took the forbidden fruit was the tree of knowledge. Shakespeare himself, in what many consider to be his last work, *The Tempest* (1661), questioned the new emerging science in England through the disasters that he attributed to the character of Prospero. This character is capable of shaping the forces of nature at will, not because he has special powers—like the magician Merlin in the Arthurian novel, who is a shaman or druid—but because he has learned the secrets of nature from books. In the work it is clear that it is Prospero's passion for study—he is the representation of the Greek *logos*—that leads him, among other things, to enslave the native, Caliban. At the end of the play, Prospero regrets having acquired this knowledge.

Playwrights' hatred of science became more evident in the twentieth century. A clear case in point is that of the poet Bertolt Brecht (1898–1956), with his three versions of his work on the life of Galileo. In the first, Galileo is a hero in the face of the Inquisition, despite the direction that the actor who plays it cannot be ethereal and graceful, 'but fat and ugly'. In the third version of the work, Brecht induces a 'Luciferian' personality in Galileo. In the prologue to this play, Brecht attacks physics:

> Respected public of the way called Broad –
> Tonight we invite you to step on board
> A world of curves and measurements, where you'll descry
> The newborn physics in their infancy.
> Here you will see the life of the great Galileo Galilei,
> The law of falling bodies versus GRATIAS DEI
> Science's fight versus the rulers, which we stage
> At the beginning of a brand-new age.
> Here you'll see science in its blooming youth
> Also its first compromises with the truth.
> It too must eat, and quickly gets prostrated

Takes the wrong road, is violated –
Once Nature's master, now it is no more
Than just another cheap commercial whore.
The Good, so far, has not been turned to goods
But already there's something nasty in the woods
Which cuts it off from reaching the majority
So it won't relieve, but aggravate their poverty.
We think such sights are relevant today
The new age is so quick to pass away.
We hope you'll lend a charitable ear
To what we say, since otherwise we fear
If you won't learn from Galileo's experience
The Bomb might make a personal appearance.
(From Brecht's *Arbeits journal*, entry for 1 December 1945)

His enormous literary talent is obvious, but also evident is Brecht's intellectual cretinism in blaming Galileo for the atomic bomb simply because he was responsible for the scientific method that drove the science of physics. It was not because he gave the political order and had the resources to develop the bomb, as did US President Franklin Roosevelt (a graduate of Harvard Law School and Columbia Law School); or because he gave the order to launch it on Japan, as did Roosevelt's successor, Harry Truman (also with a law degree, this time from the University of Kansas). If you look closely at Einstein's or Oppenheimer's diaries, you realize that if the US presidents had been physicists instead of lawyers the atomic bomb would never have been developed and, above all, that it would never have been dropped. Curiously, history and historians—especially literary political scientists—have acquitted Roosevelt (who happens to be one of the greatest presidents of all time) but not Oppenheimer. Einstein's letter to Roosevelt, warning him of the danger of the Nazis having the bomb, is also criticized; but the final decision was made by Roosevelt.

Bertolt Brecht's vision of science is contradictory because, in his work, he forces the protagonist—Galileo—to defend a totalizing and persuasive vision of scientific discourse in the face of the search for truth. For example, when the character of Galileo says:

GALILEO   Yes, I believe in the gentle power of reason, of common sense, over men. They cannot resist it in the long run. No man can watch for long and see how I—he lets fall a stone from his hand to the floor-drop a stone, and then say: 'It does not fall.' No man is capable of that. The temptation offered by such a proof is too great. Most succumb to it, and in the long run. Thinking is one of the greatest pleasures of the human race.

In other words, rationality is a negative thing, because it is just another persuasive tool. At another point in the play, the character of Galileo replies: 'One of the chief causes of poverty in science is usually imaginary wealth. The aim of science is not to open a door to infinite wisdom, but to set a limit to infinite error.'

From Aristophanes to Brecht, scientists have been presented as the anti-hero in Western literature, and this vision has obviously been transferred to the mainstream entertainment industry. The media culture and glamour, essentially from cinema, have only updated the Western literary tradition, opposing it to natural science and the search for knowledge. However, in recent years this whole tradition has been seasoned with the postmodern thinking acquired in the faculties of literature, art and communication.

One of the most topical aspects of both the twenty-first century and the late twentieth century was the rediscovery of the Gothic novel; that is, the triumph of sentimentalism. In this sense, the Gothic genre is the dark side of the rational eighteenth-century Enlightenment. In Gothic, the plot is always similar: virtuous heroes face powerful villains who pursue them relentlessly, because in their innocence they have discovered a truth that can erode the power of their persecutor. This plot is known to us because it has gone from the novel to the theatre and the cinema and television. From novels like *The Lord of the Rings* (J. R. R. Tolkien) to the *X-Files* series. According to the English literature teacher Sara Martín, 'the common denominator is always the threat suffered by someone who was innocent in principle but also privileged' (Martín 2006, 33).[1]

The literary critic David Punter in his book *The Literature of Terror: A history of gothic fictions from 1765 to the present day* (Longman 1980) states that Gothic is, par excellence, the narrative mode of the middle classes. It seems that the bourgeoisie defends the rationalist serenity of realism and naturalism but, in fact, in his opinion, in the Gothic it is venting its fear of losing its privileged status. The scientist Victor Frankenstein, the doctor Henry Jekyll and the protagonist of *Dracula* (the novel by Bram Stoker 1897), Jonathan Harker, represent us all in the classic tales of their own metamorphosis into monstrosity. This can also be seen in the work of the American writer and journalist, Edgar Allan Poe (1809–1849). The Western bourgeoisie relies on science and technology, but produces literary or cinematographic works that consider it a tool for creating monsters.

In a short chapter like this, one can't delve into all the films that destroy the image of science, but one can perhaps show a trend that began in the early days

---

[1] Sara Martin. (2006). *Expediente X. En honor a la verdad.* ('X-Files. In honor of the Truth'). Madrid: Alberto Santos.

of cinema and that is becoming more consolidated every day. I am not talking about the blatant conceptual errors about science that appear in the scripts. They are irrelevant in the face of the outrage that Western cinema perpetrates on science as an activity and of the scientist as a profession.

The first major film to offer a negative image of scientific development is *Metropolis* (Lang 1926), which clearly denounces the risks of technology by showing the first robot in the history of cinema. However, there is an earlier one, *The Cabinet of Dr. Caligari* (German: 'Das Cabinet des Dr. Caligari') (Wiene 1919), which shows the perversions of a psychiatrist. In 1936, Charles Chaplin shot *Modern Times,* which, although more focused on the criticism of the industrial processes of the then incipient capitalism, is still a denunciation of technology. This idea underlies later films such as *Master of the World* (Witney 1961) or the well-known *Blade Runner* (Scott 1982).

It also can be seen in *Brazil* (William 1985) and in *AI Artificial Intelligence* (Spielberg 2001). On television, it can be traced in the television series *Eureka* (Cosby and Paglia 2006, 2007). This negative image effect of science and technology is obviously multiplied when appearing in mainstream films, such as in the entire *Matrix* saga (Wachowsky and Wachowsky 1999).

The idea is that, in plots, Western reason always fails. This is not a failure of science, as in Frankenstein-style dystopian science fiction, but a triumph of barbarism that is encouraged by great villains with Gothic characteristics who use science as the ideal tool to facilitate the end of world order and the arrival of chaos. This perception, according to some authors, has increased after the attacks of 9/11. Professor Sara Martín, who has analysed all the seasons of *X-Files*, points out:

> *X-Files* evolved between 1993 and 2001, from the relative optimism that allows
> him to play at imagining enemies, to the devastating pessimism of his end,
> marked by the climate of the nation that has allowed itself to be carried away by
> terrorism towards political fascism, and that in reality itself is irrefutable proof
> of the triumph of the irrational. (Martín 2006, 34)

This phenomenon is extraordinarily dangerous when, in addition to being a huge success with adult audiences, the film is aimed at children or adolescents. This could be the case with *ET* (Spielberg 1982), in which scientists wanted to dissect the adorable alien, or *Jurassic Park* (Spielberg 1993), in which a mad scientist reproduces dinosaurs, with all the dangers to humanity that this implies.

These are just a few examples. But in all of them lies the Viktor Frankenstein prototype of versions of Mary Shelley's novel—from *Bride of Frankenstein* (Whale 1935) to *Mary Shelley's Frankenstein* by Kenneth Branagh (1994), through *Dr. Jekyll and Mr. Hyde* (Fleming 1941), *The Nutty Professor* (Lewis 1963) and *Mary Reilly* (Frears 1995). This Frankenstein archetype always shows a man playing at creating something that he doesn't know how will impact on the world.

The novel by the poet and playwright Mary Shelley (1797–1851), originally entitled *Frankenstein, or the Modern Prometheus,* was published in 1818. It is considered to be the first work of science fiction. In fact, it started from an unscientific reaction by the English poet when she witnessed—or at least heard about—the experiments being carried out on the properties of the newly discovered electricity, one of which, very common at the time, placed electrodes on a corpse. Obviously, as the current passed through it the corpse moved.

That movement of the corpse caused fear among the audience and some—those with an unreasonable mind—thought that electricity could resurrect the dead: hence the plot of the novel. In it, you can see some elements that are far from trivial, since they are repeated in our media culture. The first is the figure of Prometheus, the Greek god who taught humans how to generate fire; that is, the search for knowledge and technology are contrary to what the gods want. The myth of Prometheus had been widely discussed in the West as a representation of the dangers of natural science and technology, from Aeschylus (who wrote a play on the subject) to Lord Byron, a contemporary of Shelley, who analysed it in an essay.

The other interesting aspect is the representation of the result produced by science as the essence of evil. Many scholars believe that Shelley's description of Frankenstein is inspired by the character of Satan in the poem *Paradise Lost,* a classic of world literature by the great English poet John Milton (1608–1674).

Paradise Lost (1667) was inspired by the passage from the biblical narrative (another work of Western literature) that holds that God forbade Adam and Eve to pick a fruit from the tree of knowledge; that is, according to the Bible, they were forbidden to attain the same knowledge as God. The character of Satan in Milton's *Paradise Lost* questions the heavenly dictatorship, or the fact that knowledge is not available to everyone. God banishes him to hell and Satan holds that 'he would rather reign in hell than serve in heaven'.

Frankenstein is, above all, an allegory written by a poetess, not someone with scientific training, about the perversion that can be brought about by scientific development: man's search for the science of divine power, as opposed to man's search for literature, whose greatest aspiration is to praise—whether in literature, painting or architecture—the power of God or myth, while at the same time combating the Satan of science. In Mary Shelley's novel, the creature who is the result of the experiment, thus produced by science, rebels against its creator, the scientist. This contains a subliminal message about the punishment that awaits those who prefer science to myth. Above all, it is a warning to the rest of society to obstruct scientists from doing their job, otherwise it is society that suffers the consequences, just as, following Eve's decision to try the apple of the tree of science, the human species cannot be immortal but must suffer pain and death.

Viktor Frankenstein was not originally a mad scientist to whom the cinema has accustomed us since the James Whale version, but a young man passionate about chemistry:

> Natural philosophy in the genius that has regulated my fate…When I returned home my first care was to procure the whole works of this author [Cornelius Agrippa], and afterwards of Paracelsus and Albertus Magnus. I read and studied the wild fancies of these writers with delight; they appeared to me treasures known to few besides myself. I have described myself as always having been imbued with a fervent longing to penetrate the secrets of nature…
>
> He began his lecture by a recapitulation of the history of chemistry and the various improvements made by different men of learning, pronouncing with fervour the names of the most distinguished discoverers. He then took a cursory view of the present state of the science and explained many of its elementary terms. After having made a few preparatory experiments, he concluded with a panegyric upon modern chemistry, the terms of which I shall never forget: 'The ancient teachers of this science,' said he, 'promised impossibilities and performed nothing. The modern masters promise very little; they know that metals cannot be transmuted and that the elixir of life is a chimera but these philosophers, whose hands seem only made to dabble in dirt, and their eyes to pore over the microscope or crucible, have indeed performed miracles. They penetrate into the recesses of nature and show how she works in her hiding-places. They ascend into the heavens; they have discovered how the blood circulates, and the nature of the air we breathe. They have acquired new and almost unlimited powers; they can command the thunders of heaven, mimic the earthquake, and even mock the invisible world with its own shadows. (Frankenstein: Shelley 1818: 71–88)[2]

He breaks with the system and creates his creature in his laboratory. This is the prototype of the nineteenth century, the lone scientist who could not be sustained by today's science because equipment is needed. That is why the scripts or novels of the twentieth and twenty-first centuries (Michael Crichton's *Jurassic Park*, for example) incorporate large and obscure corporations where groups of scientists work to sustain a globalizing phase of capitalism.

Over time, more features have been added to the Frankenstein archetype. During the 1930s and 1940s, scientists in films matched the common pattern of middle-aged White men, often insane and evil. It is well established that

---

[2]M. Shelley. (1818). *Frankenstein or the Modern Prometheus*. Barcelona: Book Trade Edition (2016).

most fiction films depict scientists as a mad archetype: 'Without doubt, Dr. Frankenstein is better known today than any other scientists, living or dead,'[3] writes George Basalla. He adds, 'Movies and television portray scientists as frequently foolish, inept or even villainous'.

In the 1950s, under the influence of comics, cinema added a new attribute to scientists: they wanted to rule the world and have it at their feet. As early as 1978, the journal *Science*, published by the American Association for the Advancement of Science (AAAS), sounded the alarm in an article entitled 'The media: The image of the scientist is bad'.[4] It claimed: 'The popular image of scientists is remarkably bad, and the mass media must bear a great deal of the responsibility' (Maugh 1978).

In general, and with a few exceptions—*The Nutty Professor* (Shadyac 1996; Segal 2000)—in the cinema the scientist continued to be a White male, during the following decades, and has a new characteristic: an obsession with obtaining the information that he wants at the price that he wants. Amoral rather than immoral, nothing can stop him. He never allows human feelings or sympathies to stand in his way. This insensitivity manifests itself in various ways. So, even if the scientist has a family, he denies it.

Very often, scientists are represented as single or widowed and, in any case, rarely seem to have any sentimental commitments or sexual relations. In this sense, there is an obvious difference from literature teachers, who are always represented by filmmakers as either extremely committed to their students (*The Dead Poets Society*, Weir 1989) or experiencing great passion (*Shadowlands*, Attenborough 1993). 'The audience wants to hear their beautiful little daughter or assistant say that the scientist is married to the test tubes, and has no time to socialize,' filmmakers admitted at the AAAS meeting.

Another of the most recurrent themes of cinema is scientists' capacity for large-scale destruction. Susan Sontag[5] made a magnificent analysis of this issue in American cinema from 1915 to the late 1970s. She concluded that this capacity for destruction has increased as the years go by. In the horror movies of the 1930s, scientists demonstrated an ability to destroy only a person or, at worst, a small village. But soon the cinema showed that scientists had the power to destroy big cities, countries, the entire planet and, finally, many planets. In reality, all versions now coexist, from the Frankenstein archetypes to *Forbidden Planet* (Wilcox 1956), which reflects on how the misuse of science destroyed our planet.

---

[3]G. Basalla. (1976). 'Pop Science: the depiction of science in popular culture.' In G. Holton and W. A. Blanpied (eds.), *Science and its Public: The changing relationship*. Boston: Reidel.

[4]Thomas H. Maugh II. (1978). 'The media: The image of the scientist is bad.' *Science*, 200 (7 April), 37.

[5]Susan Sontag. (1977). 'The imagination of disaster.' In D. Denby (ed.), *Awake in the Dark: An anthology of American film criticism, 1915 to the present*, pp. 263–278. New York: Vintage Press.

More recently, in the film *The Core* (Amiel 2003) the Earth is about to be destroyed by military scientists' experiments in the Earth's core. But since the series *Space: 1999* (1975–1977), it has been told how a scientist develops a method of propulsion of gigantic spaceships that, for reasons that are not clear to the spectator, fails and, as a consequence, destroys several inhabited planets. The novelty at the end of the twentieth century and the beginning of the twenty-first is that, in addition, rational thought is attacked and pseudoscience and irrationality are strengthened.

From the *Harry Potter* films (2001–2011) to *X-Files* (first a television series from 1993 to 2002, although Fox announced that it would broadcast the eleventh season in 2018; then as a Bowman film 1998; and Carter 2008), in which, for the first time, scientific and magical explanation are put on an equal footing and the latter wins. Others go even further: they extol the values of witches and wizards—*Charmed* (1996–2006)—to series such as *Lost* (2004–2010), whose first episode of the second season, 'Man of Science, Man of Faith', indicates that there is a serious conflict with science as a way of accessing knowledge. This approach is destructive of science. The portrayal is of the scientist as a negative stereotype, together with something much worse: scientific thinking as a fairy tale—in Feyerabend—that will never help us to understand the world. From science as 'evil' and as 'wrong', the solution is 'the good', according to cinema and television, lying in magic and esotericism.

X-Files is also a good example of how mass television series criticize the current system of Western science. In the episode 'Synchrony' (4.19), the scientists involved work at a university and rely on a demanding scholarship system that rewards productivity; that is, discoveries for immediate application. The protagonists of the episode are researchers who fiercely compete for scarce resources by accusing each other of manipulating data to impress the committees that award the grants. The view presented by the episode is that in the profession advancement is through personal contacts (as if it were not so in journalism or cinema) and, above all, the episode exhibits an idea that is repeated in many scientific films: science and technology advance in the framework of capitalist corporations, at a multinational level and with few scruples. This is seen, for example, in the episodes 'F. Emasculata' (2.22) and 'Brand X' (7.18). The view is that scientists are naive and well-meaning, yet their experiments have unpredictable consequences and negative implications. The image given by this and almost all the series is that 'universities in every corner of the planet are increasingly dependent on private money which, logically, finances only the experiments that interest them' (Martín 2006, 281).

Although a research team is now a star in screenplays, television series or novels, 'fiction does continue to offer modern substitutes for Frankenstein, imagining them as brilliant characters but with a dark side that drags their team of young sorcerer's apprentices and, if necessary, the entire planet to perdition' (Martín 2006, 281). Examples abound, from *The Fly*'s Seth Brundle (David Cronenberg 1986) to *Hollow Man*'s Sebastian Caine (Paul Verhoeven 2000).

In contrast to the fictions of the mid-twentieth century, at the end of the twentieth century, and especially in the twenty-first century, the character of the mad scientist takes a backseat (since he may be an interesting archetype), and it is now presented as yet another victim of the excessive ambition of the corporation that he works for. 'It may not even appear at all in the forefront or in the background, giving all the prominence to corporate executives and thus underlining the idea that we are all substitutable in the capitalist environment, including the great scientific brains, which today seem to abound more than ever' (Martín 2006, 281–282).

Scientists and Invading Aliens

An interesting issue that we cannot dwell on is the relationship between scientists and invading aliens. The most common archetype of an alien is that of a cold and relentlessly hostile organism that wants to destroy the world with its flying saucers and ray weapons; that is, the alien is the antithesis of the human. In cinema, the scripts make it clear that they always want to manipulate human minds through remote control. The vision of 1950s cinema is that the mad scientist is always eager to make contact with the alien invader.

It is the military characters that dispense wisdom and clearly understand the threat, and they try to destroy these creatures. It's funny how cinema reflects the opposite of what happens in reality. Usually it is the scientists who make the discoveries and it is the military, as happened with the atomic bomb, that uses them without compunction, risking destroying the world. Scientists in the Cold War openly confronted the military to stop them from using nuclear energy, especially those who had collaborated in its development. The best-known case is that of Einstein, yet it can be said that most nuclear scientists became pacifists in the Cold War. However, the cinema represented the opposite scenario. Scientists were the crazy ones who did not care about destroying the planet and the military figures were the ones who controlled them and saved humanity.

Sontag interprets this paradox in the context of the Cold War, in which many real and intellectual scientists were accused of subversion. In this sense, cinema is a propaganda tool with political and military power that tries to discredit scientists because in real life cinema is not usually submissive. In the

Cold War, the role of the scientist in society was all-important, so politicians and the military could not allow public opinion to be so fascinated by science intellectuals and, as a consequence, to have more moral influence on the system than themselves.

Albert Einstein and Bertrand Russell created the 'Pugwash Movement' in 1953, which was joined by many scientists, many of whom (such as Einstein himself) had contributed to a greater or lesser extent to the development of nuclear science and technology. The movement not only denounced the dangers of the atomic war but openly criticized the bloc policy defended in the Soviet Union and the United States. Hence the strategy of American politicians, in collusion with some of the filmmakers who were willing to do so, to develop scripts that cast doubt on scientists' reputation, so they might be seen as subversive. In this case, it would be possible to see that, perhaps, the knight who defends us in the age of science has other masters.

Another anti-scientific view is given in the film *A Beautiful Mind* (Ron Howard 2001), which suggests that a strong inclination towards science and mathematics can lead to schizophrenic behaviour and, ultimately, confinement in a psychiatric institution. The film was Oscar-winning for Russell Crowe, and the publicity emphasized that the story had a real basis. And it was true: it was the story of mathematician John Forbes Nash, who was a mathematical genius: he was awarded the Nobel Prize in Economics for his extraordinary work on differential equations with partial derivatives that would be used, for example, in game theory. But his mind fell apart: one day (as told in the film), Nash walked into the staffroom with a copy of the *New York Times*, claiming that a short article contained a message in code, addressed exclusively to him, from an alien civilization. It is also told how he was moving away from his fellow men and that he saw human relations as mathematical theorems. However, there are thousands of brilliant mathematicians, even better than Nash, who have not had these kinds of problems. The film seems to have been created by resentful people who were not taught mathematics in high school and therefore are using cinema in revenge.

Physicist Spencer Weart, a researcher on the History of Science, proposed in the prestigious journal *Physics Today* that scientists should publicly rebel against these negative presentations of their work in the media. He suggested that the stories of real scientists should be told 'to demonstrate their commitment to the pursuit of knowledge for the sake of a civilizing and humanitarian mission'.[6]

In the AAAS and in different scientific fields around the world, it has been proposed that scientists should unite, just as ethnic, religious or sexual minorities (or majorities) have done, so that cinema does not represent them as

---

[6]Spencer Weart. (1988). 'The physicist as mad scientist.' *Physics Today* (June, p. 37).

stereotypes. In this sense, the scientists point out, such lobbies have managed to ensure that today's cinema is 'exquisitely careful to be politically correct' and they suggest that the same should be done for the group of scientists.

Harry Potter: The Consecration of Magic in Literature, Film and Television

While there is no doubt that the cinema prefers magic to science, one of the most interesting facets of the relationship between science and magic and cinema in today's media culture is that of Harry Potter phenomenon. This is a story about the adventures of some students, especially its protagonist, Harry Potter, at a school of magic. First it appeared as a novel (with several more in the years following the first in 1997), then each of the books was made into a film, and both versions constitute an unprecedented media phenomenon.

The relationship with science in this series of books and films, now constantly shown on television, is observed because the teachings that children learn to be magicians are taught in a formal school with standardized subjects and subjects, which gives it the appearance of formal science education. That is possibly one of the keys to its success. There is no precedent for successful literary or cinematographic works of fiction in the West to treat magic as a conventional teaching in a college or university, in the same way as science is taught. In the face of 'boring' subjects such as chemistry, taught in 'normal' schools, Harry Potter and his friends study the subject of 'potions'. Faced with boring experiments that try to find out the causes of the facts rationally and scientifically, they use a magic wand to modify reality quickly, without interest in the mechanisms of that change. By contrast to the archetypal mad scientist who wants to destroy the world, Harry Potter and his magician friends want to save it. They and their magic are the heroes.

The author of the saga is the British writer J. K. Rowling. Her nationality could explain why the architecture of the magic school is the classic British Gothic, typical of its oldest colleges and universities. This circumstance has been criticized, because the film's décor copies too much from one of the 'cathedrals' of modern science: the University of Cambridge, possibly the largest secular temple of science.

The first book was entitled *Harry Potter and the Philosopher's Stone*, published in 1997 and translated into 63 languages (including Latin): 400 million copies have been sold. The film version of the book was released in 2001, and is the third most profitable film in the history of cinema. It received three Oscar nominations.

A cultured adult can clearly establish that it is pure fiction, but does a child or teenager? Apart from being set in a great British university (whose architecture is recognized throughout the world) and the fact that the teaching of magic is so similar to traditional scientific teaching discredits science, at the same time increasing the prestige of magic and sorcery, there is another problem with this first volume: the term 'the philosopher's stone'.

The word philosophy means true knowledge. In fact, in English, a doctorate, regardless of the subject, is known as a Ph.D. (Doctor of Philosophy). This has provoked controversy in the English-speaking world, because it could give the impression that magic leads to true knowledge just as science does. And although 'philosopher's stone' refers to the substance that could supposedly transmute metals into gold, according to the alchemists, who were not scientists in the modern sense but more like magicians, it is also important to note that alchemical knowledge is the basis of modern chemistry. 'The research carried out by the alchemists [to obtain the philosopher's stone] led to the acquisition of a good deal of genuine chemical knowledge and, of course, to the establishment of chemistry as a science', states the *Oxford Encyclopaedia of Philosophy* (edited by Ted Honderich).[7]

All this led the British publisher to suggest that the writer change the title, which she refused to do. However, in the face of criticism in academic circles, the American edition did change the original title to *Harry Potter and the Sorcerer's Stone*, making it clear that it was about magic and sorcery. The film had to be released in Great Britain under the same title as the book *Harry Potter and the Philosopher's Stone* (Chris Columbus 2001), but in the United States it was *The Sorcerer's Stone*. As a cinematic anecdote, it is curious that, during the filming, every time the term 'philosopher's stone' appeared in the dialogue, two scenes had to be filmed: one with 'philosopher's stone' and the other with 'sorcerer's stone' for the US version. This gives an idea of the extent to which the term 'philosopher's stone' was controversial and unacceptable in a book such as this.

In French, it was entitled *Harry Potter à l'école des sorciers* ('Harry Potter at the school of sorcerers'). In Spain, the book was published and the film was released with no problem with the *Philosopher's Stone* version. Of course, no university or academy protested, or, at least, that protest did not reach the media, which obviously did not notice it either.

Harry Potter is a media phenomenon that is likely to leave its mark on the children of the West in the twenty-first century. How this updated passion for magic books will affect future generations of scientists has not yet been studied. Many of those who read the first version of the book (1997) have

---

[7]Ted Honderich. (1995). *The Oxford Companion to Philosophy*, p. 661. Oxford: Oxford University Press.

already completed their university studies. Many others, who read it now, are at the age to choose between science and literature. This idea of magic as a fun and academic discipline, similar to or at the same level as a scientific one, has continued to be reinforced over time, from 1997 to 2011, both by the different print versions and, above all, because of the greater impact of the cinematographic versions, which are also broadcast on television.

This phenomenon of the success of Harry Potter and, by extension, of the author, J. K. Rowling, with young people in the twenty-first century is significantly greater than other British authors who were successful among teenagers and whose books were also taken to the cinema and television with great success between the 1950s and 1970s. This comparison would lead us to the fact that the scientific and rational method is in decline, while magic and sorcery are on the rise.

Thus, for example, Rowling would be equivalent in many respects to another author of British juvenile fiction, Arthur Ransome, creator of the *Swallows and Amazons*. However, all of Ransome's books praise rationality and the scientific-deductive method. The same goes for another great British writer of light fiction, Agatha Christie (1890–1976), with her detectives Miss Marple and Hercules Poirot. Many literary critics have pointed out that Rowling is inspired by Christie, yet Christie's books at least use rationality and the scientific-deductive method.

I mention them, regarding bridging the abysmal gulf between the use of the scientific method and the use of magic to explain reality. Another British author, the Scot Sir Arthur Conan Doyle (1858–1930), was a doctor and writer who shaped the character of the most famous detective of all time: Sherlock Holmes. His answer, 'Elementary, dear Watson', was the culmination of a logical deduction after applying the scientific method to solve the most varied enigmas. The first novel was published in 1892, and it was undoubtedly a great step forward for young people to intuit the whole chain of events, from fact to consequence—that is, I repeat, using the scientific method. Holmes is the archetype of a character who practises both the rational-logical method and scepticism about religion.

The author of *Harry Potter* (whose fortune was estimated by the BBC in 2003 to be higher than that of the Queen of England) was born in 1965. From my point of view, her generation has already coexisted with much anti-scientific cinema. The new generation will also do so with magic, celebrating the Middle Ages in the court of King Arthur and the magician Merlin (clearly Rowling's inspiration) more than the seventeenth to twentieth centuries, and their spectacular development of British and world science, a science whose findings and methods inspired the best and most successful British writers for teenagers from the nineteenth to mid-twentieth centuries. This modification to the subject matter in youth literature, from rationality to magic, is also symptomatic of a certain decline in science.

Is There Such a Thing as a Pro-science Film?

The critic David Robinson stated that the most enduring image of the scientist on the screen is not Marie Curie/Greer Garson looking vaguely through her microscope, but that of Peter Sellers in *Dr. Strangelove: Or, How I Learned to Stop Worrying and Love the Bomb,* laughing and clucking, prey to manic delight at the prospect of the advent of the holocaust that he himself has planned. If this is true, I would at least like to mention some of those films that, according to film scholars, do give a positive image of science. Even if, from my point of view, these films fail to offer a real and appropriate image.

I will begin with the one that Robinson mentions: *Madame Curie* (Mervyn LeRoy 1943). First of all, I must say that Hollywood made this film not because of its interest in spreading the life of a scientist or science. Madame Curie had become a public figure (beyond the scientific realm) after the publication of her biography in 1937, written by her daughter. Before that, in 1921, Curie had made an American tour and appeared in the newspapers especially because she was a woman and a Nobel Prize winner. This is relevant because, being a remarkable scientist, her contribution to the history of science is less important than the huge number of books and films that have been made about her. For example, fewer films have been made about Newton than about Marie Curie, although Newton was clearly immensely more important than Madame Curie, as a scientist. This is relevant, because it is an example of the fact that the scientific criterion is not the same as the cinematographic or journalistic one.

Marie Curie was in life what we would call today a media personality. In addition to being a scientist, she was involved in society: during the First World War she went out with an ambulance to take X-rays of the wounded. She then participated in the newly created League of Nations. She was contemporaneous with famous scientists such as Rutherford, Planck and Bohr, and she appeared more in the media, basically because of the perennial characteristic of news: she had rarity value. A woman scientist was strange, and this should not be underestimated, especially because she literally gave her life for science: she died of leukaemia caused by the radioactivity that she studied, not knowing that it was dangerous and, indeed, her laboratory notebooks are still so radioactive that they are kept in a lead chamber. Her story also has merit as she worked under very harsh conditions: she had to undertake her investigations in a cold greenhouse because, according to the rules of the day, a woman could not work alongside the other scientists—all men—as they would be 'distracted'.

Everything is important, yet also very novel, that is, literary; therefore it is a good reference point for both cinema and journalism. However, these circumstances are irrelevant to science, although they sometimes contaminate it. In fact, historians of science such as Gribbin are still wondering how it is possible that he was twice awarded the Nobel Prize—for physics in 1903 and chemistry in 1911—for practically the same work.

A fact that gives an idea of Curie's (non-scientific) media projection was the extent of the repercussions in the press of the time at her supposed romance with the great French physicist, Paul Langevin. Marie Curie was a widow—her husband had died in 1906—and Langevin was married with children. Langevin's wife, Emma Jeanne Desfosses, intercepted letters from her husband sent to her by a woman in love, Marie Curie, suggesting that he divorce his wife. Another of the letters said: 'My dear Paul, yesterday evening and night I spent thinking of you, of the hours we had spent together, of which I have kept a delightful memory. I still see your good, tender eyes, your charming smile, and I only think of the moment when I will find again the sweetness of your presence.'[8]

Without delay, Desfosses sent the letters to the newspaper and they were published in *Le Journal,* one of the largest newspapers in Paris. The news appeared on the front page with a photograph of Marie Curie and the following headline: 'A love story: Madame Curie and Professor Langevin.' For a subtitle they had: 'The fires of radium'. The body of the news item began: 'A fire has been kindled in the heart of one of the scientists who was so devoted to his studies; and this scientist's wife and children are weeping.' The scandal broke on 4 November 1911, just one day after a highly important and famous scientific meeting—the Solvay Congress—that Curie and Langevin were attending, together with the best scientists of the day, in Brussels.

However, the day after the scandal broke another Paris newspaper, *Le Petit,* instead of explaining why Curie and Langevin were not in their respective homes, which would have been easy because the Solvay congress had received media coverage, preferred that reality should not a spoil good story. Under the headline 'A romance in a laboratory' subtitled 'The affair of Mrs. Curie and Mr. Langevin', the item ran: 'We knew about it several months ago...[9] We would have continued to keep it a secret if the rumour had not been spread yesterday, when the two actors in this story had fled, one leaving his home, his wife and children, the other renouncing her books, her laboratory and her glory.' In any

---

[8]Cited in Susan Quinn. (1995). *Marie Curie. A Life,* p. 261. New York: Simon and Schuster.

[9]This is a turn of phrase often used by a newspaper that has not obtained exclusive coverage yet, given the importance of the information, feels that it should be front page news the next day.

event, there was a pistol duel between the newspaper editor (Emma Jeanne's brother) and his brother-in-law, Langevin. The scandal of a widow of a scientist of national glory, such as Pierre Curie, being involved with a younger man,[10] married with children and who had been a scholarship holder for her husband, was such that Madame Curie almost had to leave France.

I don't recount these details out of morbid curiosity but because I think they are important in a person's life. And, of course, they are cinematographic. However, Hollywood refused to include them; it wanted the Joan of Arc archetype, an asexual woman concentrating exclusively on science who does not experience passions like other human beings. Among other reasons, this is because it is the belief in the social sciences that the public can understand a writer or a painter having a passionate and exciting life, but cannot conceive of this in a scientist. And that's a bad image for science because, obviously, science is a human activity, not divine or robotic, and a scientist can be just as passionate about his or her feelings as a writer. In fact, the Curie-Langevin story continued: Madame Curie's eldest daughter, Irène, married Langevin's favourite scholar, Frederic Joliot, who was very handsome according to the chronicles of the time. Evidently, in laboratories just as in any other place where there are humans, the strongest passions can be ignited.

But you don't imagine that. The most widespread archetype is that of asexual scientists who are overly focused on their work. This is suggested in the *CSI (Crime Scene Investigation)* television series, for example. It is curious that if one compares the lives of the characters in two current, world-famous fiction series, *Urgencies*, which portrays doctors; and *Ally McBeal (1997–2002)*—or *How to Get Away with Murder* (2014–2018), which is about lawyers, the scriptwriters give lawyers a more intense life than the doctors: they spend entire episodes screaming and stressed out. The homelife of the rational and brilliant *Dr. House*, in the series of his name, is a total disaster, yet the characters in the series on lawyers dress better, have nicer houses, go to more expensive restaurants and, in addition, experience intense passions. It's as if being a literature-based writer means having glamour and being a science teacher means losing it.

Therefore, the film *Madame Curie* portrays a false archetype of science. In my opinion, even the role of women in science is falsified in the film, since brilliant women with a scientific career don't usually enjoy Marie Curie's success. In fact, I think that if I wanted to make a film that portrayed a role model for women in science, I would show the life of Rosalind Franklin. Today she is almost unknown but, like Marie Curie, Franklin also died from science: from cancer caused by her exposure to X-rays. Her studies are more important than Curie's, because Franklin was the first to obtain the structure of DNA, the

---

[10]The age difference was only five years. Madame Curie was born in 1867 and Langevin in 1872.

molecule that explains life. The history of science is unanimous in recognizing the transcendental and pioneering contribution of Franklin's research to the discovery of the structure of DNA, yet it was two men, Francis and Crick, who 'stole' her results and took the Nobel Prize. It should be noted that she died before the Nobel Prize was awarded. However, the sexist atmosphere that surrounded science at the time, which made her suffer so much as a woman, represents a more realistic portrait of science. But loser-centred scripts don't attract the audience's interest, either.

Cinema has some other archetypes to represent scientists. One of them is that of the scientist as a magician, as can be seen in the Russian film *Michurin*[11] (Alexander Dovjenko 1948). There is also the scientist as an artist, in *The Man in the White Suit*[12] (Alexander Mackendrick 1951). However, these archetypes and their interpretations are in the minority, and are not worth stopping for.

Cinema and Scientific Thought as 'Another Idea'

While *Madame Curie* is a film in which, according to film experts, a relatively positive image of science is presented, another important film about the role of science is *Inherit the Wind* (Stanley Kramer 1960). This tells about something as interesting as the importance of scientific thought in the face of religious fanaticism. Scholars of scientific cinema use it as an example of cinema in favour of science, but I don't agree. As in *Madame Curie*, in *Inherit the Wind* journalism and science come together. The film is based on fact: the Scopes 'monkey' trial held in the state of Tennessee in 1925, involving a young professor who had been dismissed from his job for teaching Darwin's evolutionary doctrines. If Marie Curie was media fodder through the rarity of being a woman and a scientist, in this case it was a far more important attribute: conflict.

A struggle between religion and science will always be more interesting from a journalistic point of view than scientific theory itself, because conflict makes people take sides, and that sells newspapers, while scientific theory alone does not generate media interest. This trial, called 'the trial of the century' by the press at the time, aroused so much passion that in Dayton, the small town where it was held, a landing strip had to be set up for the planes that arrived and left with the filmed images to be broadcast in cinema newscasts (television did not yet exist).

---

[11]Vance Kepley, Jr. (1980). 'The scientist as magician: Dovzhenko's *Michurin* and the Lysenko cult.' *Journal of Popular Film and Television*, 8 (2), 19–26.
[12]Robert A. Jones. (1998). 'The scientist as artist: A study of *The Man in the White Suit*, and some related British film comedies of the post-war period (1945–1970).' *Public Understanding of Science*, 7, 135–147.

The trial was followed not only by the American but by the European press. The Western Union phone company had to install 10 new phone lines. Associated Press covered the information with two special 24-h hotlines, and the trial was broadcast live on the radio via Chicago.

There are numerous studies of this trial, the proceedings of which have been reproduced and are kept by all US law schools. Interesting analyses have been published comparing the actual statements with those that later appeared in the film, speculating on the reason for the changes.

From the point of view of media culture, science and its decline, I would point out that almost all the experts on this trial agree that the film was almost pure invention. The maxim of the greatest sensationalist journalist of all time, William Hearst, who made the phrase famous, is true: 'Don't let reality spoil your good news: invent reality to make it news.' Knowing the history, we will also learn about the science–media relations.

Dayton was a small town with two main streets, a courthouse and several Protestant churches. Its way of life was agriculture. And both evolutionists and anti-evolutionists saw in the trial a chance for the development for the town, as it would become talked about across the country. Actually, science was the least of it. What is relevant is that this represented an issue that could attract media attention, because Darwin meant elevated passions, then as now.[13] If we compare this affair with Galileo's trial and condemnation by the Inquisition (Roman Catholics) in 1633 for his support of heliocentrism (and atomism), that sentence implied that Galileo retracted his scientific theories, which were based on facts and mathematics while those of the Inquisition were in the Bible. That these two cases have some similarity is a good way to see whether civilization has changed in the course of a few hundred years.

However, I would like the reader to reflect that, while it is true that science was used only as an excuse, to draw attention to the town, the important thing was that, thanks to science and the sensationalism of the media at the time—as in the Curie-Langevin case—the public had access to knowledge on scientists that would otherwise have gone unnoticed. All this is most interesting, showing that the phenomenon of the popularization of science and its relationship with journalism and culture cannot be analysed from a simple view on whether these media controversies are beneficial or harmful to science. The answer is not easy. I warn readers that science–media relations describe a phenomenon so rich and with so many ramifications that, in my view, it is one of the most fascinating areas of study in the twenty-first century.

---

[13]In 1999, the Kansas State Board of Education approved a ban on teaching Darwin's theory in schools, and there was a film of the same title—*The Legacy of the Wind*—made for television. The main character is played by Jack Lemmon, who received a Golden Globe for his performance.

But let's get back to the movie. Based on a play, the actors included Spencer Tracy and Gene Kelly. An analysis of the content of the film suggests that the script is favourable to scientists. At one point, a protagonist says: 'In a child's ability to master the multiplication table, there is more holiness than all your shouted hosannas and holy of holies.' But the stance on the defence of Darwin and scientific method shown in the film was not adopted because these are right; in the script, the teacher should be allowed to teach Darwin in school because 'the right to be different must be defended', in the words of Spencer Tracy, playing the defence lawyer. That is to say, the film opts more for a position of justification of science as freedom of expression than as defence of science as the truth. In fact, at the end there is a scene in which Tracy, who supposedly defends science, leaves court with a copy of *The Origin of Species* under one arm and the Bible in the other. I am no expert on this type of interpretation, but in my opinion this suggests that the two are equally valid, and that is unscientific.

A phrase in the film caused a certain scandal when, in one of the central sequences, Spencer Tracy storms: 'An idea is a greater monument than a cathedral. And the advance of man's knowledge is a greater miracle than any sticks being turned into snakes or a parting of the waters'.

Scientists as Film Advisors

We've seen how there's no way that cinema can treat science well, even when it tries to. Many believe that this problem would be solved if the cinema were to accept scientific advisors. I have my reservations, because the problem is that media culture is unscientific and its forms are contrary to science. Counselling never works. The relations between scientists and filmmakers have always been conflictive, especially because of the absence of a journalistic or cinematographic culture among scientists, also because filmmakers are not interested in science unless as a pretext for an argument. They don't believe in it any more than they believe in their fiction scripts. Scientific truth is not necessary for a film to succeed. The situation is therefore as follows: filmmakers do not know about science; and scientists do not know about cinema, because they are antagonistic worlds.

To illustrate that scientists barely understand how cinema is made, we will choose a film unknown to the general public: *The Beginning or the End* (Norman Taurog 1947). This film was intended to tell the story of the scientific adventures of the Manhattan project, which would lead to the atomic bomb. Alberto Elena maintains that this film 'constitutes one of the richest and most complex cases of interaction between science and cinema throughout history'. And the interaction was so bad that the film was a disaster in every way. 'The scientists knew as little about what it was like to make a film as we did about nuclear physics,' said the film's producer, Samuel Marx,[14] justifying the reasons for the resounding flop, which, in his opinion, was the result of the scientists' unwillingness to acquiesce with the script. It is curious that, according to Marx, 'they had far fewer problems with the military or politicians than with scientists'. In his opinion, the scientists were not interested in historical rigour so much as the success of their characters. I remind the reader that the Manhattan project ended in launching of atomic bombs on the Japanese cities of Hiroshima and Nagasaki. This project involved scientists of the stature of Albert Einstein, Robert Oppenheimer and Niels Bohr (famous among schoolchildren around the world for his atomic model). Bohr even hired a lawyer to ensure that the script of the film did not respect the truth so much as 'strictly the vision of his client'. It seems that Hollywood became fed up with scientists.[15]

The second problem that can arise in the relationship between scientists and filmmakers can be seen in a recent work by David A. Kirby. This demonstrates the influence on science itself by having scientists working as advisors to Hollywood as a way of obtaining resources and, moreover, of promoting their own theories, which are not always scientifically correct.[16]

Kirby explains that in 1998 scientists researching so-called NEOs (near-Earth objects that can impact on Earth) received far more research funding thanks to two films: *Deep Impact* (1997) and *Armageddon* (1998), which told what would happen after an asteroid/comet impact. In fact, building on the former film's success, NASA itself launched its *Deep Impact* mission in 1999, for a space probe to impact the comet Tempel 1 to establish its composition.[17]

Kirby's work shows how politicians in both the US Congress and the British parliament questioned scientists on whether the scenario described in the films could occur, and suggested that they submit research projects for funding.

---

[14]Samuel Marx. (1996). 'The bomb movie.' *Michigan Quarterly Review*, 35, 179–190.

[15]Although this book does not deal with this subject in depth, I do think it is useful to cite an interesting bibliography on the relationship between cinema and science and, in particular, on this film: Michael J. Yavenditti. (1978). 'Atomic scientist and Hollywood: The begining or the end.' *Film and History*, 8 (4), 73–88. Also, the compilation *Nuclear War Films* (1978), ed. Jack Shaheen (Southern Illinois University Press) and another book: *Expository Science. Form and functions of popularization.* (1985)., ed. Terry Shinn and Richard Whitley. Dordrecht/Boston: Reidel.

[16]David A. Kirby. (2003). 'Consultants, fictional films, and scientific practice.' *Social Studies of Science*, 33(2), 231–268.

[17]Impact took place on 4 July 2005, the anniversary of American independence. It is no wonder that the calculations were made for that day, to obtain the greatest media revenue from the mission. A similar circumstance occurred when in 2001 NASA programmed the NEAR spacecraft to hit the asteroid Eros (the Greek god of love) on Valentine's Day.

Scientists in a marginal area of astronomy were delighted with cinema, but is it good for science?

The two films reinforced the idea that devoting considerable resources to researching the asteroid phenomenon is a preventive measure to avoid catastrophe. The argument in the script was that if you knew soon enough that they were going to strike, you could try to destroy the asteroids with nuclear weapons before they reached our planet. This theory is much discussed in the scientific community, yet this was the view of scientists who acted as consultants to the films. Carolyn and Eugene Shoemaker, pioneers in astrogeology and discoverers of the comet Shoemaker-Levi, advised on *Deep Impact*, while Joseph Allen and Ivan Bekey, federal government advisors, worked on *Armageddon*.

Jane Gregory and Steve Miller,[18] renowned researchers in the social communication of science, after investigating earlier work on the popularization of science concluded, interestingly, that 'popularization is essentially an act of persuasion' (Gregory and Miller 1998, 85).

However, the clearest case of the use of cinema as a tool for the propagation of a scientific theory that in no way represents the consensus of the research community is that of *Jurassic Park* (Steven Spielberg 1993) and its two sequels, *The Lost World* (Steven Spielberg 1997) and *Jurassic Park III* (Joe Johnston 2001). The scientific advisor to these films was scientist Jack Horner, a well-known proponent of a theory, much criticized by some palaeontologists, which suggests that birds come from dinosaurs and not from other branches of reptiles. In the script of *Jurassic Park,* the character Alan Grant (the palaeontologist) reproduces exactly Horner's theories. I'll describe the scene, because I think it's all a treatise on persuasion. In the foreground, Grant begins his explanation while a complete fossil of a velociraptor is shown on a computer monitor, and both he and his assistant are in shot. The script reads:

GRANT: Look at the bones of this crescent-shaped joint. There's no doubt that these guys [the velociraptors] knew how to fly. (The assistant investigator laughs complicitly). It's true. It's true. Dinosaurs may have more to do with today's birds than with reptiles (at this point, the camera focuses the velociraptor image on the computer as a shot). Look at how the pubic bone transforms, just like it does in birds. Observe the vertebrae filled with air pockets and hollows, as also happens in birds. Even the word raptor itself means 'bird of prey'.

It is clear to the viewer that this is the true theory and that the other—with more scientific support at the time—is false. With the help of palaeontologist

---

[18] Jane Gregory and Steve Miller. (1998). *Science in Public: Communication, Culture and Credibility.* New York: Plenum Trade.

Horner, the animation team turns a dinosaur into a bird. However, to make matters worse, many other scenes reinforce the theory. In one, where a herd of gallimimus dinosaurs flee from a *Tyrannosaurus rex*, the character Grant exclaims in amazement: 'Look: they move with uniform changes in direction, just as flocks of birds do when they flee from a predator.

The film also looks at Horner's controversial hypothesis that dinosaurs were warm-blooded animals. And there are no other versions shown. All of this prompted the scientific community to react. Imagine how scientists critical of Horner's theories felt when they weren't called by Hollywood. Harvard University's famous palaeontologist, Stephen Jay Gould, criticized the film in the *New York Times Review of Books*.[19] Gould acknowledged, however, that the interpretations that appear were not 'scientific blunders' but scientific theories that compete with each other, and science does not yet have enough evidence to determine which is true. But giving just one is unscientific.

From my point of view, this issue of scientists as film consultants can be seen in the same way as a 'glass half full or half empty'. On the one hand, it boosts many discussions in the public domain. It is true that, if this practice continues, American scientists will be better able to expand their theories, given the supremacy of American cinema over others. On the other hand, the role of choosing which theories are disseminated to the general public would not lie with scientists but film directors and producers, who choose the version that best suits the dramatic record of the script, not one with more scientific evidence. And this is not healthy for science.

Filmmakers would also have the ability to select lines of research that are cinematically viable, which may result in increased funding for such research at the expense of others that are perhaps more scientific yet their results and methods less cinematic. The fact that these circumstances may occur is, in my view, another symptom of the decline of science and the rise of media culture.

However, many filmmakers present a smoother version of the issue. They suggest that, thanks to such films, science reaches an audience that might otherwise not be interested. And, although there are some crazy scientists in *Jurassic Park* who are resurrecting dinosaurs from the DNA of the mosquitoes that sucked their blood (and were fortunately trapped in amber), some children might become interested in science through such films. The problem is that there are no empirical data to prove this hypothesis of the filmmakers.

---

[19] Stephen Jay Gould. (1993). 'Dinomania.' *New York Times Review of Books*, 12 August, pp. 51–56.

Television Fiction Deepens the Discredit of the Scientist's Image

So far we've talked about cinema. However, everything that we have described can be applied to television, bearing in mind that most films are shown on television and that it has a much larger audience. Yet it is precisely because of this larger audience that it is so worrying that television fiction series are taking a step further than cinema in their battle to bury science and promote irrational thinking. One of the clearest cases is the *X-Files* (premiered in 1993, and with nine seasons), a television series with worldwide distribution in which science is represented as an 'intellectual corset' with the function of oppressing the true interpretations that, let us not forget, are related with irrationality, such as extra-terrestrial experiences or myths. Of course, it is not the first series or film to deal with esoteric or paranormal issues. It is true that in both the Harry Potter books and films there is an attack on scientific knowledge by equating the teaching system of magic and science. However, in the *X-Files*, the degradation of science and rational thought takes a much more sinister step: from its first series, of such success and scope, it portrays scientific theory and mystical versions as competing and equal.

The series is inspired by the Gothic novel and reproduces the idea of literary and cinematographic pairing of Holmes and Watson, but in a modern version with FBI detectives. At the end of the nineteenth century Arthur Conan Doyle (who was a writer, and also a doctor) had in Sherlock Holmes the potential of science to solve enigmas; at the end of the twentieth century the idea in *X-Files* was just the opposite. In the Fox–Mulder series, we have a gullible parapsychologist and a sceptical doctor with a background in physics. The cases that reach them are from FBI files classified with an 'X' as unexplained because they are, as they say in the script of the series, on the 'edge of reality'. The episodes range from extra-terrestrial abductions to genetic mutations or extrasensory phenomena. When, at the end of the episode, the 'X' of unexplained is resolved, it is the unscientific argument that wins. The creator of *X-Files,* Chris Carter, stated that 'Mulder and Scully have satisfied our need not only to reconcile rational unbelief with irrational credulity but also powerful fantasies about love and work.

Numerous scientists[20] have screamed their heads off. One of the most famous, Richard Dawkins, denounced this series as 'the latest example of a television programme in which science continues to be mocked by the idea that science is incapable of providing satisfactory explanations for extraordinary phenomena, as well as another example of the continuing efforts of television

---

[20]R. Dawkins. (1998). 'Science, delusion and the appetitive for wonder.' *Skeptical Inquirer*, 22, 28 r–35.

broadcasters to subordinate scientific and rational thought to mystical spec-ulation' (Dawkins 1998). Some television critics complained that 'in *X-Files* science was thrown out the window and the paranormal reigned supreme' (Goode 2002).[21]

If, at this point, the reader is not convinced that scientists are on one side and literary culture and mainstream media on the other, I would like to clarify that during its nine years of broadcasting this series, despite numerous letters from scientists to withdraw it, has received 17 Emmy awards, granted since 1949 by the American Academy of Television Arts and Sciences. The most painful thing is that many of the Emmys have been for 'the best drama series of the season'. The *X-Files* has also received five Golden Globes (three of them for the best series of the year), the awards given since 1944 by the foreign press accredited in Hollywood. However, I reiterate, the danger is that it puts science on an equal footing with magic and superstition.

Nor do gender studies agree on what Scully's character is supposed to be: Scully the scientist and Mulder the defender of the irrational and paranormal. What is clear is that Mulder's vision always prevails over that of his fellow scientist, who can rarely prove her irrational partner wrong. The feminization of the rational and of science in Scully's character should be read not as a step forward for feminism but the opposite, as an example of misogyny. Lisa Parks, Professor of Comparative Media Studies at MIT, points out that in *X-Files* scientific rationalism is open to questioning and criticism only when it is articulated through the female body. If Scully were a man and Mulder a woman, science would surely win the game (Park 1996).[22]

Television has also reproduced the parameters of cinema with respect to the archetype of the unfeeling scientist. In the hit series and worldwide distribu-tion *Star Trek*, which began its first season in 1966, Leonard Nimoy played the *Enterprise*'s science officer, Dr. Spock. Well, Dr. Spock came from the planet Vulcan, where, according to the script, inhabitants were characterized by hav-ing no emotions. For many scholars, Dr. Spock was the perfect archetype of scientist: rational and heartless. *Star Trek* has had six series and 10 films, making it the longest science fiction saga in the history of film and television.

In a cult series such as *The Big Bang Theory* (2007–2018), a group of physics friends—Sheldon, Leonard, Howard and Raj—who work at the California Institute of Technology (Caltech) are presented as brilliant in science but real misfits, far removed from the problems of today's society and, above all, with problems relating to other people, especially girls. The most 'normal' protago-

---

[21] Jane Goode. (2002). 'Why was the *X-Files* so appealing?' *Skeptical Inquirer,* September–October 2002.
[22] Lisa Parks. (1996). 'Special agent or monstrosity? Finding the feminine in *The X-Files.'* In Lavery, Hague and Cartwright (eds.). *Deny all Knowledge: Reading X-Files.* London: Faber & Faber.

nist of the whole comedy is, obviously—it could not be otherwise, in a product of media culture—the aspiring actress, Penny.

The series presents the equations that appear onscreen on the set as real, and these are advised by physicists. However, the problem lies not in the set but in the psychological profiles, lifestyle and even clothes attributed to the protagonists. The paradigm continues: if you are rational and like science, you will have a lonely and unhappy life: the protagonists in *Dr. House* (2004–2012) and in the entire sequel to the series *CSI (Crime Scene Investigation): CSI Miami* (2002–2012), *CSI New York* (2004–2014), appear to lead pathetic lives. The main character of the latter, Gil Grisson, is an entomologist who discovers a pattern of crimes through science. In return, he has a solitary life and complications in social relations: with couples, bosses and other protagonists.

In the sequel, *Bones* (2005–2014), the scientific protagonist, Dr. Brennan, also lacks a life partner and is described as a perfectionist, obsessed with the scientific method and reluctant to socialize. Her FBI partner, Seeley Booth, has been a sportsman, has natural empathy for everyone and calls Dr. Brennan's team 'brains'. A similar case is that of the extraordinary *Breaking Bad* (2008–2013), considered to be one of the best series of all time,[23] in which a brilliant but luckless chemistry professor, Walter White, decides to go over to the dark side and make loads of money manufacturing methamphetamine and other recreational drugs of better quality than his competitors without chemistry studies can produce.

When you read the profiles of science-loving characters on the television series' websites, you can draw the following conclusion either because they say so directly in the character descriptions or because they appear in their behaviours. All of them—from Sheldon in *The Big Bang Theory,* to Dr. House, Grisson in *CSI,* Dr. Brennan in *Bones* or the chemist in *Breaking Bad*—seem to suffer from Asperger's syndrome, a psychiatric disorder that complicates mixing in society and drives away happiness.

Sorcery to Save the World?

From the Age of Enlightenment, but especially during the nineteenth century and until the mid-twentieth century, science was perceived as the best way to explain reality and to fight against the superstitions and magic that had harmed humanity so badly. It was Francis Bacon's vision of science as progress. It was associated with freedom and prosperity in the Protestant countries—which

---

[23]M. Ryan. (2012). '*Breaking Bad*': Five reasons it's one of TV's all-time greats.' *Huffington Post,* http://www.huffingtonpost.com/maureen-ryan/breaking-bad-greatest-show_b_1665640.html (accessed May 2014).

took the new science as one of their hallmarks—in the face of the irrationality of the countries that admitted the Counter-Reformation, as practised in Spanish mysticism.

In 1565, Spain's St. Teresa (1515–1582) published *Book of Life*, in which she described her 'ecstasy' with an angel:

> I saw in his hand a long spear of gold, and at the iron's point there seemed to be a little fire. He appeared to me to be thrusting it at times into my heart, and to pierce my very entrails; when he drew it out, he seemed to draw them out also, and to leave me all on fire with a great love of God. The pain was so great, that it made me moan; and yet so surpassing was the sweetness of this excessive pain, that I could not wish to be rid of it. The soul is satisfied now with nothing less than God. The pain is not bodily, but spiritual; though the body has its share in it. It is a caressing of love so sweet which now takes place between the soul and God, that I pray God of His goodness to make him experience it who may think that I am lying. (St. Teresa 1565)[24]

This pure mysticism is also a form of culture, and can inspire works such as Bernini's impressive sculpture 'The Ecstasy of St. Teresa', one of the masterpieces of the Italian Baroque. The influence of mysticism in literature can even be traced to the magical realism of the twentieth century, whose greatest exponent is the Colombian Nobel Prize winner, Gabriel García Márquez. Around the same time, the Englishman Francis Bacon (1561–1626) laid the foundations of empiricism and the phases of the scientific method: observation, induction, hypothesis, hypothesis experimentation, demonstration or refutation (antithesis) of the hypothesis and scientific thesis or theory:

> Men have sought to make a world from their own conception and to draw from their own minds all the material which they employed, but if, instead of doing so, they had consulted experience and observation, they would have the facts and not opinions to reason about, and might have ultimately arrived at the knowledge of the laws which govern the material world. (Bacon 1620)[25]

The humanities, in their reduced vision of knowledge, have always felt more comfortable and fascinated by St. Teresa's writings, Bernini's sculpture of her 'ecstasy' and the literature of magical realism than by applying Bacon's method to discover how nature and the material world work. In the middle

---

[24] St. Teresa (1565). The *Book of my Life.*Teresa of Ávila (1515–1582) is one of the most beloved of the Catholic saints. In 1562, during the era of the Spanish Inquisition, Teresa sat down to write an account of the mystical experiences for which she had become famous. The result was this book, one of the great classics of spiritual autobiography.

[25] Francis Bacon. (1620). *Novum Organun Scientiarum.*

of the twentieth century, it was thought that the humanities would decline, because without science they could not explain the world. Snow himself, as I have already quoted, referred in his book *The Two Cultures* to the fact that no one who is moderately intelligent would want to study the culture of literature. However, what Snow didn't count on was that those with a humanities background would use two products of physics and chemistry—film and television—to take revenge.

Now, with our media culture, which stems largely from the humanities of theological and mythological reverence, we are beginning to go the other way: may the scientist die and the sorcerer live! Because, while until now film and television have portrayed scientists as heartless, heartless, and as people who endanger others and the entire planet to meet their personal needs, they are beginning now to represent sorcerers and magicians in a new way. Besides *Harry Potter*, which I have already mentioned, and *X-Files,* which put the pseudosciences on the same level as science, there is an even worse example that I would like to end this chapter with the *Charmed* series (1998–2006, produced by Aaron Spelling, with a reboot in 2018). Through each innocent episode, we see that while the actors who represent scientists are rarely attractive and may even be physically repugnant (as in *Back to the Future, The Nutty Professor* or Dr. Spock himself), sorcerers are extraordinarily attractive. The three sister witches in the film and the sorcerers who accompany them in each chapter are good looking. On the website for this series created by Aaron Spelling's production company, one of the most important of all time, you can read the following:

> *Charmed* chronicles the conflicts and the love among three vastly different sisters and their need to come together to fulfill an ancient witchcraft prophecy. The Halliwell sisters were always different, not just from the world around them, but also from each other. Though bound together by blood and destiny, these twenty-something women could not have been more dissimilar. Yet the discovery that the three sisters were powerful good witches, foretold as the Charmed Ones, created a bond that reached far beyond petty sisterly grudges, and they banded together to keep their otherworldly witchcraft a secret and vanquish all comers. After seven years of protecting the innocent, demon fighting and generally ridding the world of evil, the Halliwell sisters have begun a search for a different destiny—living normal lives free of fighting demons. At the end of last season, the sisters appeared to sacrifice themselves to rid the world of the evil Zankou.[26]

That is to say, in our media culture the scientists try to awaken monsters or evil, while the sorcerers try to free us from them. According to this view,

---

[26] http://www.tv.com/charmed/show.

scientists try to separate themselves from society and become different and isolated people, while sorcerers want to lead a normal life. While scientists have no objection to endangering the planet to slake their thirst for knowledge or power, magicians and sorcerers are able to sacrifice themselves and may die to save the world from danger. With this background, how can anyone still wonder why our science departments are empty? However, in 2006, *Charmed's* website, which has many fan clubs, collapsed when it was rumoured that the series was entering its final season.

Perhaps this new concept of the twenty-first century should be further explored: magic and superstition as possessing something with a greater scope than science, even with regard to how the physical world is explained.

The 'Despised' Science

The amount of scientific information on television has significantly decreased in Spain, for example. Another trend is apparent: advance payments. In the digital terrestrial television model, there are several channels are broadcasting exclusively tarot and magic: one of several is Astrocanaltv. Never before has anything like this been seen on Spanish television. No one—neither politicians nor scientists—has protested. What is shown on cinema and television is the only view of the world to which the vast majority of public opinion has access. Science is taught in schools, but not what it is like to be a science professional. And even if schools did teach it, they could never compete with the media culture. Young people despise the scientific professions, and this will have serious consequences for the economies of those countries where there are few jobs.

Writer Alberto Vázquez Figueroa, who was a correspondent during the time of Africa's decolonization, predicted that poverty would never be lifted there because its universities were filled with lawyers rather than engineers or scientists. Time has proved him right. Africa is still in misery, while China is experiencing unstoppable economic growth, which—I believe as a result—coincides with its spectacular increase in the number of its graduates in science and engineering. Cinema and the Western media culture have barely entered China.

The Western world can accept the loss of its economic leadership, but not its scientific and rational thinking. In September 2010, *Nature's* editorial, entitled 'Science Scorned', observed: 'The anti-science strain pervading the Right wing in the United States is the last thing the country needs in a time of economic challenge', that anti-science was taking over the political ideology of the American Right wing.[27] The editorial was highly relevant and proved to be far-seeing, since in 2016 Donald Trump, who represented this trend, was elected as president. I think it is interesting to reproduce it in its entirety:

'The four corners of deceit: government, academia, science and media. Those institutions are now corrupt and exist by virtue of deceit. That's how they promulgate themselves; it is how they prosper.' It is tempting to laugh off this and other rhetoric broadcast by Rush Limbaugh, a conservative US radio host, but Limbaugh and similar voices are no laughing matter. There is a growing anti-science streak on the American Right that could have tangible societal and political impacts on many fronts – including regulation of environmental and other issues and stem-cell research. Take the surprise ousting last week of Lisa Murkowski, the incumbent Republican senator for Alaska, by political unknown Joe Miller in the Republican primary for the 2 November midterm congressional elections. Miller, who is backed by the conservative 'Tea Party movement', called his opponent's acknowledgement of the reality of global warming 'exhibit 'A' for why she needs to go'. 'The country's future crucially depends on education, science and technology.' The Right-wing populism that is flourishing in the current climate of economic insecurity echoes many traditional conservative themes, such as opposition to taxes, regulation and immigration. But the Tea Party and its cheerleaders, who include Limbaugh, Fox News television host Glenn Beck and Sarah Palin (who famously decried fruit fly research as a waste of public money), are also tapping an age-old US political impulse – a suspicion of elites and expertise. Denialism over global warming has become a scientific cause célèbre within the movement. Limbaugh, for instance, who has told his listeners that 'science has become a home for displaced socialists and communists', has called climate change science 'the biggest scam in the history of the world'. The Tea Party's leanings encompass religious opposition to Darwinian evolution and to stem-cell and embryo research – which Beck has equated with eugenics. The movement is also averse to science-based regulation, which it sees as an excuse for intrusive government. Under the administration of George W. Bush, science in policy had already taken knocks from both neglect and ideology. Yet President Barack Obama's promise to 'restore science to its rightful place' seems to have linked science to liberal politics, making it even more of a target of the Right. US citizens face economic problems that are all too real, and the country's future crucially depends on education, science and technology as it faces increasing competition from China and other emerging science powers. Last month's recall of hundreds of millions of US eggs because of the risk of salmonella poisoning, and the *Deepwater Horizon* oil spill, are timely reminders of why the US government needs to serve the people better by developing and enforcing improved science-based regulations. Yet the public often buys into anti-science, anti-regulation agendas that are orchestrated by business interests

[27]'Science scorned', Editorial. *Nature*, 467, September 2010.

and their sponsored think tanks and front groups. In the current poisoned political atmosphere, the defenders of science have few easy remedies. Reassuringly, polls continue to show that the overwhelming majority of the US public sees science as a force for good, and the anti-science rumblings may be ephemeral. As educators, scientists should redouble their efforts to promote rationalism, scholarship and critical thought among the young, and engage with both the media and politicians to help illuminate the pressing science-based issues of our time. (*Nature,* 467, 133, 9 September 2010)

The image with which opinion leaders such as Rush Limbaugh indoctrinate anti-science followers on television comes from the media representation of scientific thought. In turn, this comes from university teaching on Western literature and philosophy from which science is barred. The philosophers and sociologists of science have become the Jesuits and Dominicans who had so much power in the Spanish Inquisition. They made our kings mad about religion and outlawed science: just look at the difference between Philip II's Spain and Queen Elizabeth I's England. The problem is more serious than scientists think, and it is their responsibility to fight this crisis that threatens to destroy Western culture. This culture is based not only on Homer, but on Euclid and Archimedes.

9

Publish or Perish

Science, as well as method, is above all communication, with all the positive and negative aspects that this entails. When I am asked how I was able to study two disciplines as different as chemistry and journalism, I always answer: Why are they so different, if they both aspire to the same objective, to seek the truth and make it public? And, in addition, both professions demand great curiosity.

Curiosity was a mortal sin in the West until the eighteenth century. The Bible contained all the answers and, if they were not in it, then the question was not worth asking. Science historian Philip Ball, in his book *Curiosity. How Science becomes Interested in Everything*,[1] points out that the history of science is the history of curiosity, and that the transgressive aspect of this quality is a permanent theme in Christian theology. 'Time and again the student of the Bible is warned to respect the limits of enquiry and to be wary of too much learning (Ball 2012, 11). Basically, this is the same idea as held by many intellectuals of literature-based subjects and the social sciences, who are still against scientific progress today: 'The secret things belong to the Lord our God', it is proclaimed in Deuteronomy. Solomon cautioned: 'with much wisdom comes much sorrow; the more knowledge, the more grief.' Then he said: 'Do not pry into things too hard for you. Or examine what is beyond your reach.... What the Lord keeps secret is no concern of yours. Do not busy yourself with matters that are beyond you.'

Ball points out in his book that this aversion to curiosity as an impulse to know more than is good for you was not started by the Christian world, since Socrates is attributed to saying, 'We should not concern ourselves with things above'. According to Ball, however, Christianity established its robust moral

---

[1] Phillip Ball. (2012) *Curiosity. How science become interesting in everything*. London: Random House.

© Springer Nature Switzerland AG 2019
C. Elías, *Science on the Ropes*, https://doi.org/10.1007/978-3-030-12978-1_9

basis (Ball 2012, 13). Augustine's injunctions were repeated by the twelfth-century Cistercian theologian, St. Bernard of Clairvaux, for whom curiosity was 'the beginning of all sin'. According to him, Lucifer fell due to curiosity, as he had turned his attention to something that he coveted unlawfully and had the presumption to believe that he could gain. As a result, St Bernard said: 'The seraphim set a limit to impudent and imprudent curiosity. No longer may you, Satan, investigate the mysteries of heaven or uncover the secrets of the Church on Earth.'

To obtain the secrets of the heavens and the Earth—in short, of nature—has been humanity's greatest intellectual achievement, because it has had to fight against its own cultural tradition. In honour of the truth, it has to be said that many religious groups, especially the Dominicans, have contributed to scientific developments, from Thomas Aquinas in Oxford, Albertus Magnus in Cologne, Roger Bacon to William of Ockham, among others.

The Bible does not explain how matter is made up, why leaves fall in autumn or why, when a liquid is heated up in a saucepan, there comes a time when steam lifts the lid. These questions were absurd to the clergy (who dominated the universities) until the seventeenth century. And they were the guardians of culture. A question as foolish as why steam produces motion is the origin of thermodynamics and the industrial revolution. How matter is constituted is the origin of physics, chemistry and even biology. But these questions remain irrelevant to the literature-based or social science 'intellectuals' in the twenty-first century, who still believe that the history of film or mythology is more relevant to their university curricula than chemistry, biology or thermodynamics. They still regard science or technology as in opposition to the ideas of Renaissance humanists. This is as if the subjects hadn't been devised by humans to explain who we are and where we're going. But the scientific explanation, as the historian Margaret Jacob rightly points out in her magnificent book *The Cultural Meaning of the Scientific Revolution*,[2] is what sets Western culture apart from the rest:

> The science that became an integral part of our culture by the middle of the eighteenth century rested on certain philosophical and mathematical innovations that in very general terms can be dated precisely, from the publication in 1543 by Copernicus of his *De revolutionibus orbium coelestium* ('On the Revolutions of the Heavenly Orbs') until the publication in 1687 of Newton's *Principia* (*Philosophiae Naturalis Principia Mathematica*). This science was a very different science from that we found in other cultures (now or then), and it was largely on the actual observation of visible bodies in motion in the heavens and on Earth. It requires that the results of observations be described largely according to mechanical

[2] Margaret Jacob. (1988). *The Cultural Meaning of the Scientific Revolution*. New York: Knopf.

principles (that is, contact action between bodies) as well as mathematically. (Jacob 1988, 3)

All this is true, but the great triumph of modern science is its publication and public discussion. In fact, as Jacob refers, the milestones are not when the findings are discovered (Copernicus waited until the last months of his life to publish) so much as when they are published.

Journalists also need to be curious to ask questions that, in principle, do not seem relevant, as they can change the worldview. And, in both professions, unless what is researched is published, it is useless. Possibly there were deeply interesting discoveries in ancient Egypt or among mediaeval alchemists, but only a few have reached us. One of the great findings of modern science is to publish the results and, above all, to ensure that what is published is true and, at the same time, to encourage debate on these findings, whether books, journal articles, personal blogs on the internet or whatever. The principle of making knowledge public has been fundamental to the economic and cultural rise of the West. In her book, Jacob devotes a whole chapter, 'Scientific education and industrialization in Continental Europe', to describing the results publication cycle, the teaching of these results to the population and the industrialization of these countries (and their consequent economic potential).

One of the products of this system is the scientific journal. It is a complex scene that includes science, communication and business capitalism. We are witnessing an increase in the number of articles and the need to publish them, even though the articles are not very good.

There is another component: scientific journals are an important business since the authors (the scientists) not only never charge but often actually have to pay to be published. This element is also reaching the humanities. There is a great deal of debate about this publishing system and whether it may be damaging to science.

This debate runs constantly, and not only among academics: it has even reached the media. An interesting analysis was published in June 2017 by *The Guardian* under the title 'Is the staggeringly profitable business of scientific publishing bad for science?'[3]

The way to make money from a scientific article looks very similar, except that scientific publishers manage to duck most of the actual costs. Scientists create work under their own direction – funded largely by governments – and give it to publishers for free; the publisher pays scientific editors who judge whether the work is worth publishing and check its grammar, but the bulk of the editorial

---

[3] Stephen Buranyi. (*The Guardian*, 2017). https://www.theguardian.com/science/2017/jun/27/profitable-business-scientific-publishing-bad-for-science.

burden – checking the scientific validity and evaluating the experiments, a process known as peer review – is done by working scientists on a volunteer basis. The publishers then sell the product back to government-funded institutional and university libraries, to be read by scientists – who, in a collective sense, created the product in the first place. It is as if the New Yorker or the Economist demanded that journalists write and edit each other's work for free, and asked the government to foot the bill. (Buranyi 2017)

In an even harder-hitting article in the same newspaper in 2011, journals had been directly identified as one of the serious problems of science. Under the title 'Publish-or-perish: Peer review and the corruption of science',[4] the journalist observed that 'Pressure on scientists to publish has led to a situation where any paper, however bad, can now be printed in a journal that claims to be peer reviewed'. He stressed the serious issue behind it all: the lack of reviewers to undertake the work free of charge. Because that's what it's all about: working for free, for the love of science:

> Peer review is the process that decides whether your work gets published in an academic journal. It doesn't work very well any more, mainly as a result of the enormous number of papers that are being published (an estimated 1.3 million papers in 23,750 journals in 2006). There simply aren't enough competent people to do the job. The overwhelming effect of the huge (and unpaid) effort that is put into reviewing papers is to maintain a status hierarchy of journals. Any paper, however bad, can now get published in a journal that claims to be peer reviewed. The blame for this sad situation lies with the people who have imposed a publish-or-perish culture, namely research funders and senior people in universities. To have 'written' 800 papers is regarded as something to boast about rather than being rather shameful. University PR departments encourage exaggerated claims, and hard-pressed authors go along with them. (Colquhoun 2011)

The report describes some examples not only of bad scientific work that has been published, but also of clearly unscientific work or pseudosciences such as acupuncture, and recalled that 'The US National Library of Medicine indexes 39 journals that deal with alternative medicine. They are all "peer reviewed", but rarely publish anything worth reading. The peer review for a journal on homeopathy is, presumably, done largely by other believers in magic. If that were not the case, these journals would soon vanish'.

Richard Smith, 13 years the editor of the *British Medical Journal* (*BMJ*) and author of a book describing his experience, *The Trouble with Medical Journals* (Royal Society of Medicine Press 2006) describes his experience in this regard:

---

[4]David Colquhoun. (*The Guardian*, 2011). https://www.theguardian.com/science/2011/sep/05/publish-perish-peer-review-science.

When I first arrived at the BMJ in 1979 we probably didn't use more than a hundred reviews altogether, and there was no database of reviewers. Initially I would have to ask the senior editors to suggest somebody. The *BMJ* when I left had a database that included several thousand reviewers from all over the world, most of whom the editors didn't know. (…) One result that emerges of all studies on who makes the best reviewers is that those under 40 do better than those over 40, even after allowing for the fact that younger people often have more time. (Smith 2006, 86)

And what would be the solution? In 2018, *Nature,* along with *Science and Cell,* is the most prestigious in the world and an article in it may secure a tenured position in a university or funding for scientific projects. It launched a portal to which scientists could upload their papers and ask readers to review them. All for free. It was a failure. There were two major problems. First of all, there were few people willing to spend their time reviewing articles (in the usual journals, even if you aren't paid being a reviewer may help to achieve certain academic positions). But above all, there was a fundamental problem that shows what the academy is like: obviously, in an open system, the process could not be anonymous. And if it wasn't anonymous, no one wanted to undertake a review and thus get into trouble: the senior scientists didn't want to criticize their colleagues and, of course, the younger people weren't about to criticize their seniors, who would be needed to recommend them for better jobs in universities and research centres. Few in the academy were willing, if they were to be identified.

Not only did they fail to put themselves forward: almost everyone refused to participate when asked directly. There was an attempt to make them anonymous reviewers, but the same problem came back: there was no time. Hundreds of journals and hundreds of thousands of articles are published every year. Many with the pompous prefix 'The International Journal of…', as he liked to call his journals, were owned by Robert Maxwell (1923–1991), the powerful editor of Pergamon Corporation. According to *The Guardian*'s report, 'by 1959, Pergamon was publishing 40 journals; six years later it would publish 150. This put Maxwell well ahead of the competition'. By 1960, Maxwell had taken to being driven in a chauffeured Rolls-Royce, and moved his home and the Pergamon operation from London to the palatial Headington Hill Hall estate in Oxford, also home to the British book publishing house, Blackwell's. Scientific societies, such as the British Society of Rheology, seeing the writing on the wall, began letting Pergamon take over their journals for a small, regular fee. (Buranyi 2017). Maxwell regarded scientific journals as 'a perpetual financing machine'.

It is interesting to note that although Maxwell is known as one of the first communication entrepreneurs to make scientific journals profitable, he is known better as the sensationalist media entrepreneur who exercised power. In journalism history classes, it is common to recall Maxwell's war with the other great media mogul, Rupert Murdoch, over control of the British tabloid, *News of the World*. Murdoch finally secured it (in 1969), but Maxwell invested his profits in his scientific journal publishers (all of them prestigious in their fields) and the purchase of the other British tabloid, *Daily Mirror*. Maxwell was an unscrupulous businessman who used the sensationalist press to vilify those who opposed him. He died in strange circumstances in the Canary Islands in 1991. He saw the enormous potential of scientific journals at all times: during the Cold War he founded numerous journals on nuclear physics, then saw the growth of the social sciences and dedicated himself to the founding of social and humanities journals. He also managed to get all scientists (including Soviet scientists during the Cold War) to publish in English, and invented the concept of 'international visibility'. Even the Japanese scientists gave him their journals for publication in English. Finally, the multinational Elsevier bought Pergamon Press. In 2016, Elsevier published 420,000 papers, after receiving 1.5 m submissions.

But what I most like to tell my science journalism students is the fascinating relationship between the man who saw science journals as sound business and a means of global visibility and, at the same time, the owner of one of the most sensationalist newspapers in the world. He bought the *Daily Mirror* with the money he had made from the communication of scientific results in scientific journals, supposedly the most rigorous type of communication of all. This tabloid was his most precious jewel and the one with which he wielded true power; in fact, many consider him to be Citizen Kane of the twentieth century.

The journals weren't always so highly relevant. Until the nineteenth century (and even in the twentieth century), scientists also wrote books, and the copyright on these books was a complementary source of income. From Copernicus to Galileo, from Dalton, Humboldt or Darwin, the great history of science has been told in books, many of which, such as Galileo's *Dialogue on the World's Two Greatest Systems* (1632), Newton's *Principia Mathematica* (1687), Humboldt's *Cosmos* (1845) and Darwin's *The Origin of Species* (1859), were authentic bestsellers that also radically changed our thinking. Perhaps the last great science book to appear is *The Nature of the Chemical Bond* (1939), by the great chemist Linus Pauling. This is a brilliant work that applies quantum mechanics to chemistry, and it is why this science went from being the ugly duckling (especially when compared to physics) to a swan, as currently, as chemistry is the central science that explains everything. Pauling was awarded the Nobel Prize in chemistry for this work (he also received the Nobel Peace Prize), and most of it was published in that book (a kind of research treatise and textbook) and for which he achieved fame and, I imagine, quite a fortune from the publisher through a copyright. Within a few years, *The Nature of Chemical Bonds* had been quoted thousands of times; it is possibly one of the most cited works in the history of science.

Although the books on physics or biology are the best known, chemistry has also had success: from John Dalton and his *A New System of Chemical Philosophy* (1808, 1827), in which he recovers the notion of matter from the Greeks or Romans (Democritus or Lucretius) with the vision of new experimental physics, to the textbook that Mendeleev wrote for his students, *Principles of Chemistry* (1869). This is where the famous periodic table of the elements appeared for the first time. Geology as a science owes a great deal to Alfred Wegener (1880–1930), who proposed that the continents in the geological past were united in a supercontinent named Pangaea, which later would have been broken up by continental drift. His book *Entstehung der Kontinente und Ozeane* (1915) ('The Formation of Continents and Oceans') was the first to propose plate tectonics. And let us not forget medicine, from Vesalius and his *De Humani Corporis Fabrica* ('On the Structure of the Human Body') (1543) to 'The Interpretation of Dreams' (*Die Traumdeutung* 1900) by the psychiatrist Sigmund Freud. Would these disruptive works have passed the filter of the reviewers of scientific journals? We don't know. Einstein's theory of relativity was published in *Annalen der Physik* in 1915 in a famous article '*Die Feldgleichungen der Gravitation*, a compendium where he presented the system of 10 mathematical equations that were to change the world of physics and, with it, technology and society. He had no co-author.

The Effect of the Media on the Impact Index: Media Science

The vanity of some scientists can also affect the public communication of science and even science itself. I do not criticize vanity, because often this 'sin' is a good stimulant for intellectual production. I have always believed that vanity is another aspect that unites the scientist and the journalist. Because, let's be honest, journalists aren't motivated by altruism towards society so much as for being on the front page of the newspaper or in Prime Time news and showing off. What happens is that, usually, along the way, you can do a good service to society.

The scientist also has his vanity. He is motivated to publish in *Nature* to boast about it and, ultimately, to achieve the glory or immortality. And, normally, to achieve this, he has to undertake good science, which society also benefits from. I therefore believe that professional vanity is not necessarily detrimental to journalism or science, although it must be said that companies or academia rely on this vanity to pay low salaries (or even no salary) to both scientists and journalists.

The man who for many, including the writer of this book, is the greatest scientist in history, Newton, was so vain that he published his optics research long after he discovered it. The reason was not to spend time reviewing it for methodological flaws. It was due to something much more human: he did not want to refer the earlier contribution of his great rival, Hooke, in his research. So, he waited for Hooke to die. That way, Hooke couldn't object if there was no mention. But science lost nothing by it. Besides, time puts everyone in their place, and today Hooke's contributions to the study of light are known and valued to the fullest extent possible.

But of course, in Newton's time the mass media did not dominate society. In our current media culture, if we combine all these elements a new phenomenon may appear that could harm science and contribute to its decline: 'media science', or science designed to be in the media. The origin of the phenomenon lies in a curious fact: scientific articles published by scientific journals that, in addition, are massively disseminated by the media are cited much more by scientists themselves than those simply published in a prestigious journal.

This circumstance was demonstrated in 1991 by a study in the *New England Journal of Medicine*.[5] Its authors (Phillips et al.) analysed articles published in that journal in 1978 and 1979 to compare those that then appeared in the *New York Times* with those that did not. They noted that, in the year following their publication, the studies that had appeared in the US newspaper were cited 72.8% more than those that did not. Most importantly, this significant citation gap persisted for at least 10 years after their appearance in the newspaper, and we know what citation and impact indices mean to a scientist's academic career.

Phillips et al.'s study included a well-crafted check to assess the journalistically reasonable assumption that the *New York Times* was providing information on only the most significant scientific findings, which would obviously have been cited more often. But Phillips and his colleagues were lucky, or they had designed the methodology extremely well, because the *New York Times* had suffered a three-month strike (in 1997) during the period studied. During that strike, the newspaper had produced issues that remained as a 'record edition',

[5] David P. Phillips, Elliot Kanter, Bridget Bednarczyk and Patricia Tastad. (1991). 'Importance of the lay press in the transmission of medical knowledge to the scientific community.' *New England Journal of Medicine*, 325, 1180–1183.

none of which were distributed to the public. The authors of the paper were able to analyse this 'record issue', thus it was clear to them which articles the *New York Times* had considered worthy of coverage yet had not actually been distributed.

The study showed that the articles that were worthy of publication yet were unpublished because of the strike did not result in an increase in the number of citations in the subsequent year. Thus, it was demonstrated that a scientific article that is mentioned in the quality press is cited 78.2% more, regardless of the quality of the research. These results are in line with another study that proved that 60% of Wisconsin Medical School members hear about new scientific discoveries through the media.[6] Other research (in 1970) gave an even higher percentage for North Carolina physicists: 89% of them reported in a survey that they learned about new scientific advances that are outside their field from the media.[7] And another study (in 2014) shows 'that public communication, such as interactions with reporters and being mentioned on Twitter, can contribute to a scholar's scientific impact. Most importantly, being mentioned on Twitter amplifies the effect of interactions with journalists and other non-scientists on the scholar's scientific impact (Liang et. al. 2014)'.[8]

There is, therefore, another implication: it is scientists themselves, not journalists, who are most interested in disseminating their research work in the quality press and, of course, on television or on social networks, and this is for professional promotion. This means that there is a relationship between scientific journals and the mass media besides the figure of Robert Maxwell who, as I have said, used the benefits of scientific journals to invest in sensationalist newspapers; there are relations exclusively of communication flow between them.

As the impact index of a journal is measured by how many times its articles are cited in other publications, if a journal is able to place articles in the press it knows that in the following year it will obtain a higher SCI index. This means that the best scientists will want to publish in the journal, so this circular dynamic will be enhanced. This is the Matthew effect: 'For to everyone who has it shall be given, and he shall have more; and from him who has not it shall be taken away even that which he has.'

The distortion occurs because only *Nature* and *Science* and some medical journals have the professional press offices that really know how journalism

---

[6]D. L. Shaw and P. van Nevel. (1967). 'The informative value of medical science news.' *Journalism Quarterly*, 44, 548.

[7]M. T. O' Keefe. (1970). 'The mass media as sources of medical information for doctors.' *Journalism Quarterly*, 47, 95–100.

[8]Liang et al. (2014). 'Building buzz (scientists) communicating science in new media environments.' *Journalism and Mass Communication Quarterly*, 91, 4, 772–791.

works. As a result, these journals achieve a higher rate of scientific impact every day and the rest are inevitably left in much lower positions. *Nature* went from 28.80 in 1998 to 32.12 in 2004 and 43.77 in 2017. One of the causes may be the huge explosion of scientific journals and, of course, of published articles. In 2014, the San Francisco statement 'Ending the tyranny of the impact factor' was published in *Nature Cell Biology*, in which scientists considered that the impact[9] should be evaluated in a more holistic way, and not just the impact factor. If this happens in the natural sciences, it is obvious that the social sciences and the humanities should likewise have metrics that are not the impact of articles, instead perhaps a social impact factor (an appearance in the media, generation of social discussion, citizen mobilization, etc.).

From my point of view—and I say this from my experience as a science journalist—*Nature* and *Science* (and also *NASA*) do well in communicating scientific results to the mass media (NASA space agency is not a journal but a world hegemonic scientific source'.[10] Someone may argue: So what? Let the rest of the journals get their act together and perform like *Science* and *Nature*. However at this stage it is no longer easy, because the media and scientists consider these to be the best, therefore the authors of the best work will try to be published in them, because they know that it is an indisputable sign of quality. As the best work appears in them, journalists may be confident when selecting their press releases. In order for the reader to understand the scope of what these two journals mean at the moment, in the international ranking[11] of the 500 best universities in the world, developed since 2004 by the Shanghai Institute of Education, the criteria for evaluating universities are: the number of Nobel laureates who teach (or have taught) in them; the number of Nobel Prizes that have been trained in them; and, finally, the number of articles published by their professors in *Nature* and *Science*. In other words, you have to publish in these journals to get into the ranking.

It should be clarified that these journals are generalist and that their main objective is, of course, to publish the best science, as well as having the best impact index. This means that sometimes the prevailing criteria for news items are to select those that yield good press releases, not scientific articles. I am aware that it is a risky hypothesis that I propose here, so I will try to illustrate it with some examples. I have selected several years of random tasting to show that the phenomenon is not accidental, and relatively recent.

---

[9] http://www.nature.com/articles/ncb2905 (*Nature Cell Biology* 16, 1, editorial).

[10] Carlos Elías (2011). 'The future of journalism in the online public sphere: When journalistic sources become mass media in their own right.' *Estudios sobre el Mensaje Periodístico*, 16, 45–58.

[11] In this ranking, the first university in the world is Harvard. The best European university is University of Cambridge and the best Spanish is the Autonomous University of Madrid, ranked 150th in the world and 60th in Europe.

The first example could be the article published by *Nature* on 4 January 1996 on the analgesic effects of myrrh.[12] The article was of minimal scientific relevance, but it was published in the week that the festival of the Three Wise Men is celebrated. The *Nature* study had a typically descriptive title for the scientific articles—'Analgesic effects of myrrh'—yet the press release ran, 'Why the Three Wise Men gave myrrh'. Obviously, the item was picked up by many media, because it was so perfectly suited to the week's news, first by the newspapers and then, as is often the case in science journalism, it was reported by radio and television.

In 1998, *Nature Neuroscience* had a surprising story about the 'cells of the human soul', published, among others, by the *Sunday Times* and the Spanish newspaper *El Mundo*.[13] Just one month earlier, on 13 February, *Nature* had published a study with the following conclusion: 'The preference in kissing couples for turning the head to the right.'[14] However, that was just the conclusion of *Nature*'s press office, because the actual article, which was in the category of 'short communication' not a full paper, was entitled: 'Adult persistence of head-turning asymmetry'.[15]

However, the press office was not entirely to blame for the misrepresentation. The research methodology quoted in the scientific article itself stated: 'I observed kissing couples in public places (international airports, large railway station, beaches and parks) in the United States, Germany and Turkey…'. This is absolutely terrible, because the image that the scientist is offering to society is that he is really a voyeur, a peeping Tom with a science pathology, the archetypal mad scientist with no affections that the cinema transmits. The perception of a young people—or not so young—who hear the news—is that real people kiss for real, but scientists are only there to watch others kiss, not themselves to kiss. That's bad enough but, at worst, they could equate a scientist with a sexually deranged man.

Even if that research were serious, I believe that I would not have included it in an academic journal. But not only was it included, but it was the star item selected by the press office (and by the media worldwide) for that week's issue (421). Among other topics, this contained research on autoimmunity, quantum gravity, an analysis of protein movements in cells, the origin of carnivores in Madagascar, the role of interleukin-23 in brain inflammation and the effects of European colonization on the coral of the Australian Great Barrier. Normally,

---

[12] Piero Dolara et al. (1996). 'Analgesic effects of myrrh.' *Nature*, 397, 29–29.

[13] *El Mundo*, 12 March 2003, 34.

[14] Published and disseminated by the EFE Agency. Database key: ela/jm/euo (United Kingdom-Science. 13.02.2003. time: 09.47. category: company.

[15] Onur Güntürkün. (2003). 'Human behaviour: Adult persistence of head-turning asymmetry.' *Nature*. 421, 711.

there is room for just one science story a day in a newspaper or newsletter, yet the one that took that space was 'kissing on the right'. Quite possibly, that researcher will have better opportunities for the next 10 years and secure more funding for competitive projects. This is a perverse effect by science whose origin is media culture.

Whose fault is it? The journalists, who simply follow what researchers or scientist assure them, that using these articles will help with the CVs and prestige among their colleagues? Recall that this widespread coverage of irrelevant topics and their increasing citations after they are published in the press will lead many scientists in future to study side issues in science, as these are much more newsworthy than basic science. Hundreds of scientists around the world who research the analgesic effects of various substances will mention in their references the one on myrrh published by *Nature*, so that an article that, in principle, is irrelevant will become important and its authors will gain merit under the current system of evaluation of researchers that prevails around the world.

The editors of dominant journals themselves acknowledge that, because of this dictatorship, they often publish research that is not scientifically but methodologically relevant. In fact, Richard Smith in his memoirs[16] about his 25 years of experience—13 as director—with one of the most prestigious medical journals, the *British Medical Journal*, acknowledges that, many times, it is not the strictly scientific criteria that prevail:

> I remember a debate over a paper we published in our Christmas issue, an issue traditionally devoted to slightly strange, wacky and amusing material. A long-term follow-up of a population in Wales showed that men aged 40 who had 50 orgasm a year lived longer than those who had fewer.[17] There would inevitably be doubts over the data. How honest are people about their sex lives? Perhaps the facts that the men had more orgasms was simply a maker of other characteristic that made them likely to live longer. Or maybe healthier men were capable of more orgasms. This study was neither medically nor scientifically important, but it would interest our readers a get a lots of media coverage. I decided that we would publish the study - and it did get lots of coverage. (Smith 2006, 181)

Due to the increasing competitiveness to which they are subjected and contaminated by our journalistic culture, journalists will cause other scientists to desist from investigating tedious questions with little prospect of becoming news, even though they are relevant from the point of view of the scientific

[16]Richard Smith, (2006). *The Trouble with Medical Journals*. London: Royal Society of Medicine Press.
[17]G. Davey Smith, S. Frankel and J. Yarnell. (1997). 'Sex and death: Are they related? Finding from the Caerphilli cohort study.' *British Medical Journal*, 315, 1641–1644.

corpus, and will direct their studies to collateral effects of them. In the coming years, we will see the findings of 'brainy' studies published in 'prestigious' journals on what the impact of a meteorite on Earth would mean, the chemical substance responsible for falling in love, the star that guided the Three Wise Men, the happiness equation, the possible presence of anti-cancer components in the most mundane of foods, from wine and grapes to olive oil or chocolates. They will all become suitable topics for research, rejecting any tedious and, above all, less media-friendly subjects. (It would be interesting to know if the 'Ig Nobel Prize'—science to make people laugh and think—harms or benefits the image of science in society. In my opinion, the image of their work and what science does that these scientists share is not good, yet the award is published more in the media than the real Nobel awards.).

Under the current system of evaluation, scientists who investigate side issues are valued more highly than their colleagues who have chosen to address relevant but less newsworthy issues. The latter, disappointed by low expectations of seeing their results will be published in the media, abandon their investigations for others that are more effective from a journalistic point of view. Because, even if they are not appointed to new roles, they will at least be recognized by other colleagues and by society as a whole. The extent to which world science is being harmed by these phenomena that have arisen in recent years is beyond the scope of this book, but I believe that it should be studied in a more serious way. Perhaps the solution would be to value less the impact index and to look for other systems to evaluate the performance of scientists. For example, solving an enigma.

Do Scientists Give Out an Absurd Picture of Their Profession?

It is often the scientists themselves who, with their 'childish' behaviour, generate a ridiculous image of science that transcends and is embedded in public opinion, even if the scientists do not believe it. The clearest example of this was the meeting of the International Astronomical Union (IAU) in Prague in August 2006. The image given out by physicists was pathetic: 'Scientists vote today if Pluto is a planet,' was the headline across all the newspapers on Thursday, 24 August. The discussion was absolutely absurd, but all the media pounced on it. Some readers may wish to reprimand me daring, in previous pages, to complain about Kuhn' theory that science is only an agreement between scientists at any given time, since this headline states that the scientists were indeed acting as if that science is voted on.

Of course, Kuhn is wrong. And, of course, scientific truths are not fashions, as he maintains, but certainties that are empirically demonstrated and, moreover, with which reliable predictions can be made. A scientific proposal or law is never either is voted on or consensual in an assembly by vote. In fact, at the above-mentioned congress, the physicists were voting on the internal rules for classifying asteroids. Their vote was no more relevant to science than if they were choosing the criteria to use to sort the postcards sent to them by their scholars. But the image that the eminent congressmen unconsciously gave out was that science is not a truth but an agreement that is voted for by the majority.

After the vote Pluto ceased to be a planet, according to the criteria adopted. All the television news programmes told the same story: from now on, the science books will have to be thrown away. They're no good anymore. They're already out of date. From the list of planets that teachers made us learn, scientists have now excluded Pluto. A chatterbox geek on the radio assured the world that he had been suspended from class for forgetting Pluto in a test several years earlier, and now asked the audience to phone in to vote on whether teachers should consider passing him.

Who hired the UAI press office, and why did they opt for the easy way out rather than just closing down those discussions that are no more scientific than how to order a research project filing cabinet? And the worst question: did physicists realize that with this news they were besmirching the credibility of science around the world? Possibly not, because few physicists have read Kuhn, so are unaware of the significance of the phrase 'scientists vote'. Were journalists in the press office aware that by leaking the news to the media they were comparing a vote on internal rules of procedure to voting on a physical law? Of course they didn't. Journalists have studied Kuhn but not Newton, and believe that everything in science is voted on: whether Pluto is a planet or whether the law of gravity is valid. Astronomers didn't realize that with those headlines science will lose vocations. A young person who hears the news on television might wonder why they should study something as hard as physics if it's worth nothing. People might vote for something else, they will think.

The fact that all of this is terrible for science, and that the discussions lasted so long, only gave the journalists more ammunition. I reproduce here the editorial of the newspaper El *Mundo*, which explains this idea perfectly:

Millions of textbooks around the world will become obsolete today if the proposal of the International Astrophysical Union (IAU) is successful. The academic discussion about the extension of planetary status to other minor bodies has ended up defending one of the nine existing bodies. (…) But beyond the arbitrariness of the decision, what is surprising is the Byzantine discussion that has given rise to a mere question of nomenclature in the scientific community. The Prague Congress has debated it so vehemently that it has at times reminded us of the Council of Constantinople, which tried to elucidate the sex of angels. (*El Mundo*, 24 June 2006, 3)

Obsolete books? A book on law—and the entire career histories of those who studied, under Franco's dictatorship, for example—or my book about trash TV may be obsolete. But a physics book never becomes obsolete. Newton or Galileo are not, and will not become, obsolete. But I am interested in the final part of the editorial: 'the physicists recalled the Council of Constantinople, which tried to elucidate the gender of angels'. I absolutely agree with that sentence, as that was the image that the scientists had projected. If the first sentence was the reason why vocations were ruined (why study something that is so easily made redundant), the second was the reason why scientists lost social support and subsidies, and why society should finance extravagant whims, as public opinion thinks this attitude is irresponsible. There was spectacular mockery about this scientific event. For example, the UAI defined planets—or so it appeared in the media—as 'round, influential objects, capable of dominating their environment, that never stop spinning and may even possess living beings'. That definition did not come from a journalist. Someone from UAI must have given it to her. But, as one radio talk show host put it so well: 'It sounds like a politician's definition.' I didn't know that physicists were inspired by politics to make definitions.

Similar mockery sparked another scientific investigation among journalists, with great success: obtaining the mathematical equation that describes how to flip a potato omelette in a frying pan without breaking it.[18] The website of Cadena Ser, the most popular radio station in Spain, and which does not usually include much information on science, used this as its major news item that day: 'The formula establishes that the angular speed at which the tortilla must shoot out of the frying pan is equal to the square root of the product of pi (3.1416) by gravity, divided by 4r, with r being the distance between the elbow and the center of the tortilla'. And there were cooks and housewives on all the television channels, saying that said they didn't need to know mathematics to turn an omelette over, laughing out loud when the journalist quoted the formula. Scientists were portrayed by the media as following 'an absurd profession'.

A very interesting example of this theory is the research of two young astrophysicists from Johns Hopkins University: Ivan Baldry and Karl Glazebrook. At the annual astrophysics meeting in January 2002 they presented a paper on background radiation from galaxies. In a footnote, they put forward the idea

---

[18] It appeared in the media on 8 March 2003.

that by adding together all the radiation from the 200,000 known galaxies in the universe the resulting colour could be considered to be the colour of the universe. And that colour was…. turquoise to aquamarine, they concluded. They did it as a joke, as they put it, after huge media coverage. I'm not so sure about that, though. It might have been a game, but they wrote it in a serious context.

A journalist who attended the meeting published the news and it went around the world on all the media. From the *New York Times* to the BBC. From Australia to Spain. The response from physicists around the world was spectacular, showing that they hear about findings through the media. They all began to measure the same thing, as if it were the essence of physics. However, it turned out that the discoverers of the 'colour of the universe' had not taken into account a rare physical effect called the 'white dot', the point at which light appears white, which depends on the environment in which it is contemplated. If this effect is taken into account, the colour is no longer turquoise but…. beige. The impact of this rectification, made two months later (in March 2002), also had worldwide coverage.

Were the two young astrophysicists doing any more serious research on astrophysics? Possibly, yes, but all over the world—in both academia and the media—they were known for that 'colour of the universe' finding. Within three months they appeared on our screens twice across the world, thus strengthening the so-called media reinforcement factor. They weren't anonymous anymore. People who had never noticed them looked at their resumes on Google to see what they were doing. It should be noted that they were both relatively young and it will be necessary to investigate in future whether their careers were enhanced by this episode.

However, here, as in the case of Pluto's nomenclature, the image conveyed to public opinion is that scientists are dilettantes whom society has to finance. Their whims for very expensive toys—observatories, space telescopes—are in order to have fun. And I insist: the culprit behind this image is often scientists themselves, who play with the media as if it were harmless. In such cases, it seems that they believe that they should act publicly as the cinema sets down that scientists should act: like extravagant madmen. Science is a serious business. I always wonder why all science museums have to be fun and for children to play in, while art museums are serious and educational.

The Impact Index or the Lesser of Two Evils

So far, we have talked about how the media can modify the impact index by increasing it dramatically. But…. would the impact index be a good method of evaluating science, even if there were no such media contamination possible? Sir Winston Churchill, the best politician in history according to a BBC poll of the British, argued that democracy is the least bad of all political systems. And he said that one of his strengths is that he constantly favours the search for faults and negative talk. I think similarly about the system for the evaluation of science and scientists through the index of citations (SCI): it is the least bad, yet has serious flaws that can be harmful to scientific progress in the long run, if not corrected.

What no one can deny is that this system copies the one developed by the media culture to measure audience size. According to this method, the best television or newspaper is simply the one that has the largest audience, not the one that offers the best content. But audience systems do not quantify whether the viewer likes the programme or, after watching it, thinks that it is disappointing; they only measure the number of viewers, which is why the dictatorship is irrational. Something similar happens with the SCI. We have talked about how one of the consequences of this dictatorship is that it forces scientists to publish a great deal and, as a result, their working conditions have worsened since a few years ago, which many also describe as a decline in science that explains the lack of vocations.

But I believe that the method itself can be improved and, above all, criticized. Because, as some experts in methodology argue, 'what began as a specific and very specific way of understanding the 'impact' of scientific publications (without reading them) has ended up being the way of evaluating scientific contributions.'[19] The problem is that, in addition to the one already described involving the appearances of their authors in the media, there are other ways to heighten the impact of a scientist, a journal or an article without them necessarily being either scientifically relevant or interesting. Bibliometric experts are critical of the fact that, for example, the impact factor is calculated by taking into account the total number of citations (which may be in minor journals), regardless of the type of journal in which they are published. This encourages repeated citation of groups of scientists, parrot-fashion, to enhance the prestige of that group and its publications.

Authors will respond indignantly to this accusation by claiming that the reviews of their work are carried out by 'peers or equals' and that in order to increase objectivity its evaluators are anonymous. However, it has been shown

---

[19]V. Pelechano. (2000). *Psicología sistemática de la personalidad.* Barcelona: Ariel.

that this 'peer or equal' evaluation often provides conflicting results (one might praise the work and another criticize it), in which case the editor of the journal has the casting vote, and that it is proven to have a bias towards certain theories or research.

Normally, the more prestigious the journal, the more conservative it is; the more conservative it is, the more the editor avoids publishing theories that contradict the dominant paradigm. For example, researchers Ader and Cohen found it difficult to publish their experiment and a new medical specialty was initiated: psychoneuroimmunology. Once the work was published in a smaller journal on this topic and circulated among the scientific community, the authors were able to disseminate subsequent work in the prestigious *Science*, *The Lancet* and *Annual Review of Psychology*.

In addition, it should not be forgotten that the head of a journal is interested in having the article for publication rankings and for another reason: for him or her, scientists who ask for their work to be published are far from anonymous. A 'tornado effect of quotes' can be seen, appearing on the horizon, so it is clear when a famous or much-cited scientist is boosting citations for the journal. Thus they will prefer good work by a famous scientist to a work, although it might be evaluated as better, by someone unknown. But, as the history of science has shown time and again, the great advances come from unknown young researchers.

Another perversion of impact index-related journals is that they publish only positive research results. This is another major difference between science and literature, and it may be the cause of the current decline in the former. In science, you can spend many years working yet, if you don't achieve results, it's work and lost time, because you can't publish it. In sociology, economics, communication or education studies, if you carry out a survey, a content analysis or interviews with the subjects of the study, you will always have some results. If you write an analysis or an interpretation of the work of a writer or a painter, you will achieve something. It is true that it is often debated whether the explanatory mechanism is invalid or influenced by causal inferences. Therefore, many articles in the social sciences do not explain causes but provide data. Although they're not much use to the world, they do serve to get a paper published. Even in medicine, it's the same. If you give a new treatment to a patient, there are two possibilities: they will either be cured or not cured. Both cases can be published. You never waste your time. One of my works in my doctoral thesis in journalism was, as I mentioned, to quantify and identify the scientific sources that appear in the press. Whatever the outcome, publication was assured.

That didn't happen for me in my first line of research in chemistry. I had to synthesize and crystallize new molecules (manganese (III) fluorides with nitrogenous bases). If I could achieve this and obtain crystals, I could establish its molecular structure and measure the magnetic and optical properties of these new molecules to analyse in them a well-known and interesting structural effect called Jahn-Teller distortion. But it was not easy to obtain molecules that had not been produced before. It was as if everything had already been 'invented'. While I achieved synthesis, I did not achieve crystallization.

One day I achieved both, but the worst thing that could happen to a scientist happened: I found out that this molecule had already been published by a French group just a month before… How much time and money (reagents, use of instruments, etc.) had been wasted! I could have written a treatise on my failed attempt and why it didn't produce results, but that wouldn't interest any journal. I had spent all my interest and care on it. I did not seek to fail, as Popper erroneously suggests that the scientist intends. I wanted to get it right or die, but those molecules didn't appear and, I thought, I could go years without positive results. How unfair is the life of a scientist!

However, I believe that in science, in certain cases, it is just as important to know the reasons why a phenomenon cannot occur as to know the reasons why a phenomenon does. Publishing negative results or unsuccessful attempts in journals can help science. And that work, I thought, should be valued. When I commented on this in one of the debates at the LSE, someone confirmed that, indeed, failures in science are just as important as successes and pointed out that, perhaps, journals do not publish failed attempts because scientists seek to discredit Popper (and his idea that the principle that drives scientists is to search where science fails). I don't think so. Very few scientists know of Popper or Kuhn, and they take these theories into account even less, not even to discredit them. For a scientist, the discredit of philosophers of postmodern science is already obvious.

I believe that journals are uninterested in unsuccessful attempts because such articles are not cited and therefore cannot increase the rate of citation and impact. However, some rightly argue that negative quotes from research do also count towards impact assessment. Richard Smith suggests in his memoirs on his years as head of the *British Medical Journal* that scientific journals publish only positive results because they seek to please the media. It will never be news that an AIDS vaccine has not been discovered, Smith's right about that. However, this does not discredit my hypothesis about impact assessment, because, as I mentioned, journal publishers want their articles to appear in the media to increase the impact rate.

At a dinner party with scientists from Imperial College I slipped into the conversation my idea that journals should publish these failed attempts. I had been invited by Joao Cabral, a physics professor at that institution whom I had met through a journalist friend. Joao regularly invited his friends, roommates and doctoral students to his home. As head of one of the Imperial student residences, he lived in a magnificent apartment on the top floor, from which the best of London could be seen. The residence and Imperial College are in one of the most exclusive areas of the British capital: South Kensington, where the Natural History Museum and the Victoria and Albert Museum are located. This London model is often repeated: good universities must be close to major museums and in the most exclusive areas.

At Imperial, as in the LSE, students and teachers of all nationalities live together. And at that dinner there were Portuguese, Spanish, American, Chinese, German and British scientists. Joao belonged to the chemical engineering department, so most of those present were physicists, chemists or chemical engineers. The discussion began with how hard the life of a scientist is and how well I had done by dedicating myself to journalism, where, according to them, 'anything goes'. I commented that the life of scientists is hard because they have designed it this way and suggested that, to improve it, they could regard a failed result as a relevant publication. Then, whatever work they did was not lost work, and neither would they would have such psychological pressure to achieve results. Everyone looked at me strangely. They thought it was madness. I pointed out how interesting these failed results would be for science and that leaving them unpublished was due to the contamination of our media culture. I didn't think to even mention Popper.

It's been years since I've met so many scientists together for lunch. Their perception of science is radically different from the approach taken by those from the humanities. They had no doubts about the inconvenience stemming from my idea, and used such pragmatic arguments that they disarmed me: if journals were to publish failed attempts, scientists would waste time reading articles that led to nothing. It is better for them to read what has been achieved and take the research forward from that point, they commented. No one wants to know why you failed and to look for another line that also leads to failure. Furthermore, in their opinion, my suggestion could pervert the citation system, as articles on failed attempts would be cited and thus provide their authors with unworthy recognition. Science is something serious that really solves issues, they repeated. In some cases, it may be worth publishing a judgement, but it shouldn't be the norm. It's not enough to try to solve the riddle; you have to solve it. Above all, they argued that science, in order to progress, must enhance effective resolution, hence its true worth relative to other disciplines, they insisted. Science is as hard as nature in its selection of species, pointed out others. If you reward academically the one who tries yet doesn't succeed, why try hard? It is hard and terrible for scientists, but it is beneficial to science. Scientists' working conditions are irrelevant compared to the welfare of science, they stressed.

All the opinions were so convergent, regardless of their country of origin, that I had to give up and admit that this was the general feeling. I talked about working conditions, lost time and frustration, and how well philosophers or sociologists, who publish any result, live in comparison. Let's not mention how in cultural studies they don't even need results but follow their brotherhoods of pedants. The scientists at the dinner party said that you always have to try, because you might achieve. 'It is no good unless you solve the problem,' they insisted. I was once again dazzled by this positivism on the part of scientists, which I had not experienced for years. It was as if Bertrand Russell were right: 'Being a scientist is the profession that leads you to happiness,' the philosopher said.

I don't know if I agree with Russell on this, but I must admit that the fact that the scientific journals are so favourable to a positive result and despise a failed attempt is good for the cause. In my case, for example, I was able to obtain at least one molecule. I finally obtained six, and I published three articles in impact journals. If my rulings had been operational, I wouldn't have made the effort. Science won. I almost lost my health and my interest, but a scientist is easily replaced. My effort was worth it so that others would no longer have to think about it. They can devote themselves to other molecules and other problems. Mine are already achieved, and anyone can obtain them by following the steps in my articles.

However, I do agree that this leads to harsh working conditions and uncertainty about the future, and this chases away vocations in fields such as chemistry, physics and mathematics. The gap between effort and outcomes when raising, detecting and analysing a problem and solving it is, in most cases, an abyss: Who would want to dedicate themselves to risky issues, to go to the frontiers of knowledge, if they do not have the certainty of a positive result?

On the other hand, anyone can predict that, obviously, if I (and the six other contributors who, on average, author each article) had not synthesized and studied those molecules, others would have achieved it with exactly the same result. Such is the science of the ungrateful. Others will see it as the greatness of science. Perhaps they are also right, but this dynamic may explain another cause of loss of vocations. This book, for example, can only be written by me. Hundreds may appear on the same subject. Most of them will be better than this one, but no one will do it exactly like me. If it were not published, no other approach would ever appear that takes these approaches. And that gives a uniqueness and potentiality that modern science does not give.

But the system of scientific journals has further perversions that discourage vocations. For example, the reviewers of a paper may be worse scientists than the ones who are being reviewed. Since the reviewers are unknown to the author, they can take on a good study, and, I repeat, its authorship is unknown to the reviewers. As the bibliometrics expert Buela-Casal points out:

> the anonymity of reviewers facilitates intentional, disproportionate or cruel criticism. If the reviewers are specialists in an area of work, it is because they have carried out research and publications in that area. If a paper undergoes critical review, or the results totally or partially contradict its work, the reviewer is more likely to reject the paper, which is facilitated by the anonymity.[20]

For example, an anonymous reviewer went so far as to write in the evaluation of an investigation by the president of the American Psychological Association himself, Robert Sternberg, that the text appeared to be written by a 'charlatan lawyer', and equated the investigation to that of a first-year college student. Sternberg decided to make this verdict public, and the rest of society heard what defines today's science.

Journals and Politicians

I don't want this book to present a view opposed to impact journals. The fact that they may discourage vocations among brilliant candidates does not mean that the system should disappear. In fact, I stress that modern science has advanced thanks to these journals, and that they are a tradition to be defended to the utmost. The publication of science in journals (and now on the internet) is essential, because science is the heritage of humanity, not of scientists. It doesn't belong to any particular country. Science must go beyond geographical boundaries, which are always arbitrary and irrational. Everyone, regardless of where they come from, has the right to receive information about science and to publish, if they deserve it, their ideas and discoveries in any journal in the world. I would just point out that the systems of publication and evaluation of scientists must be seriously reviewed in order to ensure that,

---

[20] Gualberto Buela-Casal. (2003). 'Evaluación de la calidad de los artículos y de las revistas científicas: propuesta del factor de impacto ponderado y de un índice de calidad.' ('Evaluation of the quality of scientific articles and journals: proposal of a weighted impact factor and a quality index.'). *Psicothema*, 15, 1, 23–35.

in some way, the strong impact that media culture has on today's society does not damage scientific journals and science itself.

I believe that science needs independent journals focused on scientific development, not on the media. Above all, they must not succumb to the power of politicians. In this sense, in my view one of the episodes that most clearly describes the decline of science in the twenty-first century occurred in February 2004. The President of the United States, George W. Bush, banned the publication of scientific articles in US journals from Cuba, Iran, Libya and Sudan. It is worrying that something was attempted during the Iraq war that never happened during the Cold War, as the Soviet scientists could still publish in American journals (thanks, among others, to Robert Maxwell). But of course, during the Cold War, science was considered important. In the twenty-first century, it's not so much valued.

The measure that we were talking about was part of a US Treasury Secretary's rule that the publication of scientific articles from these countries 'violated the trade embargo'. One of the scientific articles that suffered from politicians' (and frightened scientists') paranoia was 'Physical-chemical changes in the hull of corn grains during their alkaline cooking', submitted to the *Journal of Agricultural and Food Chemistry* by a Cuban researcher. The editor of the journal at the time, James N. Sieber, ordered the Cuban researcher to withdraw the article. 'The American Society of Chemistry (ACS) has reluctantly established a moratorium on the publication of articles from Cuba, Iran, Iraq, Libya and Sudan. Failure to do so would put ACS, including its directors and staff, at risk of serious civil and criminal penalties', Sieber wrote in a letter to the Cuban scientist.

Rarely, in my opinion, had science sunk so low. The issue, surprisingly—or perhaps not so surprisingly—went unnoticed by the world, either journalistic or intellectual. But for me it was revealing that we were taking a step backwards in civilization. Galileo went to the stake. What risk was James exposed to? N. Sieber? A fine? That's why the editor of this journal stopped publishing a scientific idea that, in view of the title, did not appear even remotely dangerous.

Some independent scientific associations protested. Both the AAAS (editor of *Science*) and the American Physical Society opposed Bush's ban. The argument, from my point of view, was curious, because they did not argue that this ban undermined one of the foundations of the history and tradition of science but instead appealed to a media principle: they said that it went against the first amendment to the American Constitution, the one that speaks of freedom of speech and of the press. Apparently, legally, in the West nothing can protect the tradition of science but a tradition from the media culture.

I estimate that there could be about 10,000 scientific journals in the United States at that time. The vast majority of their chiefs looked the other way. What should have been a monumental scandal barely made the news. It certainly received far less publicity than Pluto's expulsion from the list of planets or the polemic on the colour of the universe. Philip Campbell, head of *Nature*, said: 'We see no reason whatsoever to reject publications from these countries, but we are giving each other legal advice.' And he qualified that *Nature* is, of course, British.

Scientific Journals and the Imposition of an Information Agenda

Everything is so perfect. The press office is so professional. If there really is, as I have said, an ideal of symbiosis between science, journalism and dissemination, why talk about *Nature* in a book about the causes of the decline of science in media culture? First of all, it is because the spread of science is dying due to the success of *Nature*. Because the supply is so good (as with *Science*), journalists around the world select news only from these sources, depriving researchers of the opportunity to publish in the media. The Matthew effect occurs again: the scientist who appears in *Nature* is much quoted because many people read it and, in addition, the number of quotations of the work is increased because the author enjoys much media coverage, at the expense of those who publish in other journals, who are seldom quoted.

The layman will ask himself: why do the press offices of other journals or universities not work as well as *Nature*'s? Possibly many will ask that. But *Nature* every Thursday and *Science* every Friday have managed to impose what in journalism is called a news agenda. Normally, there is just one science story a day in a newspaper or television newsreel (if you're lucky). What journalist will risk choosing a local university's item that has been published in a journal when he knows that all the newspapers in the world (literally: from the *New York Times* to the smallest regional newspaper in the United States, Japan, India and Australia) and all the television channels will publish the item that is in *Nature* news. This is a piece of news that, possibly, because of the tough selection criteria of this journal, will be better than the one sent in by the local university, since otherwise the local researcher would have sent it to *Nature*.

All this explains why it is very difficult to break this dynamic of imposing the agenda by eclipsing the rest of the scientific results. In fact, during my time as a science journalist for *El Mundo* it was practically impossible for me to publish anything other than what was broadcast on Thursdays by *Nature* (and on Fridays by *Science*). It was a huge risk not to follow them, because both my bosses and my readers would not failed to understand why I was withholding the *Nature* news in favour of a piece with lesser scientific scope.

'And that's bad for science? Science is universal and *Nature* is perhaps what has the greatest scientific relevance. It is negative for the local researcher, who is mediocre, and has not been able to publish in *Nature*, but not for science,' they told me at a talk at the LSE on the causes of the decline in science.

I recognize that here I have a journalistic streak: I think it's bad to leave to the exclusive discretion of *Nature* and *Science* what the world needs to know about science. In my view, an excessive global focus on a single news item may distort the scientific process itself and the dissemination itself. Globalization, applied to science, makes a scientific finding more interesting than, for example, an international summit. And neither science nor scientists are prepared for such situations of extreme media stress.

A clear example can be seen if we describe the whole media attention surrounding one of the most relevant scientific news of recent years: the cloning of Dolly the sheep. In this case, both *Nature*'s summary and the press office agreed that it was 'very relevant'. Its publication in *Nature* on 27 February 1997, in addition to provoking a broad debate in society on cloning, meant a study on the way in which the news should be communicated, with the aim of obtaining a controlled, but at the same time planetary, media impact. This may look like an oxymoron. The communication strategy, the impact caused by the news, as well as the treatment in the media, can be considered a paradigmatic case of the phenomenon of social communication of science that we have already seen in this book. Many people criticized scientists for denaturing the world. The archetype of the unscrupulous scientist was repeated. The most pedestrian opinion-formers, with no scientific training, filled whole pages of newspapers, while those who really knew about the subject were frightened and reclined in their laboratories, waiting for the storm to break. However, here I am interested in analysing the effect that this media stress had on Dolly's own responsible scientists and, by extension, on science itself. Ian Wilmut, one of Dolly's managers, said in an interview, 'when you know that *Nature* is going to include your research in a press release, you prefer to prepare another one with more information yourself'.[21]

The first effect was the destruction of the research team. Led by Ian Wilmut, the team from the Roslin Institute in Edinburgh, Scotland, had worked closely with the communications advisors of PPL Therapeutics, the company that had funded and collaborated on the cloning project. Thus, when *Nature* confirmed

---

[21] Interview by Octavi López with Ian Wilmut, published in *Quark: Ciencia, Medicina, Comunicación y Cultura*, 14, 79–81, January–March, 1999.

the publication of the article, the communication team had only 10 days to plan the media strategy. Journalists from the *Nature* and PPL Therapeutics press office chose Ian Wilmut as their main spokesperson. The media culture needs heroes or villains, and always individual protagonists, never collective ones. They decided to transfer two specialists from London to Edinburgh to advise them on how to appear on television. The result was obvious: the chosen spokesperson became a media star and his brilliance continues.

Dolly's scientists knew that *Nature* would include the research in its weekly press release distributed on Fridays, which would be embargoed until the following Wednesday and published on Thursday. However, two calls on Saturday night alerted them that the *Observer* would publish the story the next day, so the news came out three days ahead of schedule. In just one week they handled more than two thousand phone calls, spoke to nearly a hundred journalists and granted Dolly access to 16 film crews and more than fifty photographers from around the world.

How did Wilmut take this media impact? Interestingly, one of his criticisms in the aforementioned interview is not directed against the media but the system of dissemination to the media of the results of the work published by *Nature*. Wilmut describes *Nature*'s and other journals' system of keeping the research secret until they published it as 'unfortunate':

> This strategy – explains Wilmut – implies a decrease in the pace of discussion of your work. It also magnifies and amplifies things when information comes in, so you get an exaggerated response from the media. The process would be improved if the investigations could be discussed more openly. Then, once you have the article, it would be elaborated in detail for *Nature*, for example. But in order to preserve the position of exclusivity, some magazines would not allow it. (*Quark* 14, 80)

I find Wilmut's reflection most interesting since, although *Nature* defines itself as the great defender of science, something that I do not deny, it is obvious that as it is a private company it is logical that, like other means of communication, it prefers an exclusive to losing out in the interests of greater rigour in the work or encouraging scientists' debate on specific issues. *Nature* acts fiercely by threatening to withdraw an article from publication if authors do not respect its rule that the research that the journal publishes cannot have been published anywhere else. In this way, the publication of a scientific result becomes what we call in journalism an absolute (exclusive) scoop. This boosts the multiplier effect in the media, as it is known that they have an unpublished news story that nobody knew anything about before. As Wilmut says, the scientific debate—so necessary in the scientific method itself—loses out.

Regarding the extraordinary impact of the news of Dolly's cloning, *Nature*'s 1997 editor-in-chief, Phillip Campbell, said in an interview[22] that neither the editorial staff at *Nature* nor the reviewers' opinion of the paper predicted the impact that the news would have. 'In fact', explains Campbell, 'a similar work, although with fetal cells, had already been published a year earlier and went almost unnoticed.'

But of course, Campbell forgot to point out that in Dolly's case the PPL Therapeutics communications consultants, who had paid for the research, wanted free media advertising. To this end, they had drawn up a strategy that produced good results: to ask the scientists to talk about the possibility of cloning human cells, something that is not even mentioned as a remote possibility in the work published in *Nature*.

But news without conflict does not sell, and to occupy the front pages the scientists, advised by the journalists, agreed to talk about the human repercussions. In fact, Wilmut acknowledges in the interview that the pressure from the media had positive results, since the direct benefits of advertising included that 'it has been easier to get money to start a company that takes advantage of the technique used to obtain Dolly'.

Scientists as Media Stars

One does not need to be an expert in the history of science to know that, apart from perhaps in mathematics or theoretical physics, from the twentieth century onwards few discoveries are the product of a single person. This is what is called the professionalization—which some call the Taylorization—of science. However, since one of the idiosyncrasies of the media is that attention cannot be focused on multiple protagonists, one of the scientists is usually chosen to be the spokesperson. If the news is important—as in the case with Dolly or, as we will see later, with the Higgs boson—the spokesperson becomes a star scientist, with all the benefit and harm that this entails. As *Nature* sends its communiqué to the world, one scientist becomes a world media star—is that good for science?

During my stay in London from 2005 to 2006, I had the opportunity to follow in the press the trial against Ian Wilmut for 'taking over Dolly's cloning work and not letting her collaborators have their share of fame'. The trial, which itself deserves a book or a doctoral thesis, was fascinating because it gathered together the mechanisms by which science works in this twenty-first

---

[22] Interview with Phillip Campbell by José Luis de la Serna, head of the *El Mundo* Health Supplement. *El Mundo* (7 May 1998, pp. 7–8).

century. During the trial, Wilmut had to admit that the others did most of the work, although he was the one who had the fame. This case is not like that of the South Korean scientist Hwang Woo-suk, who falsified the data, because Dolly was real. However, both are connected as they demonstrate the great competitiveness and aggressiveness in today's science, often due to the contamination caused by media culture.

The trial was held in Edinburgh in March 2006. In Great Britain, it not only had considerable media coverage but was widely commented upon in academic circles, as I saw in the LSE staffroom. In my opinion, it shows how hard and ungrateful is science when the line of research does not depend on just one person but a group. The lawsuit originated because Wilmut's collaborator, cell biologist Keith Campbell,[23] felt that Wilmut had mistreated him professionally. Wilmut had to admit, on the basis of the evidence provided by Campbell, that 'at least 66% of the credit for creating Dolly went to Campbell'. However, Campbell was mentioned last in the *Nature* article while Wilmut was shown as the first author.

The second author of that article, Angelika Schieke, who at the time was a doctoral student, told *The Scientist* that Wilmut had taken the lead because he personally asked for it at a previous meeting and the rest of the co-authors granted it to him, despite the fact that this leading role did not reflect his contribution at all. While the trial against Wilmut was taking place, two of the technicians who worked hard to manipulate the hundreds of eggs and cells whose final result was Dolly, also complained in the press. The technicians felt that Dolly's merit was not really a new interpretation of a natural process or a pure scientific finding, but rather a technical application. Even the technique had already been described. Dolly's great achievement, as the technicians at the trial stated, was to repeat a known technique to hundreds of samples until one sample turned out well. In this sense, they considered that their work had not been recognized either. Overall, the research team complained that in the media portrayal of Dolly it seemed like Wilmut had had the idea—which was not true—and that he had worked all day in the lab as if he were a modern Marie Curie, which they claimed at the trial was not the case either. What was his merit, selecting technicians and fellows and finding money? The debate in British teaching staffrooms was electric.

All this leads us to consider that there is no clear protagonist in current scientific results and that, many times, researchers use the media to attach importance to themselves that they do not possess scientifically. However, media fame works for the scientist's own modern projection. In fact, of Dolly's team only Wilmut was awarded the prestigious German Paul Erlich award. Wilmut

---

[23] In 2013, Campbell killed himself in drunken fury.

was also appointed as director of the new Centre for Regenerative Medicine at the University of Edinburgh, one of the few to be granted permission by the British government to clone human embryos. That is to say, the media effect worked in favour of Wilmut, not the scientist who had worked most on the project and who contributed 66% of all the work, Campbell.

Is this usual? It has always happened, but the media culture is conducive to it. In fact, before leaving this area of investigation, I will relate how in the same country and at the same time a similar case arose. It was at Newcastle University, the first centre to obtain permission to clone human embryos. There, in 2005, Serbian scientist Miodrag Stojkovic left the university after learning in the media that his boss, Alison Murdoch, had hastily presented at a press conference the results of their investigation into producing first human embryo, on which they were working together. The worst thing for Stojkovic was that at that press conference Murdoch took much of the credit when, in the opinion of the Serb, he was the one who had contributed the most.

After this type of controversy, many team leaders, seeing the danger, took Wilmut's or Murdoch's side to defend their case in the media. They clarified that, although they might not undertake all the work, they direct and define it, therefore they should take all the honour. The junior scientists in both cases replied in the press that their work had been carried out 'in spite of the bad management of their bosses'.

The controversy grew, with young British researchers arguing that in twenty-first century science there are only two kinds of bosses: those who recruit only mediocre people so that they don't become a threat, and those who prefer bright people but promote the career of mediocre scientists lurking in the shadows. 'When they squeeze our talent dry' said the young scientists, 'they throw us out and look for new, innocent blood.' In any case, these controversies—deployed with good journalistic sense by the press—show that today's science, unlike that practised a century ago, is not only highly competitive but extremely unfair when it comes to rewarding merit because, among other reasons, it is so difficult to establish where it lies.

From Co-author to Contributor

What this controversy brought to light was that today's scientists are immersed in an increasingly perverse phenomenon called 'the transition from co-authorship to simple contribution', and this has encouraged some to seek media support. I have already said that, in the early days of the history of science, scientists were entirely responsible for research. It was an individual work—the work of the genius—just as now it would be that of an actor, a writer, a painter, a footballer or a journalist. You can work as a team, but genius is outstanding and the role of each person is always clear: in the case of cinema, for example, the director is not the same as the scriptwriter or the make-up artist. The work together is valued, but also the individualities and the role of each one is perfectly detailed. And there are films that receive an Oscar for make-up or costumes, not for best actor or director.

But in science that doesn't happen. Since the twentieth century, science has gone from individuality to collaboration between several people in order to carry out research. For example, the publication of the molecular structure of DNA in 1953—a date that many propose as the date of the last great scientific finding—had only two authors: Crick and Watson (the latter was a 25-year-old postdoctoral fellow). Since then (and especially since the 1970s and 1980s), the trend in the natural sciences—and also in medical sciences—has been a significant increase in the number of authors per article. Even in the social sciences or the humanities it has increased, and in these areas this trend is truly worrying. It is a way of increasing considerably not only the scientific production of each scientist, but also his or her production of scientific publications, which is what is measured in media culture. In fact, scientists of the late twentieth and early twenty-first centuries are no longer even co-authors. Another step has been downgraded, and they are now referred to by science policy experts as 'mere taxpayers'. Richard Smith's book makes this clear in the title of his chapter 9: 'The death of the author and the birth of contributor'.

This system generates many injustices and perverts the whole system. One of the worst results is the ghost author. Smith recalls an interesting anecdote: 'I did, however, recently meet a professor who told me that he sat next to a woman on a plane—in business class—who boasted that she had published more papers in major medical journals, including *BMJ*, that any other living author. Yet her name, she said, never appeared. She was a champion ghostwriter and employed by pharmaceutical companies' (Smith 2006, 116).

It is not surprising to find works with more than twenty names, but the average is 10. The system has favoured a process in which there is a buy-sell-sell-sell-exchange of firms, and different groups include researchers who have done nothing in return for other favours. For example, it is common in countries with little or medium scientific impact, such as Spain, for research groups to include major global scientists among their contributors simply to make the first cut of the major journals. This is because high-impact journals, such as *Nature*, can receive an average of ten thousand articles a year and obviously not all can be evaluated.

It has been shown that the 'first cut', which is made by the journal's staff rather than the evaluators, has a number of biases in favour of scientists from prestigious universities or scientifically powerful countries. The way to overcome it may be to include prestigious names in your articles, even if their contribution is minimal or zero. This endorses the article. Other times, the exchange of names is done for more mundane reasons, such as a thank you for allowing a scholarship holder to stay.

This perversion is in everyone's interest: the big scientist from powerful countries has more and more publications and is 'bigger', and the small ones can aspire to publish in better journals. Having many publications and being cited are the parameters that measure a good scientist in the age of media culture, as I have already made clear. But journal publishing under the 'taxpayer' system does not adequately reward or evaluate individual talent and effort. And what's worse, it burns the brilliant and empowers the mediocre.

Richard Smith, director of the British Medical Journal for 13 years, says in his book that science cannot work well without methods to clarify what part of the work each contributor has done in each article. Smith proposes that journals should include a list of the co-authors with their biographies and what they have contributed to each study. This relationship should be developed and agreed upon by the authors of the paper themselves (Smith 2006, 111–118).

But this idea does not convince the great world leaders in science, who are very comfortable with the system of contributions as it is the one that has provided them with that leadership. However, it discourages talented young people from engaging in it. The current scientist does not feel that the research is his own. He or she produces scientific literature on which his or her name is written, but many times has not even written the article, or even a part of it. In fact, it is becoming increasingly common for powerful research groups to hire specialized writers—young doctors—to whom the scientists pass the data.

But a ghostwriter is a mere automaton, since their contribution is supposed to be null and void and therefore their name does not appear. They write in an impersonal style, without emotion or literary elegance. And, most seriously for the development of science, this scientific literature, written by mercenaries who have not obtained the data, does not even provide the consensual or personal point of view of each researcher. As long as we have one more paper, many researchers think, it doesn't matter what is published, because the evaluation is quantitative, not qualitative. And many suggestive interpretations remain only in the mind of the scientist.

Richard Horton, editor of the medical journal *The Lancet*, carried out a very interesting study[24] in this sense, inspired by the maxim of the French philosopher, Simone Weill, who maintained that 'all statements that begin with "we" are lies'. Horton selected 10 scientific articles published in *The Lancet* in 2000, written by a set of 54 co-authors, giving an average of more than five per published study. Interested in the weaknesses, interpretations and implications of these studies, he wrote to all of the contributors and asked them to write a personal report. He found significant disagreement among them on the same study, yet that diversity of interpretation did not appear in the articles. In addition, all the weaknesses in the study methodology (referred to by the different contributors) were removed from the final article. If this is unfortunate in disciplines such as medicine, it is tragic in the social sciences or humanities, because it eradicates any dissident thinking. Moreover, the effect of the narrator's talent, so necessary in disciplines that claim to call for social mobilization, also disappears.

The 5,000 Authors of the Higgs Boson Paper

In May 2015, *Nature* addressed an issue with an article entitled 'Physics paper sets record with more than 5,000 authors'.[25] It stated that 'a physics paper with 5,154 authors has—as far as anyone knows—broken the record for the largest number of contributors to a single research article. Only the first nine pages in the 33-page article, published on 14 May in *Physical Review Letters*,[26] describes the research itself—including references. The other 24 pages list the authors and their institutions. The article in question is on the work of the Large Hadron Collider team on a more precise estimate of the size of the Higgs boson.

Robert Garisto, an editor of *Physical Review Letters*, says in *Nature* that publishing the paper presented challenges above and beyond the already Sisyphean task of dealing with teams that have thousands of members: 'The biggest problem was merging the author lists from two collaborations with their own slightly different styles,' Garisto says. 'I was impressed at how well the pair of huge collaborations worked together in responding to referee and editorial comments', he added. Too big to print? 'Every author's name will also appear in the print version of the *Physical Review Letters* paper', said Garisto to *Nature*. By

---

[24]Richard Horton. (2002). 'The hidden research papers.' *Journal of American Medical Association*, 287, 2775–2778.

[25]Davide Castelvecchi. (2015). *Nature News*. https://www.nature.com/news/physics-paper-sets-record-with-more-than-5-000-authors-1.17567.

[26]G. Aad et al. (ATLAS Collaboration, CMS Collaboration) *Phys. Rev. Lett.*, 114, 191803 (2015).

contrast, the 2,700-odd author list for a *Nature* paper on rare particle decay, published on 15 May,[27] will not appear in the June print version but will be available only online. Some biologists were upset about a genomics paper in 2015 with more than a thousand authors[28] but, as *Nature* news remembered, physicists have long been accustomed to 'hyperauthorship' (a term credited to information scientist Blaise Cronin at Indiana University Bloomington).[29] An article published in 2008 about the CMS experiment at the LHC5, before the machine started colliding protons, became the first paper to top 3,000 authors, according to Christopher King, editorial manager of Thomson Reuters ScienceWatch. The paper that announced the ATLAS team's observation of the Higgs particle in 2012 had 2,932 authors, of whom 21 were listed as deceased. [30]

The Nobel Prize for the Higgs boson was won in 2013 by Peter Higgs, who did not do any experiments but predicted it theoretically in 1964 in an article in *Physics Review Letters* entitled 'Broken Symmetries and the Masses of Gauge Bosons',[31] and most importantly it bore only his name. But of course, this was the 1960s, not the second decade of the twenty-first century.

These circumstances force many scientists to become media stars in order to differentiate themselves or obtain some kind of recognition, even though they often do not deserve it. But from the point of view of the decline of science and, above all, of vocations, what is relevant is that these discussions about authorship, ingratitude and intellectual exploitation experienced in the scientific environment are often aired by the media. It is fair, convenient and healthy that it should be so. But it can provoke a refusal by bright and hard-working young people to dedicate themselves to a world that is perceived as unjust and arbitrary. A lawyer, an actor or journalist can succeed on their own merits, but a scientist depends on the bosses and the money that they provide. In the working environment developed by science since the final decades of the twentieth century and the beginning of the twenty-first century, the process has become so Taylorized that a brilliant young person will never be happy. Newton would never have survived in a science department today.

[27] CMS Collaboration and LHCb Collaboration *Nature* http://dx.doi.org/10.1038/nature14474 (2015).

[28] W. Leung et al. *Genes Genomes Genet*, 5, 719–740 (2015).

[29] B. Cronin. *JASIST* 52, 558–569 (2001).

[30] ATLAS Collaboration. *Phys. Lett. B*, 716, 1–29 (2012).

[31] Peter W. Higgs. (1964). 'Broken symmetries and the masses of gauge bosons.' *Phys. Rev. Lett.* 13, 508. https://journals.aps.org/prl/abstract/10.1103/PhysRevLett.13.508.

10

Science in the Digital Society

The building is located in one of the most central and exclusive areas of London, in Carlton House Terrace, close to Piccadilly, St. James' Park and Buckingham Palace, Trafalgar Square and the British Prime Minister's residence, Downing Street. The exterior is elegant and in the style of its neighbouring buildings, all Victorian mansions. A porch supported by two columns gives access to the interior, which combines ancient elements—pictures, books, instruments—with modern decoration. The whole, in my opinion, offers an image of both simplicity and sophistication.

It did not look like an opulent institution, but appeared to be a wealthy one. Nothing seemed decadent; on the contrary, its walls were strong, vigorous and even fresh. After spending time reflecting on the decline of science, it wasn't what I had expected to find when I went to the Royal Society, the world's most prestigious scientific institution. The clash was the more brutal because I was going with an image in my mind of Spain, where the headquarters of its equivalent, the Royal Academy of Sciences (founded in 1834), is a rather abandoned building, located, moreover, on Calle Valverde in Madrid, until a few years ago one of the most derelict areas of the Spanish capital, where, for example, a significant percentage of street prostitution and sex shops is concentrated.

The Royal Society, on the other hand, is in one of London's most glamorous neighbourhoods, which—if only metaphorically—further widened the scientific chasm between Spain and Great Britain. The prestige of the Royal Society is not only guaranteed by its antiquity—it was founded in 1662, when science was taking its first steps—nor by the fact that one of its presidents was the most important scientist of all times, Isaac Newton. Above all, it is because from its foundation he believed that the best initiative for science to flourish

© Springer Nature Switzerland AG 2019
C. Elías, *Science on the Ropes*, https://doi.org/10.1007/978-3-030-12978-1_10

was to combine intellectual and methodological rigour with vocation, not only to inform all humanity of its results but to yield them to humanity itself.

His well-known and influential motto, *Nullius in verba*, something like 'in nobody's words', refers to the need to obtain empirical evidence in order for knowledge to advance, totally denouncing the 'criterion of authority' that scholars and a large part of the academy in Latin countries maintained. The Royal Society's philosophy of rejecting the 'criterion of hierarchical authority (political, academic, religious or military)' and acting only on the basis of empirical evidence is what has scientifically and culturally promoted English-speaking countries above Latin countries.

On the walls hang portraits of Royal Society presidents. At a glance I found William Bragg's, Ernest Rutherford's, Dalton's and Humphry Davy's. 'Almost all the scientists I studied in high school are here,' I thought. If we had to find a synonym for science or scientific progress, it would be the Royal Society and, in general, British and North American scientific societies. The prize was to have English as the universal language of science, but it must be recognized that these countries scientists have always worked with great rigour. And, at the present time, they are at the forefront of the dissemination of science. Therefore, the professional fact that without doubt I am most proud is that my first research[1] was of sufficient quality to be published in one of the journals of one of the Royal Society's current branches, the Royal Society of Chemistry.[2]

The Royal Society is not just a building. It is an active forum for the discussion of ideas throughout the year, to take a position in the scientific community on the problems that affect the world. In this regard, it highlights the campaign that is being promoted among the media and politicians to raise their awareness of the real danger posed by climate change. For someone who loves science, it is always a real privilege to attend one of its seminars.

Many times, the Royal Society deals with well-established issues, but at other times the approaches are bold, controversial and multidisciplinary. The best part is that discussions are public, and most of them free of charge—as are the main British museums—and the speakers are usually the best in the world in their speciality. They invite you to tea (and even coffee) with pastries and, to attend, they usually only require registration in advance through their website. I chose a controversial seminar which, in addition to seeing how a debate at the Royal Society worked would perhaps give me ideas for my research into the

---

[1] Pedro Núñez, Carlos Elías, Jesús Fuentes; Xavier Solans, Alain Tressaud, M. Carmen Marco de Lucas and Fernando Rodríguez. (1997). 'Synthesis, structure and polarized optical spectroscopy of two new fluoromanganese (III) complexes.' *Journal of the Royal Society of Chemistry (Dalton Transactions)*, 22, 4335–4340).

[2] The Royal Society of Chemistry is the world's foremost organization in chemistry. With over 54,000 members (in 2018), it is the result of the merger of, among others, the Chemical Society, founded in 1841, and the Royal Institute of Chemistry, founded in 1877. The origins of both lie in the Royal Society.

decline of science. I was not looking for any more data on the decline because, in my opinion, I already had enough. My idea was to find a radically new approach that would inspire or reaffirm different interpretations. I registered for and attended 'Social Intelligence: From Brain to Culture'.[3]

The hypothesis that was debated for two intense days was that the attribute responsible for the evolution of intelligence had been not so much a genetic mutation (although something might have influenced it) as group life; that is, the existence of society. The topic focused on primate zoology, ornithology, oceanography, psychology, anthropology, ethnology and palaeontology, among other perspectives. The scientists explained how, for example, birds and mammals have the ability to learn by observing others. Even invertebrates have that ability, according to one of the speakers.

Professor Andrew Whiten of St. Andrews University explained how orangutans and chimpanzees transfer skills from one generation to the next. Using projections of orangutans in the jungle that took you back to the beginning of time, Prof. Whiten explained how this ability to transmit knowledge led to groups of orangutans with increasingly complex cultures. In order to manage this greater cultural complexity, they needed greater capacity in their brains. That is to say, living in a group and learning more and more skills could be a factor that stimulates a greater capacity in the brain. Not everything simply had a genetic explanation.

Could learned skills—such as a delousing technique—be lost forever if the offspring who had not yet acquired them were separated and placed in groups that had not developed those skills, asked one of the attendees. Another question was even more disturbing: Can a species involute? Can man become an ape without intelligence?

Visual Media as a Public Health Hazard

Obviously, the question has no easy answer. We know that there have been cultural regressions—for example, in the West from the fall of the Roman Empire to the Renaissance—but that did not mean a loss of intellectual capacity but of acquired cultural skills or levels. It was a time of obscurantism and irrationality, consequences of poor culture but, apart for due to hunger in the needy classes, there was no danger that the children's brains would not develop properly. The only danger that can threaten the intellectual evolution of the human species is that there is something in the environment that limits the biological capacity to develop its brain. And this danger, apart from more or

---

[3]'Social intelligence: From brain to culture,' held at the Royal Society, 22/23 May 2006.

less nonsensical education theories that can always be exploded, is present at the moment and has as its fundamental basis one of the pillars of media culture: television. And, by extension, everything audiovisual: videos on mobile phones, tablets, computers.... The danger also affects all countries and all social classes globally, so it can be considered a real threat to the species.

In August 1999, the American Academy of Pediatrics published a guide recommending that children under the age of two should never, under any circumstances, be exposed to television. It extended this ban to any type of entertainment broadcast on screens. According to the Academy of Pediatrics,[4] television 'can adversely affect brain development', adding that children of any age should never have a television in their room.

In 2004, a group of neurologists led by Dr. Christakis discovered that exposure to any type of screen image (television, video games, computer) during the critical period of synaptic development in neurons is inevitably associated with later problems of attention and concentration in school.[5] This critical period ranges from birth to seven years. When a human is born, its neural connections are not formed, thus they are assembled as the child grows. Once they reach a certain age, if they are not well connected they will never be able to do so.

For example, all children are born with a predisposition to speak, yet if they do not grow up in a language environment the neural connections for speech will not be made and they will never possess the language proficiency of another child who grew up in a speaking environment. It also happens with vision: if a child grows up in a room without light, he or she will never be able to see. This implies that if television or video games prevent the correct and optimal neuronal connection, we are talking about a real public health problem that would result in children of the television age not being able to reach the levels of brain development of their predecessors.

My high school science teachers always used to remind us of a phrase: 'Patience is the mother of science.' The alarming decline in the level of scientific knowledge of children in the West is directly related to the fact that they are not able to maintain the necessary attention and, above all, to sustain it for the time necessary to learn complex and abstract concepts, such as science. Well, Dr. Christakis' group showed that for every extra hour of television an infant received each day, the child's chance of developing a neurological disorder called ADHD was increased by 9%. Studies indicate that every year those affected

---

[4] American Academy of Pediatrics Committee on Public Education. (1991) 'Media education'. *Pediatrics*, 104, 341–343.

[5] D. A. Christakis, F. Zimmerman, D. L. Di Giusseppe and C. A. McCarty. (2004). 'Early Television Exposure and Subsequent Attentional Problems in Children.' *Pediatrics*, 113, 708–713.

by this syndrome grow exponentially worldwide and it is estimated that 5% of children in the West suffer from it to a severe degree.

Any primary or secondary school teacher with several decades of teaching experience can easily confirm that the vast majority of children in the late twentieth century and early twenty-first century suffer from this disorder, the symptoms of which are that the student is unable to remain calm and reflective at his or her desk. This is a condition, I repeat, that is essential for learning a language as complex as the scientific one. In other words, the problem may not be pedagogical but neurological—a health disorder caused by one of the channels of media culture: television.

Obviously, political, economic and, above all, media power will never acknowledge this fact, preferring to blame teachers and education systems. This is easier than banning television (or audiovisual communications in any format), a true pillar of media culture. But the scientific debate is going in a different direction. In 2007, the British psychologist Aric Sigman published a book in which is condensed all the scientific research that relates the audiovisual field to biological injuries: works that are already widely cited in English-language bibliographies and congresses.[6] According to this novel avenue of research that relates television to a deficit of neuronal connection, the education system cannot do anything, because these children cannot attain the levels of concentration and abstraction of their ancestors. Watching television is to the brain (and a future scientist) what cutting off the legs is to an athlete: one can run with a prosthesis, but one can never run for real.

> Television viewing among children under three years of age is found to have deleterious effects on mathematical ability, reading recognition and comprehension in later childhood. Along with television viewing displacing educational and play activities, it is suspected this harm may be due to the visual and auditory output from the television actually affecting the child's rapidly developing brain. (Sigman 2007b: 15)[7]

In the work of Prof. Christakis cited above, the researchers based their study on 1,278 one-year-olds and 1,345 three-year-olds. The conclusion was clear: the greater the exposure to television at this age, the more problems the children had in focusing their attention when they turned seven. This study, carried out with American children, was confirmed by a similar study and published a year later (in 2005) by a team of New Zealand scientists.[8] In that research,

[6]A. Sigman. (2007a). *Remotely Controlled. How Television is Damaging our Lives.* London: Random House.

[7]A. Sigman. (2007b). 'Visual voodoo: The biological impact of watching TV.' *Biologist*, 54, 12–17.

[8]R. J. Hancox, B. J. Milne and R. Poulton. (2005). 'Association of Television Viewing During Childhood with Poor Educational Achievement.' *Archives of Pediatric and Adolescent Medicine*, 159, 614–618.

they found that if children under the age of three watch television they will suffer a decline in their mathematical ability and in their understanding and recognition of written texts when they reach adolescence.

This study, conducted by paediatricians, was part of an ongoing follow-up (over 26 years) of 1,037 children born in New Zealand between 1972 and 1973. Every two years, between the ages of five and 15 years, the doctors themselves surveyed the children about how much television they watched. Paediatricians found that their neurological observations fit like a glove the academic results of the children: those who had seen more television in their childhood and adolescence obtained fewer grades and levels in their studies. The relationship was so clear that the director of the research team, Dr. Hancox, came forward to testify: 'The more television you watch in childhood, the more likely you are to drop out of school. Watching too little television is the best way to get to college and get a degree'.[9]

Only 7% of the children watched less than an hour of television every day, and these were the ones who achieved the best academic results in all stages of education. In this regard, the researchers suggest that the American Academy of Pediatrics' recommendation to limit 'at most' two hours of television a day in adolescence is too flexible. Television must be absolutely forbidden in childhood (as are drugs, tobacco and alcohol at that age) and viewed for a minimal time during adolescence. Other research confirmed that 'the mean of hours of television viewing during childhood was associated with symptoms of attention problems in adolescence'.[10]

Dr. Hancox qualified these results by referring to children who grew up in the 1970s and 1980s, when there were only two television channels in New Zealand. What, he wondered, alarmed, what would happen to the children of the 1990s and the twenty-first century, who had access to more channels and, above all, to computer games, whose effects were even worse?

There is an urban legend that says video games can be educational (promoted by shady business interests). It's not true. They're not even stimulants. Neurologists have compared the blood flows of children who were asked to play with video game consoles with those of a group who had to complete repetitive sums of single-digit numbers. They found that to calculate the sums the children had to use both frontal lobes of the brain (the right and the left)

---

[9] The statements are from a press release of the press office of the University of Otago about the cited paper (Hancox et al. 2005).

[10] C. E. Landhuis, R. Poulton, D. Welch, R. J Hancox. (2007). 'Does childhood television viewing lead to attention problems in adolescence? Results from a prospective longitudinal study'. *Pediatrics*, 120, 3.

almost entirely. However, to use video games it is enough to stimulate only a small part of the brain associated with vision and movement.[11]

Neurologists have also found that there are different types of care, and that they relate to different parts of the brain. Thus, television can cause different impairments and degrees of attention deficit. One of the most noticeable impairments in the brains of the younger generation is related to the ability to maintain attention for a long time. In the early days of film and television, the pace of storytelling was much slower and there were not so many sequences per minute. But today, both the use of the remote control device and the population's loss of attention due to overexposure (in many cases from a young age) to television, television, cinema or video games have resulted in a succession of information being broadcast at dizzying speed, otherwise, the youngest stop watching.

Let's not forget that audiovisual production is based on Pavlov's old experiment: images and sounds that give us pleasure are created, and we wait for more images to experience more satisfaction. We drool over what is promised next. Television and video games take this principle of 'constant gift giving' very seriously. Some research in the audiovisual field has established that attention is maintained for not much more than 15 s.[12] Therefore, it is necessary to offer constant stimuli (visual or sound or both). This is achieved thanks to the technological evolution of the sector which, at the moment, allows rapid changes of camera, depth of focus (zooms) or the incorporation of sudden strident sounds in the same sequence, all assembled to maintain constant attention. A study of the *Sesame Street* children's series reports that, over the past 26 years, the number of shots per minute has doubled.[13]

Loss of Abstraction Capacity

If the British and North American approach may be characterized by the fact that television can derange the complex system of neuronal connections in the human brain, causing an involution towards an inferior species unable to concentrate (it is only a working hypothesis), the Latin approach could achieve the same result in another way: the loss of capacity for abstraction caused by the audiovisual.

---

[11] R. Kawashima et al. (2001). Report for *World Neurology*, vol. 16, cited in A. Sigman. (2007). 'Visual Voodoo: The biological impact of watching TV.' *Biologist*, 54, 12–17.
[12] M. Rothschild, E. Thorson, B. Reeves, J. Hirsch and R. Goldstein. (1986). 'EEG activity and the processing of television commercials.' *Communication Research*, 13, 2, 182–220.
[13] A. Sigman. (2007b). Op. Cit.

Teachers of mathematics, physics and chemistry—precisely the subjects with the greatest drop in vocations throughout the world—are well aware that there has been a loss of abstraction. Children's ability to understand complex ideas has diminished, something that did not happen to their older siblings. Even those with a good and prolonged level of attention do not understand science or abstract thinking. This has been demonstrated in Germany, where there is an 'abstraction level' test for university entrance. German teachers continue to denounce the fact that it is impossible to reach the levels of a few decades ago. In Spain, too, we can quantify this by analysing the university entrance exams of the past twenty years for subjects that require this ability, such as mathematics, physics and chemistry.

Many pure science teachers blame the new education trends in the West for this decline in the capacity for abstraction. I am not going to deny this hypothesis, because it is obvious that education studies, like most social sciences, is a lax subject governed more by ideology than by pure science. In fact, there is a paradox that the more educationalists work in—or influence—an education system, the more young people suffer a poor assimilation of knowledge. Because of some 'pedagogical renewal' movements, the textbooks of the last four decades are filled with more images and less text every year: because now no one understands an experiment, a theory or a language if it is simply described in words. 'Seeing is understanding', is the new motto, disregarding Descartes' statement that suggested that the senses—especially sight—deceive and that truth can only be reached through rational and abstract thought: 'I think, therefore I am', concluded the French philosopher.

Anyone who has studied science will easily confirm that if a child learns knowledge essentially through images, he or she will not be able to develop the full potential of his or her capacity for abstraction necessary to become a scientist in the future. Maybe they can become an educator or a journalist, but not a physicist. Because what we observe with our senses says, for example, that it is the Sun that moves around the Earth, which is a lie. In other words, the truth can rarely be found through what we observe simply with our senses. The history of science shows us that the way to understand reality is through abstract rational thought. There is no real knowledge that comes from the perception of an image. That's why a physicist can always be a good journalist, but the opposite is impossible.

Modern educationalists have contributed to the deterioration of the children's capacity for abstraction, together with erroneous education laws which, mostly elaborated by intellectuals of literature-based subjects, have reduced the presence of scientific and mathematical language in education. However, from my point of view, the real culprits are the media and, in particular, television.

The dates of the generalization of television clearly coincide with the decline of science. We have already described that one of the causes may be the poor image that television portrays of scientists and a weighting towards irrational ideas.

But television has consolidated a dangerous tendency in the new communication that is contrary to rational thinking: 'I see it on television, then it exists.' But it should not be forgotten that a television image is even more manipulable and uncertain than a visual image. Considering that in developed countries, and in almost all developing countries, television is a common household medium, it could be hypothesized that, if this effect were true, most generations will have lost their capacity and interest in abstraction by the beginning of the twenty-second century.

This hypothesis of the loss of the capacity of abstraction of those who watch television has moved into the arena of its popularization after the publication of the book *Homo videns*[14] by the Italian essayist Giovanni Sartori (1924–2017), professor at the universities of Florence (Italy) and Columbia (USA) and holder of the Prince of Asturias Award for Social Sciences. That is why I consider it the Latino view of the problem. It is a complementary hypothesis to the English-speaking one, since the capacity for abstraction requires great concentration and a high and prolonged level of attention.

Although Sartori's book has been criticized, especially by the audiovisual media industry, in my opinion no criticism can dismantle his main thesis. Moreover, the empirical data that we noted in the first chapter of this book, on the drastic decline in abstract science students (physics, chemistry and mathematics) in Europe and the United States, are evidence in support of Sartori's theses. Also, since the 1980s, there has been an abandonment of the so-called hard cultural studies relating to philosophy, history, philology and literature in the new generation of students. Paradoxically, as we have seen, there has been an increase in enrolment in careers with 'soft' conceptual content, such as journalism and, above all, in film or cultural studies, courses that scarcely involve subjects requiring a capacity for abstraction or analysis of complex subjects.

Sartori points out that this loss of abstraction began in the mid-twentieth century with television, and was accentuated in the 1980s with cybernetics and the emergence of computers and multimedia technology. Television, Sartori says, allows us the deception of seeing close up what is really far away; but cybernetics allows us to create virtual realities, so that children become fonder of

---

[14]Giovanni Sartori. (2002). *Homo videns. La sociedad teledirigida* ('*Homo videns*. The remote-controlled society'). Madrid: Taurus.

simulation than seeking answers to the real world, which is what distinguishes science and rational thinking.

There are hypotheses (currently being investigated) that establish that, from a physiological point of view, the eyes of children who watch television from a very early age (before they learn to read) become used to a certain speed of reception and that, consequently, when they begin to read they become bored since the speed and focus of their eyes are different for reading than for viewing television images. The problem is that science is not learned in documentaries but through books and formulae, repeating exercises and experiments.

Sartori proposes the concept of 'video-children'. In her book he says that statistics indicate that television has replaced maternal care and has become the new nanny and primary school teacher. Watching television is more fun, entertaining and interesting for a child than attending school, so if children come into contact with television before school age they may become school phobic and, above all, may develop a tendency to respond to stimuli that have to do with show business, loud noise and sensations. Young people who watch television as children are dominated by impulses and their motto is to act before thinking, the premise of every audiovisual message.

However, learning science (or any other complex and abstract content) needs, above all, attention and concentration. The generation of television renounces the search for the causes of things, the logical chain that determines events, the reasoned and reflective sequence that leads us from one event to another. The television generation doesn't want to look for causes, it just wants—and I know this from experience with my journalism and audiovisual communication students—to present a newscast, to broadcast a football match or to comment on a reality show, and refuses to learn the scientific basis for why a camera records and why the signal is transmitted. The young people of this generation just worship the show. They love easy and fun. From an early age, they gave up the ability to ask themselves about functions that they cannot see with their eyes. They repudiate the capacity for abstraction that, for example, mathematical or chemical language implies. And they abhor those who have the faculty to understand these abstract and complex languages or concepts, denigrating them at the earliest opportunity.

The Abstraction and Development of Civilization

'What is the use of the capacity for abstraction? I think that attraction is more important,' students reproached me when I naively suggested that my university's new film studies curriculum should include subjects such as optics or electromagnetism, as well as a good foundation in physics and mathematics.

Biologists classify the human species as *Homo sapiens*, belonging to the order of primates.[15] What makes us unique? Well, from a physical and even genetic point of view, almost nothing. We share 98% of our genome with the chimpanzee. *Homo sapiens* is not the animal that sees best (its eyes only perceive a small part of the light spectrum), nor, of course, the one with the most sophisticated sense of smell: everyone who has had a dog knows this. Nor is it the best listener (the hearing systems of cetaceans, for example, are much more effective). Its organ of touch is limited compared to that in other animals and the sense of taste, which some humans have developed, is of little use since most of the substances in nature cannot be tasted, because we would die trying. Neither is *Homo sapiens* the one who runs the best, and the species cannot even fly unaided, something that can be done by much more primitive animals such as birds.

From an evolutionary point of view, the only thing that differentiates *Homo sapiens* from other animals is that the species has a larger brain. Scientists point out that this increase in size may explain the capacity for abstraction. In fact, the passage from Australopithecus to *Homo erectus* meant a 50% increase in cranial capacity and took three to four million years to evolve. The move from *Homo erectus* to *Homo sapiens* took place in just a few hundred thousand years and resulted in a 60% increase in cranial capacity!

If we take into account that the bipedal posture favoured a narrowing of the pelvis, this increased brain development meant that babies had to be born almost in a foetal state to be able to leave the birth canal. Some sociological theories suggest that the birth of such immature offspring led to coexistence in society and the division of labour as the only way to be able to care for offspring for so long. However, there are animals that are also born in this state (kangaroos) and others that take a long time to raise their offspring (elephants).

In order not to dwell on biological differences, we could conclude that mankind differs from other animals only in that the brain has the capacity to have abstract thoughts and to write them down: that is, to transmit and preserve them by means of abstract symbols. This 'strange' ability has allowed us to fly, run faster than any animal, 'see' in areas of the spectrum such as infrared and ultraviolet, swim underwater, listen and interpret ultrasound. In

---

[15] Biologically, the human species has the following classification: belongs to the Cordados filum, since they have a spine; the Vertebrates subfilum, with a segmented body; the Tetrapods superclass, with four limbs; the Mammiferous class, in which the offspring are fed by milk-producing glands; and the Primtates order, since their descendants are arboreal, with flat fingers and nails; the Hominid family, with the eyes located in the front of the head, upright and bipedal locomotion; the genus Homo, with large brain and language ability; and finally, in the species *Homo sapiens*, with prominent chin, high forehead and scarce body hair.

a word: it has allowed a species that does not excel at anything physically, and that even has an evolutionary handicap—its offspring are born helpless—to become hegemonic. And this hasn't happened before, either.

In fact, for hundreds of thousands of years, *Homo sapiens* was dominated by the rest of the animals. This is important to bear in mind, because during the two hundred thousand years that *Homo sapiens* has lived and, in particular, during the last 35,000 years that so-called *Homo sapiens* survived, the species has always been defeated by other animals.

Man's true take-off from other animals came about 35 centuries ago with the birth of writing. It had an important advance 25 centuries ago with the birth of rational thought in classical Greece, and it was consolidated four centuries ago with the birth of modern science. Everything sublime in this time, from poetry to architecture, from mathematics to philosophy, religion or the novel, was based on abstraction and one of its manifestations: symbolism. Even painting and sculpture can be defined as the path of abstraction (e.g. in the development of perspective theory), to 'imitate' reality.

Although this story may seem well known, I think it is not clear at present that we may be undermining the foundations that have made *Homo sapiens* possible over the past 35 centuries. Everyone believes that, in general, the normal process of civilization is to advance knowledge. But it doesn't have to be that way. We have already commented on how, from Roman times to the Renaissance, there was a cultural involution that lasted for more than a thousand years.

For hundreds of years then, superstition again dominated over reason in the explanation of reality. And although for anthropologists, superstition and reason are different variants of the capacity for abstraction—animals are not even thought to be superstitious—there is no doubt that rationality is more difficult to cultivate and gives greater results than superstition. Logic and, in general, rational thinking are among the greatest achievements of the capacity for abstraction.

But we cannot lose only sophisticated skills, such as rational thinking, and thus return to magic. There may even be a setback in simpler skills such as the technology that currently imbues the natural sciences. For hundreds of years, for example, the technical schemes for buildings such as the Parthenon or the Roman Colosseum were lost. It is no wonder that in the Middle Ages many regarded Greeks and Romans as 'supermen' who could never be surpassed.

The ancients possessed skills and know-how that the mediaeval population lost, and thought that these shortcomings were proof that they were intellectually inferior. Fortunately, in the Renaissance the principles of civilization were returned to as the only way to gain momentum and overcome the classics. The strategy was to study them in depth and try to imitate them. It was shown that mediaeval man had no less intellectual capacity than the Greeks; it was just a problem of acquiring the knowledge and, above all, copying the educational methods of classical Greece. Perhaps in this twenty-first century we should take that path again.

Still at the end of the eighteenth century, in 1781, when William and Caroline Herschel discovered the planet Uranus, the focus of society and the incipient media was that Uranus was the first discovered planet that the ancients had never known. There was hope for overtaking them in the knowledge of the universe, but the focus given to it by the media shows that there was still a latent feeling of intellectual inferiority with respect to the classics.

It is not the subject of this book but, according to many experts, there is currently a crisis in abstract disciplines such as painting, architecture or literature, among others. I have attended debates in which it has been said that the great geniuses of cinema are precisely those who did not watch—because it did not exist—television in their childhood. From Griffith (1875–1948), Hitchcock (1899–1980), Billy Wilder (1906–2002) or Orson Welles (1915–1985) to the recent Scorsese (1942) and Spielberg (1947), the latter who probably scarcely watched television before they learned to read although they are obviously not comparable to the former.

It is curious to note the paradox that many film and image experts think that this generation of filmmakers, who learned to read before watching television, will never be overtaken. Not watching television or watching movies until you're 15 or 16 may not only produce better scientists, but also, and this is the curious thing, better filmmakers. But here we cannot deal with this debate, which has many angles.

We have already mentioned that, in the opinion of many scientists since 1953, with the discovery of the structure of DNA, there have been no important theories in science, which is what we are talking about. We have already commented that one of the causes may be that, since natural science is an 'absolute' truth, there are fields that are being closed and it is difficult to derive new theories. Another reason could be that since 1953 the generations who have devoted themselves to science—and everything else—were raised on television.

Rational thinking is therefore very important, and this is another characteristic that distinguishes *Homo sapiens*: that it is not our ability to communicate with other individuals of our species, because that is also the case for other animals, but the ability to 'communicate with ourselves' and, above all, to try to answer the simple but wonderful question of why things happen. Why does ice cool down a cup of tea? Why does a brick wall break down over time without anyone intervening, but never does a pile of bricks join together to form a wall? In short, only mankind has the capacity to obtain the second law of thermodynamics and to reproduce it at will.

I'm Connected, So I Exist

To all this we can add the digital society, and I was especially aware of this during my stay at Harvard in 2013 and 2014. Some call Cambridge 'the Left Bank of Boston (of the Charles River)' not only because of its geographical location but also because of its progressive tradition. Many others believe it is the Athens of the twenty-first century: this small city of just 100,000 people is home to two of the world's leading universities: Harvard and the Massachusetts Institute of Technology, separated by just two metro stops on the Red Line or a beautiful walk down the Charles River, especially inspiring with the intense yellow and red in the autumn or the spectacular blossom of spring.

My faculty sponsor at Harvard, Professor Emeritus Everett Mendelsohn, was a veteran professor who had been at Harvard since 1960. One of his most interesting books is *Heat and Life: The development of the theory of animal heat* (Harvard University Press, 1964), in which he explains that the heat of living bodies (corpses are always cold) has nothing to do with a divine principle or with an 'innate heat' to life, as Galen or Hippocrates affirmed, but is only the product of chemical reactions. Mendelsohn dives into his book for the ideas that great Renaissance or Enlightenment thinkers—like Harvey, Descartes, van Helmont, or Boyle—observed about why animals (and not vegetables or minerals) have warmth while they are alive and lose it when they die. The conclusion reached by enlightened scientists would have been revolutionary: there are no special biological laws governing living systems. As late as 1828, many people thought that there were differences between the living (organic) and the inert (inorganic), until the German chemist Frederich Wöhler (1800–1882) demonstrated that he could synthesize urea (a compound of urine that is exclusive to living beings) from substances derived from minerals such as potassium cyanate and ammonium sulphate.

Heat is the product of chemical reactions, just the same in living bodies as in the inert. That is to say, the distinction that many draw between the biological and the inorganic (technological or robotic) has no basis, from a strictly physical or chemical point of view. Hence the importance of the Craig Venter experiment in 2010, manufacturing in a laboratory the complete DNA of *Mycoplasma mycoides* bacterium. This was a very simple living being, it is true, but it broke the taboo: life had been created from inert chemicals, which are used in haemoglobin as well as in nuts and bolts.

Although the rivalry between Harvard and MIT is great—though not as great as between Harvard and Yale, due to their traditional regatta—their collaboration is also intense. At Harvard, they say that the difference between them and MIT is that at MIT they are working on what will become reality in a decade's time and at Harvard they are thinking about what will happen in 50 years' time. 'But we have to feed back, and there are many spaces for collaboration,' Mendelsohn told me.

One such space is the science, technology and society group, which encourages discussion among scientists, historians, engineers and social scholars. In the academic year 2013/14 there were many lectures, but perhaps one of the most interesting was that of Sherry Turkle, Ph.D. in sociology and personality psychology from Harvard and professor of social studies in science at MIT. She is one of the greatest specialists in the interaction of technology—the inert—with living beings, especially humans. The title of her talk, 'The dystopian presented as the utopian: Does the internet lead us to forget what we know about life' was already a declaration of intent. Dystopian (or antiutopian) society is a recurrent theme in science fiction cinema, where technology degrades human beings, as opposed to the utopia of always being connected and having constant access to all information. 'I'm still excited about the technology, but we can't let it get us where we don't want to go,' she said.

Its strong point is that 'technological devices such as mobile phones change not only what we do but what we are'. In her opinion, we are increasingly connected in the same proportions as we are alone: 'Alone but connected' could translate the title of her 2011 bestseller, *Alone Together: Why We Expect More from Technology and Less From Each Other*.

Turkle studies social robots, the devices designed to keep people who are alone from feeling isolated, ranging from robots that talk to the elderly to *tamagotchi*, the digital pet that children can take care of instead of a traditional dog. 'If no one cares about us, we get hooked on the technology that does,' she said. And she is less concerned about the problems of digital identity on the internet, data theft or the right to forget than the absence of social interaction that our current technology—especially mobile phones—is causing. In Turkle's science, technology and society group's talk, she illustrated her ideas with photos in which families looked at their cell phones at breakfast and did not talk to each other. So did couples, friends, siblings in a room or people attending a business meeting. 'We are running away from conversation and replacing it with mobile messages that we can edit and retain,' she warned, and transformed the motto of the mathematician Descartes from 'I think, therefore I am' to 'I share, therefore I am'.

Finally, she observed that with technology—from the avatars in multiplayer games to the profiles that we publish on the web—we can love our 'friends, our virtual body and our designed world, but we cannot escape from reality. Turkle is the kind of American professor—very much appreciated at Harvard or MIT—who is a world celebrity. Her lecture on the TED platform, which was attended by Twitter founder Biz Stone, Amazon founder Jeff Bezos and Melinda Gates, among others, has had millions of hits; *El País* described her as a cyber-diva.[16]

The most curious thing about Turkle is that in 1995 she said the opposite: she wrote a book, *Life on the Screen,* that turned her into a technological guru. She was on the cover of *Wire* magazine, the bible of technology, and defended how enriching would be the interaction with the internet or mobile screens for human beings as far as she could see. She later said in an interview (in 2012), to explain this change of approach:

> But then I was not able to see that our real life would be cut short by our digital existence. I believed that we would get on the internet and what we would learn inside would help us to improve our life outside, that our digital experiences would enrich our real life, but always in and out of it. I did not understand that the future would consist of living constantly in symbiosis with a computer on: the mobile phone.[17]

Living Is Being Glued to the Screen

'Living is increasingly about being glued to the screen and connected to the Web', say Gilles Lipovetsky and Jean Serroy in their book *The Global Screen.*[18] One of the characteristics of digital society and virtual realities is how cyberspace changes our reality. The limiting effect is that our real reality is all about cyberspace. A survey conducted in 2010 by UNICEF and the *Ombudsman* (Spain) found that people aged 12–18 have become so familiar with new technologies that they can no longer live without them.[19] Four out of 10 of their friends are virtual: that is, they have never seen them personally, and 42% of their time—during the school day—is spent surfing the internet.

---

[16] *El País*, 25 March 2012.

[17] Bárbara Celis. (2012). 'La 'Ciberdiva' que nos pide desconectar ['The' Cyberdiva 'that asks us to disconnect'], interview with Sherry Turkle, (*El País,* 25 March 2012).

[18] G. Lipovetsky and J. Serroy. (2009). *La pantalla global* ('The Global Screen'). Barcelona: Anagrama.

[19] Presented in Madrid on 5 November 2010. Statistical sample: 3,219 students between the ages of 12 and 18. Work directed by Esperanza Ochaíta, professor at the Universidad Autónoma de Madrid.

Some 64% of these adolescents (minors) have posted private images (both their own and other's) on the internet, while 14% said that they had received sexual advances and 11% said that they had received insults or threats via the internet. Their parents cannot control what they do, because they are not digital natives: 60% of the children surveyed admitted that they surf without any adult being involved during the time that they spend online or in what they do on the internet.

The term 'digital native' was coined in 2001 by Marc Prensky[20] to explain aspects such as recent school failure and, above all, the gap between these new teenagers and their parents and teachers, whom Prensky calls digital immigrants: 'People sitting at their desks have grown up with video games and MTV, with music downloading, with a phone in their pockets and a library on their laptops, while constantly texting. They've been connected most of their lives, have little patience for reading and have little step-by-step logic,' Prensky says.

The people in charge of the study defined this generation of teenagers who prefer virtual reality to real reality as the 'bunker generation': they dig themselves deep into their rooms, move away from their real families and friends and, in a 'virtual' reality, go through a phase of their lives in which they ought to know and recognize 'real' reality. These experiences will affect them in the subsequent making of real decisions.

But something else may happen: The internet can control their lives and they may become addicted. It could be the new cyberdrug. In October 2010, the European Commission presented a study in which it stated that 41% of Spanish adolescents between the ages of 11 and 16 showed symptoms of internet dependence.[21] That same month, the University of Castilla-La Mancha held the 5th Conference on the Prevention of Drug Addiction under the slogan: 'Prevention: A key tool in the face of new technological addictions'. Psychologists say that there are similarities between the problematic use of new technologies and substance dependence. All drug addicts—including technology addicts—have the same goal: to escape from reality.

Neither the World Health Organization nor the American Psychiatric Association accepts the new disease yet. However, the reality is stubborn and many people are hooked. In Spain, drug rehabilitation organizations such as Proyecto Hombre offer specific programmes for these patients. Some newspapers talk about the subject, for instance under the headline, 'Photo of the Bunker Gener-

---

[20] Marc Prensky. (2001). 'Digital Natives, Digital inmigrants.' *On the Horizon*, 9(5).

[21] Sonia Livingstone, Carmelo Garitaonandía et al. (2010). 'EU Kids Online. A study of a sample of 23,000 surveys, conducted by professors at the London School of Economics.' In Spain, the results were published by the University of the Basque Country.

ation' (*El País*, 6 November 2010, 52) or, a few days earlier, 'The New Junkies' (*El Mundo* 27 October 2010).

Those interviewed, many of them from universities, talk about traumatic experiences from which they do not know how to move on:

> I was connected all day long, from when I got up at eight in the morning until I went to bed later than midnight; I didn't even sleep in my bed. (…) I spent all day writing on Wikipedia and watching news. It had nothing to do with gambling or pornography…. I was attracted to knowledge. The pleasure of having everything under control was what had me hooked.[22]

It is curious how, for example, in this case they believed that they had control of real reality through virtual reality, but that control made them lock themselves in their homes and, in any event, become inoperative in reality.

This type of addiction keeps you in your room without making trouble for the rest of society, so these people go unnoticed and are not dangerous, like other addicts: 'Those hooked on the Internet do not disturb, do not commit illegal acts, do not transmit diseases…. That's why it's a latent problem, but very abundant', says one of the people in charge of Proyecto Hombre who was interviewed in one of the reports cited. The internet also reinforces other addictions—such as sex, compulsive shopping or gambling—among people who have these pathologies.

Being hooked on the internet is a form of addiction that is 'sweeping Spain,' said Javier García Campayo, a psychiatrist at the Miguel Servet Hospital in Zaragoza and an expert in this type of addiction. During a press conference in Madrid he described the profile of patients with addictions other than substance abuse, and the role played by primary care doctors in their detection, diagnosis and treatment.[23] He stated that 'between 10% and 15% of the Spanish population is already addicted to cyberspace'. The United States is the country with the highest prevalence of this type of addiction (where there are already specific treatment centres), followed by Japan. In recent years, the numbers have also skyrocketed in China. 'Spain has arrived late to new technologies, so it is foreseeable that cases will increase', he added.

A few questions can tell if a you are a potential internet addict who may need to be treated. For example, while at work do you check every 15 min for new information on Facebook, spend more than three hours a day hooked up to the internet or have a child who won't leave the room yet has 500 friends on social networks? Even though you or your child may not know it, according

---

[22] *El Mundo*, 10 October 2010. Campus supplement 4, 1–3.
[23] Statements collected by many means, among others, the EFE Agency and *El Mundo* (26 January 2011).

to psychiatrist García Campayo a positive response to any of these questions may indicate that there is internet addiction.

Cases of the phenomenon known as *hikikomori* (or isolation syndrome, imprisonment), a kind of social *harakiri* that affects more than one million Japanese youths between the ages of 16 and 22, 10% of the population of this age group, have been flooding Japan since the 1990s. The first case described in the West was that of a young Spaniard who spent almost a year in his room, barely leaving, hooked on the internet and his own cyber reality. García Campayo managed to free him from his self-imposed prison and the case was described in *El Heraldo de Aragón*.[24]

Jesus (not his real name) was an only child and had been fatherless since he was 12 years old. At the age of 17 he began to withdraw to his room, as if absorbed by the internet. He dropped out of school, stopped calling his few friends and went into a kind of social suicide. Through the computer, his only link with the world, he created his own cyber reality. According to the newspaper's story, his mother left his meals at his bedroom door. For almost a year he barely left home, and if he went out it was always at night to buy computer equipment and chocolate bars from shops that stay open 24 h a day. In time, he didn't even do that: he would shop online so that he wouldn't have to leave his cyber bunker.

However, not everyone who isolates themselves via the internet can be considered sick; it can be a vital option. One of the most important social changes in digital civilization is the triumph of *Homo solitarius*. Stigmatized in the past—a single woman was repudiated by society—a solitary person is now at the centre of an emerging social and economic system. Symbolized by the multinational Ikea's slogan, 'Welcome to the independent republic of my home', home is a consecrated, autonomous space to live in without going out: from here we can compare food, work or walk (virtually) through a museum or study at university.

In the United States, the term 'cocoon' has proliferated in reference to people who voluntarily confine themselves to their home. As their website, www.cocoonzone.com, describes, these people do not wish to socialize but to achieve peace of mind and personal well-being through a new lifestyle based on retiring, like cloistered nuns, to live behind the walls of their homes, isolating themselves as much as possible from the threatening stress of the world outside the walls. A friend of mine, Álvaro Santana, another Spaniard who emigrated to the United States, a doctor of sociology from Harvard and a professor at that university, defines them better: 'Content in his digital cavern, the cocoon has become

---

[24]Nuria Casas. (2010). 'Un bunker cibernético en el dormitorio' ('A cybernetic bunker in the bedroom'). *El Heraldo de Aragón* (5 December 2010).

a connoisseur in his home paradise and, in front of the computer, acts like an earthly god, whose only physical and inescapable contact with the outside world's mortals is reduced to touching a keyboard.'[25]

Internet-Modified Brains

Neurobiologists have known for years that the brain evolves during the adult stage in the human species, not just in childhood. New connections are continually created according to our experiences and the use that we make of our brain. Exposure to the internet not only affects new generations but may be physically altering the brains of all of us who are in contact with cyberspace. Obviously, if this is so, the internet could be changing our perception of reality, not as a consequence of the virtual representation that we see on the internet and how this conditions our real behaviour but because our own perception of real reality (which is the representation of physical space by our brain) is conditioned by our neuronal system evolving to adapt to the demands of cyber reality.

In June 2010, American writer Nicholas G. Carr published an interesting book: *The Shallow: What the internet is doing to our brains.*[26] In it, Carr, advisor to the editorial board of the *Encyclopaedia Britannica* and editor of the *Harvard Business Review*, argues that the internet is changing our brains, and in a profound way. It's depriving us of our ability to concentrate. Now, we select data instead of reasoning and paying attention to the text. He believes that the internet also has an impact on our ability to think critically.

All the new services on the web—Facebook, email, news alerts—bombard us with information in small doses, creating constant interruptions that keep the brain watchful and prevent it from concentrating on other text. In this sense, our minds no longer seek data or try to draw conclusions from it; they simply pick what they need from the oceans of information with which they are surrounded.

Carr first published these ideas in January 2008 in the influential magazine *The Atlantic* in an article entitled: 'Is Google making us stupid? What the internet is doing to our brains'.[27] His point of view aroused a great deal of

---

[25]Álvaro Santana Acuña. (2010). ¿Eres un cocoon? *Truman Factor,* http://trumanfactor.com/2010/eres-un-cocoon-580.html, last updated 2017.

[26]Nicholas G. Carr. (2010). *The Shallow: What the internet is doing to our brains.* New York/London: W. W. Norton & Company.

[27]Nicholas G. Carr. (2008). 'Is Google Making Us Stupid? What the Internet is doing to our brains.' *Atlantic* (July/August). Accessed May 2018, http://www.theatlantic.com/magazine/archive/2008/07/is-google-making-us-stupid/6868/#.

controversy. Among other elements, Carr believes that part of the problem lies in the web's own intertextual essence:

> Over the past few years I've had an uncomfortable sense that someone, or some-thing, has been tinkering with my brain, remapping the neural circuitry, repro-gramming the memory. My mind isn't going – so far as I can tell – but it's changing. I'm not thinking the way I used to think. I can feel it most strongly when I'm reading. My mind would get caught up in the narrative or the turns of the argument, and I'd spend hours strolling through long stretches of prose. That's rarely the case anymore. Now my concentration often starts to drift after two or three pages. I get fidgety, lose the thread, begin looking for something else to do. I feel as if I'm always dragging my wayward brain back to the text. The deep reading that used to come naturally has become a struggle. (Carr, *The Atlantic*, 2008)

Links are a fantastic tool, but they represent constant interruptions that pluck us out of the text to take us elsewhere. Even if we don't click them, they are a distraction for our brain, a signal to remind us that we can continue reading elsewhere, that there is something else we need to know. 'It wouldn't make sense to put an end to having links on the web, but it would be better for users to have a list of links at the end of a long, in-depth text, rather than to insert them into the piece,'[28] Carr says.

Michael Rich, associate professor at Harvard Medical School and chief exec-utive officer of the Boston Center for Media and Children's Health, told the *New York Times*: 'The worry is we're raising a generation of kids in front of screens whose brains are going to be wired differently.'[29] He also considered that one cannot be stimulated all the time: 'Downtime is to the brain what sleep is to the body.' In a report entitled 'Growing Up Digital, Wired for Dis-traction', the American newspaper explained how this generation is unable to concentrate in class or find time for homework. He illustrated his story with the case of Vishal Singh, a 17-year-old student who spent 10 h a week on video games, updated his Facebook at two in the morning despite having school the next day, and was so famous for sending links to videos that his best friend called him a 'YouTube bully'. Singh described how he lives: 'I'm doing Facebook, YouTube, having a conversation or two with a friend, listening to music at the same time. I'm doing a million things at once, like a lot of people my age. Sometimes I'll say: I need to stop this and do my schoolwork, but I

---

[28] Statements to *El Mundo*, 19 September 2010. 'Cerebros antontados por internet' ('Brains stunned by the internet), Eureka supplement, p. 5.

[29] Matt Richtel. (2010). 'Growing up digital, wired for distraction'. *New York Times* (21 November 2010).

can't. If it weren't for the internet, I'd focus more on school and be doing better academically.' His Facebook timeline is proof of this:

Saturday, 11:55 pm: 'Editing, editing, editing'.

Sunday, 3:55 pm: '8+ hours of shooting, 8+ hours of editing. All for just a three-minute scene. Mind = Dead.'

Sunday, 11:00 pm: 'Fun day, finally got to spend a day relaxing... now about that homework....'

In the report, it described how the main topic of discussion among Singh's teachers was that he was one of the brightest students in high school, although he received poor grades in English (D plus) and Algebra (F). 'He's a kid caught between two worlds, one that is virtual and one with real-life demands', said David Reilly, Director of the Woodside Institute in Silicon Valley, where he studies Singh. Nonetheless, Singh was awarded an A in film critique.

Another student, Allison Miller, 14, said in the report that she sent 27,000 messages a month and that her fingers are so fast that she can even 'feel' conversations while texting. That frantic activity was responsible for her bad grades: 'I'll be reading a book for homework and I'll get a text message and pause my reading and put down the book, pick up the phone to reply to the text message, and then 20 min later realize, "Oh, I forgot to do my homework".'

The director of the centre developed a pioneering education programme that used iPod or video games to teach concepts. He claimed that his intention was to combine technical skill with deep analytical thinking, but he did not know whether he could achieve this. A member of his group, Alan Eaton, a charismatic Latin teacher, declared indignantly: 'It's a catastrophe!' He states that technology has led to a 'Balkanization of their focus and duration of stamina,' and that schools make the problem worse when they adopt the technology: 'When rock 'n' roll came about, we didn't start using it in classrooms like we're doing with technology.'

PowerPoint Makes Us Stupid

The problem is not only that students do not receive real education in their virtual worlds. Even if they are in the classroom, under the care of their teacher, the new technologies can condition their learning. The way that we express reality or our thoughts orally has changed with the internet. In 2010, a book was published in France, *The PowerPoint Thought: Inquiry into this program that makes you stupid*, written by journalist Frank Frommer.[30]

PowerPoint is the product of the interdisciplinary nature of elite American universities, capable of forming highly differentiated profiles that create trends in today's competitive world. The program was created by Bob Gaskins, an honours graduate in computer engineering, linguistics and English literature from the University of California, Berkeley. His first books and research dealt with the use of programming in the humanities (art, literature), so it was not surprising that in the early 1990s he devised a program to turn literary discourse into images.

Used by 500 million people, according to Microsoft, PowerPoint has become the irremediable crutch of most conferences. Even in a university setting, where oratory and rhetoric are the essential characteristics of a teacher, the program is used with intended didactic application. With fatal frequency, a lecturer merely repeats the phrases projected onto the monitor in the auditorium. An excellent and ironic article published in the *New Yorker* began with 'Before there were presentations, there were conversations, which were a little like presentations but used fewer bullet points, and no one had to dim the lights[31]', criticizing the fact that people had stopped speaking at meetings (business, academic, etc.) when giving presentations. For the French journalist Frommer, PowerPoint crushes the virtues of traditional rhetoric, because it forces the speaker to privilege form over content:

> Exhibition is more important than demonstration and seeks to hypnotize the audience and limit their ability to reason. Many times images are incorporated that have nothing to do with what is being said, simply as an ornament or anesthetic. The staging calls for a darkened room in which people are attentive to the screen and consume 15 slides in half an hour. When you leave the room, saturated with images, you've practically forgotten.

The biggest problem is that it alters the habits of argumentation, because speech is subject to the images on the screen. Argumentation is not based on images but on abstract processes through which a speaker leads an audience. No wonder, then, that Frommer is most concerned that PowerPoint is used in education. The essential data never reach the student, who is hypnotized by the speed of the slides. But the worst thing is that students don't learn to express themselves orally, without PowerPoint, as they lack examples to emu-

---

[30] Frank Frommer. (2010). *La Penesé Powerpoint. Enquête sur ce logiciel qui rend stupide.* París: La Dècouverte.

[31] Parker, Ian (2001). 'Absolute Powerpoint. Can a software package edit our thoughts?' *New Yorker*, https://www.newyorker.com/magazine/2001/05/28/absolute-powerpoint.

late throughout their academic life. Even among highly motivated audiences, therefore, their messages may be misunderstood.

This issue has aroused so many misgivings in advanced countries such as Switzerland that a computer engineer, Matthias Poehm, founded the Anti-PowerPoint party, on whose website[32] he states that Europe would save €350 billion if these soporific presentations were abandoned (around 30 million a day across the world) and people concentrated on real work. According to their estimates, 11% of Europe's 296 million workers use this program at least twice a week to explain their arguments. The party is not full of geeks, and was taken very seriously by the *Wall Street Journal* technology analyst Nigel Kendall, who points out that the abuse of PowerPoint is a real problem that needs to be solved.[33] The party wants Switzerland to hold a referendum on banning the use of this software, and intends to become the fourth political force in the country.

Poehm explains that he founded the party and wrote a book, *The PowerPoint Fallacy*, when one day he realized that at a meeting he couldn't express an idea orally unless he turned it into a slide. In many European countries in 2018, there are still schools and colleges that prefer to have students use PowerPoint rather than teach the use of traditional rhetoric and oratory without such a crutch.

One of the greatest critics of the PowerPoint program is Clifford Nass, a professor at Stanford University, who, as a dual graduate in mathematics and sociology, is considered one of the foremost authorities on the interaction between humans and computers. His latest book, *The Man Who Lied to his Laptop*, addresses the complex relationships between computers and humans and the ability of software to interfere with how we choose and organize information. Regarding PowerPoint, he says: 'We will end up communicating by making lists of points one through five. Without making any effort to relate one idea to another. In fact, this is how we are understanding each other, without paragraphs, without pronouns, the world condensed into slides.'

One of the surprises that we science journalists had when NASA sent us the final report on the causes of the 2003 disaster involving the space shuttle Columbia, in which seven astronauts were killed, was the page on which PowerPoint presentations were blamed. The data were analysed by Edward Tufte, professor emeritus at Yale University and world expert in statistical evidence and information design:

---

[32]http://www.anti-powerpoint-party.com/.

[33]http://blogs.wsj.com/tech-europe/2011/07/08/swiss-party-campaigns-against-powerpoint/.

As information is transmitted from one level of the organization to another immediately superior, from the people who perform the technical analyses to the middle management and finally to the management, the key arguments and quantitative supporting information are filtered over and over again. In this context, it is easy to understand why a manager would read this PowerPoint slide and not understand in the act that it referred to a life-threatening situation.[34]

From that moment on I became interested in the work of Tufte, an eminent statistician and world authority in the transmission of quantitative information, who coined the term 'chartjunk' to explain how Excel prevents people from understanding the true meaning of numerical tables. One of the examples that he uses is a medical presentation on typical survival rates in cancer patients, by type of cancer, where he demonstrates that the conclusions obtained from the presentation are the opposite to the real ones. This is terrible, not only for the doctors and their patients but for the journalists who are to report on it.

In 2004, Tufte published a controversial essay that fascinated those of us who work in information technology, 'The cognitive style of PowerPoint'[35] He selects different presentations given by various experts and shows that the slides proposed by PowerPoint drastically reduce the analytical quality of the arguments. He stresses that pre-designed templates 'weaken verbal and spatial reasoning and almost always corrupt statistical analysis'. In his opinion, the low resolution of the slides and the limitations of space and design make it necessary to reduce the contents of a presentation 'to a succession of simplifications, generalizations, ambiguous, incomplete or quasi-propagandistic phrases, in the form of advertising slogans'.

He adds that the standard format requires the separation of numerical data and text, as slides presenting statistics are not normally accompanied by the corresponding written analysis. The typography of the pre-designed templates confuses or distracts the reader, instead of improving comprehension of the text. Slides omit logical relationships, assumptions and even complete subjects, verbs and predicates. Tufte concludes that this program should never be used to explain complex information, and recommends eliminating it from lectures and, above all, from university classes.

Another curious fact: a US military officer (leader of American and NATO forces in Afghanistan), General Stanley McCrystal, publicly accused this Microsoft application—not Iraq's alleged weapons of mass destruction—for being the main enemy of the American army: 'PowerPoint makes us stupid.

---

[34] *Columbia Accident Investigation Boards. Final Report* (August 2003, p. 191).

[35] Edward R. Tufte. (2004). *The Cognitive Style of PowerPoint: Pitching out corrupts within* (2nd edn., 2006). Connecticut: Graphic Press.

When we understand that slide, we'll have won the war (in Afghanistan)'.[36] As Julian Borger, *The Guardian*'s world affairs editor, wrote:

> The diagram has an undeniable beauty. Done the right way (embroidered perhaps) it would make a lovely wall-hanging and an ideal gift for the foreign policy-maker in your life. But it is a bit of a red herring... It was designed to convey the complexity of the Afghan conflict. But the big problem with Power-Point is that it makes everything seem simple. It does not link ideas or facts in any kind of human narrative. It just lists them as bullet points, a shopping list of things to do or jobs already done.[37]

Computational Thinking in the Digital Civilization

In my book *Galileo's Selfie: The social, political and intellectual software of the twenty-first century* (Peninsula-Planeta, 2015)[38] addressed the importance of teaching mathematics and computer programming to all types of students, including journalism students. One of my doctoral students, Guillermo González Benito, feared criticism when he read his thesis, because its conclusion was too ground breaking for the stagnant European academic system. He had argued that for all professions—he studied the case of journalists—one must not only know how to use computer programs but, above all, know how to design them. That is to say, just as since the Renaissance every educated person had to know how to express his or her thoughts in written language, in the digital civilization, to be educated, one must know how to program (not just how to use Facebook or Word). He finally read it successfully in 2017, which means that the mentality is changing.

The British and North Americans call this trend computational thinking and follow the same guidelines that allowed the birth of modern science. From Galileo onwards, experiments have been designed to be mathematical: no mathematics, no science. One of the biggest problems that Galileo had with the Inquisition was that his method kept anyone who did not know mathematics from a knowledge of nature, causing intellectual vertigo among

---

[36] Simon Rogers. (2010). The McChrystal Afghanistan PowerPoint slide: Can you do any better? *The Guardian* (29 April). https://www.theguardian.com/news/datablog/2010/apr/29/mcchrystal-afghanistan-powerpoint-slide.

[37] https://www.theguardian.com/world/julian-borger-global-security-blog/2010/apr/27/afghanistan-microsoft.

[38] Carlos Elías. (2015). *El selfie de Galileo. Software social, político e intelectual del siglo XXI*. Barcelona: Península-Planeta.

philosophers who were incompetent in theorems, and they were enraged at him.

In 1623 Galileo would write a paragraph in his revolutionary book, *The Assayer*, that would hurt—and still hurts. I have already quoted it, but will do so again so that the reader may observe that it is totally topical:

> I seem to detect in Sarsi a firm belief that, in philosophizing, it is necessary to depend on the opinions of some famous author, as if our minds should remain completely sterile and barren, when not wedded to the reasonings of someone else. Perhaps he thinks that philosophy is a book of fiction written by some man, like the *Iliad*, or *Orlando Furioso*—books in which the least important thing is whether what is written there is true. Sarsi, this is not how the matter stands. Philosophy is written in this vast book, which continuously lies upon before our eyes (I mean the universe). But it cannot be understood unless you have first learned to understand the language and recognize the characters in which it is written. It is written in the language of mathematics, and the characters are triangles, circles, and other geometrical figures. Without such means, it is impossible for us humans to understand a word of it, and to be without them is to wander around in vain through a dark labyrinth. (Galileo, *The Assayer*, 1623)

This doctrine underlies the modern concept of computational thinking: the digital world can only be written in algorithmic language. It aspires to widespread literacy in the computer programming language. We must all be able to translate our questions into the language of computers, not wait for a computer specialist to solve our problems. The thesis that I supervised for my student raised common types of questions on reports in journalism, and for each he designed a small software program to achieve results and take a different approach than the usual. This was intended to show how relevant it is in journalism (a typical literary profession) that people know of computer engineering, moreover that a computer engineer who does not know journalism cannot replace a journalist.

The hypothesis can be applied to other professions. Engineers, physicists, mathematicians and even chemists and molecular biologists know programming, but not lawyers, journalists, economists, educationalists, psychologists, doctors or teachers. My student was able to carry his thesis to a good conclusion (he created an algorithm to determine certain types of messages from political sources on Twitter) because he was a computer engineer and had graduated in journalism. Unfortunately, however, Western educational systems still do not marry literature-based subjects to science, mathematics and technology. Few students enter university knowing how to program, and fewer still in the areas of humanities and social sciences.

Learning to program from an early age, while learning to read, write or add, is now essential and has become the real emerging debate in today's education, with frequent press coverage. Under the title 'Learn to program as you learn to read', the newspaper *El País* (7 March 2013) put this question to experts. Luis de Marcos, a lecturer in computer engineering at the University of Alcalá de Henares (Spain), explained his position in the aforementioned report: 'It is no use being taught tools like Word that will have disappeared or will have changed a lot when they finish their studies.' No application, if well designed, takes long to learn. The computer must be used as a tool for solving problems. What the British heritage countries call computational thinking. Because the debate is: are we digital creators or consumers of content? Do we want our children to use up their mobile data allocation in 15 min, or do we want them to develop something and share it?

In the United States, Microsoft founder Bill Gates, a mathematics student at Harvard, and Facebook founder Mark Zuckerberg, a computer science student at Harvard, agree on only one point: you have to introduce the programming language in American schools from elementary school on. To achieve this, they have joined forces—and funds—to create the Code.org Foundation. In the video presentation, Bill Gates says: 'Hi guys! Forget about being doctors, football stars or rappers. Even if they call you geeks at school, the future is in computers.' And Zuckerberg emphasizes: 'Learning to program doesn't mean wanting to know everything about computer science or being a teacher.' Above all, the Code.org Foundation aims to solve one of the major strategic problems of the United States, which we have already mentioned: to significantly increase the number of software engineers. Only a measly 2.4% of the country's graduates are IT graduates, a much smaller figure than a decade ago, despite the fact that the number of jobs that require programming skills has since doubled.

US education officials note with alarm that only one in 10 elementary schools teach a programming language. With those figures, it will be impossible for the United States to lead our modern digital civilization. The numbers in Europe (especially southern Europe) in this sense are at a third-world level: there is practically no provision for teaching programming in primary or secondary education, and only in science and engineering faculties, with fewer students every day, is a programming language learned—and not in all of them.

However, the aforementioned *El País* report detected a ray of hope: in Spain, in 2013, at the faculty of computer science of the University of Valladolid workshops were being run for children between the ages of eight and 12 years so that they could learn to program. And it was a success, though more among the boys: only three girls versus 15 boys in the group. 'It's not just that they learn to program, but that they develop a logical mind that will suit them for any facet of life,' said the instructor, Belén Palop. His colleague, Pablo Espeso, was also enthusiastic about the experience: 'They understand algorithms that we were unaware of at first in computer science; pure mathematics that they apply naturally, because it's simpler than it looks.'

Computer programming, like mathematics, music, chemistry and English, involves languages that are obviously best learned when children are young. And, often, they can no longer be developed when adult. Unfortunately, the mathematical/scientific level of Western students has fallen, according to comparative analyses of examinations over the past forty years. From my point of view, it has not all been the fault of erroneous educational laws but of the fact that the consumption—not the creation—of digital culture limits their capacity for abstraction and prolonged concentration, skills that are not necessary in a journalist, filmmaker or lawyer—vocations that have increased in popularity, despite there being no demand for this work—but essential in mathematics, science and engineering.

The Culture of Entertainment and Fame

I have left until the last part of this journey through the causes of the decline in vocations something that is perhaps the most important, encompassing everything and at the moment is strangling the future of science with a tremendous grip. This cause is the main characteristic of media culture: fame. From London to Melbourne, from New York to Bombay, from Tokyo to Buenos Aires, popularity, fame, is what dominates the world and the market. And, popularity is not acquired in laboratories, libraries or archives; it is achieved through television.

Hollywood's film industry was the first to realize the economic power of building a character's media fame by artificially creating what was then called a celebrity. It discovered that the films were watched not because they were good or bad but because of the celebrities who acted in them. Building a celebrity involved, among other actions, filtering out details of their private life. Details were often invented, aiming to feed the public with what they wanted to hear: experiences that they could never have.

The concept of glamour was strengthened: parties on wonderful yachts and fascinating palaces, of turbulent loves, of extravagant behaviour. The celebrity would be a new god, the hero whom everyone would want to imitate. The media became consolidated as a channel to transmit how the new gods lived. To further engage the audience, it was determined that in the new media culture fame would be within everyone's grasp. And the media emphasized how celebrities had come up from below. You just had to be in the right place (a Hollywood set) and at the right time: the twentieth century.

With the spread of television in the 1960s, Hollywood's culture became global. The fascination with fame, for being a celebrity and for living a life of glamour, designed in film studios, took on global connotations, and the effect was all the more intense in those countries with a significant media presence linked to a market economy. The media realize that the public aspires to an existence of luxury and glamour. And if you don't have glamour, you can at least buy newspapers and magazines and watch television shows that you can be excited about.

The media industry is so important—perhaps the most powerful, at the moment—that anyone who comes into contact with fame knows that they will become immensely rich. Famous people can turn everything that they touch into gold. That is why there is a call to sell merchandise; the famous person becomes an engine of the economy, and is blessed by the whole system. Their presence alone secures millions, for being famous, without having to work. What they sell is their own persona and existence, away from hard graft and surrounded by glamour.

Who qualifies as famous? A doctor who, after studying for eight or nine years, goes whole nights without sleep to save lives? A scientist who, in his laboratory, steals time from his rest to find a vaccine to cure diseases or a solution to alleviate climate change? A historian who dives into the archives to unravel clues to explain why we live in this kind of civilization? A firefighter or policeman who risks his life to save people? A teacher who patiently teaches children who would prefer to play? All these do work that is not only highly useful for society but responsible for its progress. But none of them will climb to fame, because they lack the one and only requirement: to appear continuously in the media.

Who meets this requirement? All those whose work is a focus for the media, regardless of whether their occupation is relevant to the advancement of society. The most important thing is that the media light is constantly on these 'workers'. And what professions are attractive to the media? Just a few: sportsmen, actors, television presenters, film directors, journalists, models, bullfighters, politicians, singers and reality television contestants.

The system is that a clever youngster who wants to succeed in the life of the twenty-first century knows that these are the only professions, not because they give him or her sure route to fame but since they at least offer a chance. No one can pretend to be famous by studying medicine or physics. If you have the body of a model or an aptitude for sport or performance, for music or bullfighting, you will do your best with it. Young people without these skills, and who are awake, will try to study journalism or audiovisual communication, the only two university degrees giving direct access to fame.

Others, with lesser interest in their studies, will do anything to become reality television contestants. This is not reprehensible; it is the system that their parents and grandparents created. A 16- to 18-year-old cannot be blamed for taking advantage of the ways in which the system in which he or she lives to achieve success. It is a law of life: young people have the right to succeed and, in the twenty-first century, this triumph is linked to media culture, just as in other times it has been related to war or, much earlier, to hunting.

For most youngsters in the twenty-first century, work will consist of effort, precarious working conditions and a lack of social recognition. Richard Lambert, Director General of the Confederation of British Industry, alarmed at the lack of scientific vocations, told the BBC that the problem is not so much young people as secondary school teachers:

> Employers are increasingly worried about the long-term decline in numbers studying A level physics, chemistry and maths, and the knock-on effect on these subjects, and engineering, at university. (…) They see, at first hand, the young people who leave school and university looking for a job, and compare them to what they need – and increasingly are looking overseas for graduates. (…) China, India, Brazil and Eastern European countries were producing hundreds of thousands of scientists and engineers every year (…) This is not a criticism of young people – they work hard to achieve the best possible grades in the system provided. But it is clear we need more specialised teachers to share their enthusiasm for science and fire the imaginations of pupils, and to persuade them to study the core individual disciplines to high levels. We must smash the stereotypes that surround science and rebrand it as desirable and exciting; a gateway to some fantastic career opportunities. But the UK risks being knocked off its perch as a world leader in science, engineering and technology. We cannot afford for this to happen.[39]

I agree with Lambert about changing the stereotypes surrounding science, but not about blaming secondary school teachers for the decline in vocations: How can a humble high school teacher can compete with our media culture? The serious media themselves take an approach that is the opposite of effort and study.

As Sartori points out, loquacity, histrionics and the magic of collective hypnotism are now the key to success. The most important value in a media culture such as the one in which we live is neither morality, nor sanctity, nor altru-

---

[39] BBC (13 August 2006). 'Schools "letting down UK science"', http://news.bbc.co.uk/1/hi/education/4780017.stm (accessed May 2018).

ism, nor humility, nor intelligence, nor even artistic ability. The main thing is to become famous, because now it is the equivalent of wreaths of glory. Years ago, one had to do something for the public good: win a war, make a discovery, devise an invention or write something important. Never again—at least while this media culture prevails—will excellence, intelligence, wisdom or even money be needed. It is enough to have an attractive figure, to know how to seduce, to create impact, to make noise to attract media attention; in short, to be an exhibitionist on the media stage.

A few chapters back, I mentioned how we believe that young people may have opted for journalism or film studies because they are easier and more entertaining than physics or mathematics. This is obvious and can be demonstrated by studying both courses of study or by comparing a university textbook on quantum physics with one on the history of cinema. However, this could give the impression that those who choose film or journalism are not good students in high school. Nothing could be further from the truth.

In this media culture, the cult of fame and the search for the smallest chance to achieve it mean that the entry requirements for studying communication or film, for example, are higher and higher, while departments of science and engineering need to lower their offers because they have barely any potential students. In Spain, the so-called 'Letizia effect', in which a middle-class girl, a good student, choses the career of journalism, became famous on television presenting the news and finally married the prince—and now she is queen—has updated the story of Cinderella, in which the new glass shoe includes journalism, film studies or media studies. The score for entering the faculties has skyrocketed, following the 'Letizia effect'.

Vital Asian Optimism

One of the aspects that surprises me most when I meet an Asian is the optimism in the face of Western, let alone European, and especially southern European, pessimism. In 2013, the dean of the faculty of politics at the National University of Singapore, the Indian intellectual Kishore Mahbubani, published a controversial book, *The Great Convergent: Asia, The West and the Logic of One Word*. This presents the Asian vision of the twenty-first century: in the face of Western sadness, in Asia everything is joy. In 2009, the *Financial Times* included Mahbubani among the intellectuals to be reckoned with in this century. Two of the most thoughtful journals in global thinking—*Foreign Policy and Prospect*—have selected him as one of the most influential thinkers.

In March 2013 he gave a lecture at the LSE that I was able to attend: listening to Asian intellectuals, I repeat, is like being on another planet. According to Mahbubani, the last two centuries of human history are an aberration: they were dominated by the West when the most advanced region on Earth has always been Asia (Columbus discovered America because he was looking for a quick route to the riches of Asia). The International Monetary Fund endorses this analysis: in 2013, the world economy was saved by Asian locomotives.

In a few years the Chinese economy will lead the world, and Asia will obviously demand a greater presence in international organizations at the expense of Europe and the United States, which will gradually see a watering down of their power, influence and way of seeing the world. Asia will impose its line of thought and idiosyncrasy: for example, the Chinese have never taken a weekly day of rest from work to pray, as prescribed by the ancient Judeo-Christian tradition, the origin of the Western leisure culture.

The new global reconfiguration benefits Asians and is devastating for Europeans: the latter cannot understand why they are no longer the centre of the planet and, above all, fear that a global world with Asian hegemony will condemn them to live worse than their parents. They long for the lost world of their ancestors, and nostalgia presides over their lives.

In 2011, an essay was published in Germany (a country with no economic crisis) describing the generation of digital natives taking their first steps as adults: *Heult doch: Über eine Generation und ihre Luxusprobleme* ('Stop whining: On a generation and its superfluous problems' 2011).[40] Its author, still in her twenties when she wrote the book, was Meredith Haaf (Munich, 1983), a brilliant—and eternal—journalism scholar who has graduated from several literary courses, philosophy and history. The book caused a howl of protest that shocked many social theorists: it was a first-person description of a those in a generation that can accumulate university degrees yet know that they will live worse than their parents and that, rather than face the real world, prefer to escape into the virtual reality offered by the internet and communication technologies in general. Such behaviour will have an impact on social mobilization.

Haaf argues that if there were a cemetery for a whole generation, the epitaph of those born between 1980 and 1990 would be the advertisement for a flat-rate mobile phone tariff: 'Talk your head off.' In the advertisement, radiant young people are talking nonstop. Phrase by phrase, their bodies shrink until you can only see men and women with huge mouths lying flat on the ground: 'They were empty, but they were still sounding off', Haaf sums up. In her

---

[40]Meredith Haaf. (2011). *Heult doch: Über eine Generation und ihre Luxusprobleme*, Piper Verlag. The references are from the Spanish edition: '*Dejad de lloriquear. Sobre una generación y sus problemas superfluous.* (Alpha Decay, 2012).

opinion, the members of her generation have one thing in common: a passion for communication:

> What that communication is about is, in principle, irrelevant. (…) We feel helpless and isolated when we lose our mobile phone or leave it at a friend's house. (…) And when it is reported in the newspapers that Chinese factory workers only manage to escape by committing suicide under the pressure of producing Apple computers or that in the war zone in the Congo, coltan, the raw material for the production of mobile phones, is extracted under criminal and inhumane conditions, young people worry and perhaps even post the article on Facebook, which, however, does not change our perception of having the fundamental right to communicate permanently. (Haaf 2012, 43–44)

Haaf believes that, unlike generations who studied in earlier times, the education system of the 1980s and 1990s—when the digital natives were trained—made speaking for the sake of it more powerful than any other ability:

> How I am and what I think at the moment is interesting, so I talk about it, what I am and what I think at the moment. This logic at the core of our expressive behaviour ultimately provided the breeding ground for the explosion of communication technologies during 1990. We were hooked from an early age, well prepared for media addiction. It is not surprising, then, that many of us have chosen the career of communication sciences, an academic field that does not consist of much more than analyzing different forms of conversation. (Haaf 2012, 62–63)

Haaf considers her generation to be highly 'individualistic' rather than 'supportive', and says that people's aim is to differentiate themselves from the masses, either through clothing brands or working their Facebook profile: the only project that must succeed is Me, Myself & I Ltd., and one's personal image must not be affected.

> My generation is in a paradoxical situation because, on the one hand, it has enjoyed a much higher economic well-being during childhood than any generation that has preceded it – and this regardless of the income band – but, on the other hand, it has, compared to previous generations, a worse chance of maintaining even that well-being. Social mobility has been considerably reduced from one generation to the next. (…) Only a few of us will be able to live above the standard of living of our parents; however, many of us have grown up with economic security that has made us extremely materially demanding, even though, without outside help, we would not be able to meet such demands. (Haaf 2012, 225)

Pessimistic Western Narrative

From my point of view, whoever regards the glass as being half full always wins. In recent centuries, after the French Revolution, the Western narrative has always been optimistic about scientific and social progress. But in the twenty-first century this optimism has turned to sadness: from Tony Judt to Meredith Haaf, everyone mourns the world in which the West and its outlook on life were dominant. The joy and optimism are now found in Asia—and in some Latin American countries such as Brazil: 'There have never been fewer people killed in wars like these,' Mahbubani said at his LSE conference, 'and global poverty has been reduced to unprecedented levels. Infant mortality has never been so low.' He added: 'In Asia there are currently 500 million people considered middle class, but by 2020 there will be 1.7 billion and by 2030 more than half of the world's population will be middle class.' Mahbubani illustrated his claims with data showing that Asia and, indeed, China are now superpowers, and there are a billion fewer hungry people than there were three decades ago. Asia exudes pure optimism.

For Mahbubani, the key to this current progress in non-Western countries is the global spread of modern science—and its technological side, the philosophy of using logic and reason, which in his opinion prevents war. Moreover, the free market has turned a country that was hungry, thirty years ago, such as communist China, into today's world superpower. It is no wonder that China is leading globalization against US President Donald Trump, who wants to return to protectionism.

In Mahbubani's opinion, the major global issues that may dilute Asia's celebrations—and which he therefore considers to be the real problems of the twenty-first century—are not unemployment, low wages, the decline of political parties, a lack of democratic representation, the power of the bank or the lack of expectations of young people, but climate change and pandemics. These two unknowns scarcely appear in surveys of emerging conflicts in the West. The problem is that Mahbubani is right, and we in the West may not be used to thinking globally that these are imposed by our culture of digital civilization.

The Singapore Model: Betting on Mathematics

Some students asked at the conference about the 'social contract', whereby the state must improve the lives of its citizens. I don't think that they study Rousseau, Hobbes or Locke in Asia. They are interested in other Western thinkers. Asia is not just the giant, China. Mahbubani lives in Singapore (one of the Asian tigers) which, in a way, can be considered a model for studying the enormous potential of Asia. A British ex-colony in China—its inhabitants speak Chinese yet study science and mathematics in English—Singapore is an island city-state (there are several) of barely 700 km^2 and a population of more than five million inhabitants; in other words, a brutal density: 7,680 inhabitants per square kilometre. But it has a human development index of 0.895 (one of the highest in the world) and its universities—it has four and is designing the fifth—are among the best in the world (the Singapore National is among the top 25 according to the *Times* supplement). Its port is one of the busiest and it is a leader in the chemical, electronic and financial industries.

Singapore has no natural resources, only human resources, so its commitment has been to training and talent. And it has had results: with Korea and Finland, it is one the three countries with the best positions in mathematics in the PISA reports. Aware of the enormous potential of mathematics for its future as a country in the digital civilization, Singapore has also developed its own method of teaching mathematics—known as the Singapore method—which makes it the country with the most mathematicians per 1,000 inhabitants. Rather than oil or territory, the country regards mathematics as their most valuable asset to enable them to prosper in the future, and sells it as such.

In some Western countries, attempts are being made to introduce such a method to improve the mathematical skills of the younger generation. But it is not working, basically because it involves much study and practice on the part of the students, an Asian idiosyncrasy that is discredited in postmodern Western pedagogy. Mathematics will always be difficult: Euclid said so, when the King of Persia asked him for an easy way to learn it: 'Majesty, there are no royal ways to mathematics'.

The Scientific Method, Key to the Progress of the Western World

During my stays in Great Britain and the United States, I have discussed with my Asian colleagues the reasons for the rise of the West in recent centuries and the reasons for its current decline. The only thing that we agree on is the advantage that Europe gained from the scientific method of studying nature. Although history depends on many perspectives, if one looks at global politics in terms of science and technology one sees that, until well into the second millennium, Europe was always behind Asia in technological innovations. We should remember that these, together with the conquest of other territories or natural resources, have been the foundation of economic development since the beginning of humanity.

The three inventions that Europe used to expand its potential from the second millennium were Chinese: the compass, with which Europeans could cross open sea; gunpowder, with which it conquered, subdued and enslaved the territories and peoples accessed with the compass; and paper, without which the printing press would not have had the influence that it did. These three Chinese innovations came to Europe through the Arabs who, in addition, enriched these contributions with Greek culture and, above all, their extraordinary numerical notation (the one that we still use). The cultural mix and the desire to change tradition benefited Europe.

Not only did the Arab countries fail to consolidate but they rejected their rich mediaeval scientific tradition. The most important astronomical, medical and mathematical centres in the world in the tenth, eleventh and twelfth centuries were Arab cities, from Cordoba to Baghdad. It is not the object of this book to delve deeper into the decline of Arab science, but from the fifteenth century, when these countries rejected science, their political and cultural decline began. Many point out that the Arab countries lacked a Renaissance—which could have unleashed a scientific-technical revolution—and an Enlightenment. Therefore, despite the rise of social networks and digital technology, it would have been difficult for democracy to flourish in countries with cultural values where religion is regarded as above reason and science. Although technology is global, the consequences of its use are local and depend on the previous scientific/technological culture.

There is heated debate about whether Greek mathematics would have evolved sufficiently if it had not been for Arabic numeration: it is clear that it would have been extremely difficult to operate mathematically with Roman numerals, for example. But the Greeks were different. Today, we know that Archimedes (killed in 3rd century BC) may have achieved a kind of differential calculus (undeveloped until the seventeenth century, with Newton and Leibniz). In 1906 a *palimpsest* (a Greek parchment, erased to copy out Christian psalms) written by Archimedes was discovered in which it can be intuited that the Greek mathematician could have advanced the notion of the infinitesimal limit. A rhetorical question frequently asked at congresses and seminars on the history of science is where we might we be now, had Rome not conquered and subdued Greek talent and had Christianity not appeared.

In this twenty-first century, the Chinese have already adopted the Arabic numerical system and are learning Latin spelling, not to read Virgil or Cervantes but for chemical formulation and mathematical theorems. Despite what we may think due to our cultural Eurocentrism, by the seventeenth century—the birth of modern science in Europe—more texts had been written in China alone than in the rest of the world combined. Its continuously produced literature—ranging from poetry to essays and even literary and political criticism—is the oldest in the world: from the Shang dynasty (sixteenth to eleventh centuries BC). China accounted for 25% of world GDP in the eighteenth century. Its decline in the nineteenth century was due not to reading romantic novels or buying impressionist paintings. It was the consequence of failing to take on science and technology, and failing to adopt scientific methods and advances in physics, chemistry, biology and engineering.

Contemporary Asians have inspected Western culture over the past 2,500 years and have left with only the scientific method (and its technological applications) to learn about the nature of matter and energy. It's strange to meet Chinese researchers around the world, as they've maybe never heard of the Beatles, the Nazi holocaust or *Gone with the Wind,* yet are scholars in microbiology or quantum mechanics. The only Western thinker who has influenced them is Karl Marx who, remember, read his doctoral thesis in 1841 on the difference between the philosophy of nature in Democritus and Epicurus. This is a brilliant and little-known work, of which we have spoken in another chapter. In it, Marx takes from the importance of the concept of the atom and matter to the theory of the celestial bodies of Democritus and its influence on Epicurus and his revolutionary phrase, 'He who despises the gods is not ungodly, but he who adheres to the idea that the multitude is formed by the gods'.

Why, then, did Europe—and the West—lead the world economy and progress in the nineteenth and twentieth centuries? In my opinion, it was not because of their literature, architecture, painting, music, politics or spiritual philosophy. Europe led because it developed the scientific method in the seventeenth century to unlock the secrets of matter and energy. A small step in its applications (already initiated by Galileo when working for the Venetian government, thus emulating Archimedes) unleashed an unprecedented set of technological innovations: from thermodynamics, which explains the steam engine (eighteenth century) of the first industrial revolution, to the chemical (nineteenth century), electronic (twentieth century) and mechanical (twentieth century) industries.

It has not been an easy road: even in the nineteenth century something as common as electricity was vigorously rejected by many intellectuals, such as the English writer and philosopher Mary Shelley and her work *Frankenstein, or the Modern Prometheus* (1818) or the Church itself. While she was terrified by experiments in which electrodes produced movements in corpses, cardinals feared that physicists would have the power of resurrection. In our twenty-first century, some are opposed to therapeutic cloning or embryonic stem-cell research.

Workers' unions and workers' rights were not won in convents, farms or millennial universities, but in science/technology-based industries in countries where religious tradition had instituted a day of rest for prayer. In our global world, with instantaneous transport (especially in the computer industry) and where technology and science are not the exclusive property of the West, we are competing with cultures in which rest is not a tradition, thus the European way of life seems to be collapsing.

Why the Western World?

There is a discussion—only in the academic, not in political or economic spheres—over why it was in Europe and not in China that science appeared. Three clues could explain it. The first was Greek thought, but not of Plato or Socrates but of the mathematicians Euclid, Democritus and Archimedes—a great inspirer of Galileo—who, it seems, either failed to arrive or insufficiently influenced China, although they were the basis of the European Renaissance. Mathematics revolutionized the approach to the most subversive area of human thought, which is not the abolition of capital or the bourgeoisie but the nature of matter—not only physical but biological—and energy. And that changed the world.

The second key is found, paradoxically, in Christianity: the founders of this monotheistic theology created a tradition of thought, influenced by Aristotle and Greek rationality, in which the material universe had order. The Christian contribution was to design an all-powerful creator who not only established moral or spiritual standards but ordered matter and energy, breathed life into it and organized all laws. The discussions of the Catholic theologians of the sixteenth century on the moment of inertia—and in general on the laws of movement—are highly illustrative in this sense, although it is not the aim of this book to go into this in depth.

Discovering the nature of matter and its order would make it easier to understand God, some mischievous people have reflected. In fact, biology or physics were important subjects in theology until the seventeenth century. When one reads the biographies of the Christian thinkers who devoted themselves to science—from Bacon to Copernicus to Galileo to Newton—one discovers that they did so in order to draw closer to God, understanding the universe which, according to Christian doctrine, He created:

This was a very powerful stimulus – perhaps the most powerful – for their efforts. So as a result of such rigorous traditions, the Church inadvertently provided a cognitive framework for generations of scientists whose work ultimately undermined the very foundations of Christianity, replacing superstition with logic and reason.[41]

And the third reason—and this makes some Chinese nervous—is a certain culture of freedom that took place in specific places and at specific times in Europe: the Venice of Galileo, and the England of Newton or Dalton. Where that freedom did not flourish, science did not prosper. The paradigmatic case was Spain: we had the same cultural heritage as Venice or England. The first two criteria—classical cultural tradition and Christianity—were more than met, as it was in Toledo that the School of Translators introduced many Greek mathematical texts into Europe. In fact, it was in Toledo, at the time of Alfonso X the Wise (1221–1284), that the astronomical tables that Copernicus used to place the Sun at the centre of the universe were compiled. These tables—the first after those of the Greek, Ptolemy (3rd century AD), whose realization can be considered as the first research project in the history of Europe and science, as I have already mentioned—were devised by Arab mathematicians and astronomers, supervised by Jews and financed by a Christian king. This multiculturalism in science and thought has been lost since the Catholic monarchs and, in my opinion, has not been recovered in Spain. Donald Trump's anti-immigration laws in the United States may have the same effect as the expulsion of Jews and Arabs in Spain: the country may suffer huge cultural decline.

Be that as it may, in this twenty-first century Asia is well aware that the cause of its loss of world hegemony was its backwardness in understanding the natural sciences—it does not find that it misses either Hobbes or Locke and their social contract. Therefore, its educational system is producing more and more mathematicians, scientists and engineers, in contrast to Europe and the rest of the West, which are saturated with social, cinematographic and literary studies, inspired by weak theories such as psychoanalysis and post-structuralist thought, and deficient, as we have already mentioned, in engineers and scientists who study matter and energy.

It is not only that the Chinese study matter and energy, but that they use these studies to approach sociology, politics, law and economics. As Plato said, in order to dedicate oneself to metaphysics (the forerunner of the humanities and social studies) one must first study physics. The Chinese already have their space station (while the West proposes to abandon theirs) and plans to

---

[41] Michael White. (2010). *Galileo. Anticristo. Una biografía* ('Galileo, Antichrist. A biography'). Barcelona: Almuzara.

have bases on the Moon by 2020. For the last 25 years, the Chinese have had presidents who are engineers.

Europe and the United States are acting as Asia did in the seventeenth and eighteenth centuries: rigidly maintaining their traditions. Many of its universities (some departments of science studies, cultural studies, literature, media studies and even sociology, history and philosophy) are acting as the Spain of the Counter-Reformation that saw in science only danger and that, in any case, did not consider it to be culture. The most frightening thing for the future of Europe has been seen on the banners waved in demonstrations by young Spaniards or French people bearing the slogan, 'We want to live like our fathers'. There is nothing more dangerous to any kind of progress than a society that aspires to live like its forebears.

But the past doesn't come back. And if it was splendid for Europe, it was because of its science and technology, its Greek mentality that was recovered in the Renaissance through studying matter, describing it in mathematical language and applying scientific method. That is, it was achieved through an extraordinary way of seeking the truth (through observation, experimentation and mathematics) that no other culture had ever done before. This is highly relevant, because Western culture is not superior to any other in terms of literature, architecture, painting, political philosophy, law, cinema, music or ethics. It may be different, but it is in no way superior to Chinese or Indian culture, or others. Considering that Western literature, religion, ethics, law or art is superior to Eastern art is unacceptable Eurocentrism and cultural imperialism. However, the way of describing the world from chemistry, physics, biology and geology with mathematical language is a hugely important contribution to humanity of European and, by extension, Western culture. No other civilization has achieved such success in describing matter, which is the same as describing the reality of the world.

That a Western creator or intellectual—whether a teacher, politician, journalist, filmmaker, sociologist, political scientist, philosopher, historian, writer, or so on—does not have a great knowledge of physics, chemistry, biology, geology and mathematics is a huge setback in Western culture. This is also a danger to their economy and welfare state, not because of the benefits of natural resources but for the scientific and technological talent applied to natural resources.

The other great Western contribution is that this scientific culture is public and has been shared with all humanity. The East has not had to invent a new chemistry or physics. When young people refuse to study science, they are turning their backs on the most important intellectual adventure ever achieved by human beings. And they also welcome fake news, magic or the alternative facts that have darkened Europe from the time that the Roman Empire succumbed to the religious fundamentalism of some Christian sects until the Renaissance, when fight began against the idea that theology could explain the world and, using texts such as *De Rerum Natura,* to explain it again from the point of view of physics.

Printed in the United States
By Bookmasters